Lecture Notes in Computer Science 4816

Commenced Publication in 1973
Founding and Former Series Editors:
Gerhard Goos, Juris Hartmanis, and Jan van Leeuwen

Bianca Falcidieno   Michela Spagnuolo
Yannis Avrithis   Ioannis Kompatsiaris
Paul Buitelaar (Eds.)

Semantic Multimedia

Second International Conference on Semantic
and Digital Media Technologies, SAMT 2007
Genoa, Italy, December 5-7, 2007
Proceedings

Springer

Volume Editors

Bianca Falcidieno
Michela Spagnuolo
Consiglio Nazionale delle Ricerche
Istituto per la Matematica Applicata e Tecnologie Informatiche
Via De Marini 6, 16149 Genova, Italy
E-mail: {bianca.falcidieno,michela.spagnuolo}@ge.imati.cnr.it

Yannis Avrithis
National Technical University of Athens
School of Electrical and Computer Engineering
Image, Video and Multimedia Systems Laboratory
9 Iroon Polytechniou Str., 157 80 Athens, Greece
E-mail: iavr@image.ntua.gr

Ioannis Kompatsiaris
Informatics and Telematics Institute
1st Km Thermi-Panorama Road, Thessaloniki, 57001 Greece
E-mail: ikom@iti.gr

Paul Buitelaar
DFKI GmbH
Language Technology Lab
Campus D3 2, Stuhlsatzenhausweg 3, 66123 Saarbruecken, Germany
E-mail: paulb@dfki.de

Library of Congress Control Number: 2007940049

CR Subject Classification (1998): H.5.1, H.4, H.5, I.2.10, I.4, I.7, C.2

LNCS Sublibrary: SL 3 – Information Systems and Application, incl. Internet/Web
and HCI

ISSN   0302-9743
ISBN   3-540-77033-X Springer Berlin Heidelberg New York
ISBN   978-3-540-77033-6 Springer Berlin Heidelberg New York

Typesetting: Camera-ready by author, data conversion by Scientific Publishing Services, Chennai, India
Printed on acid-free paper   SPIN: 12197521   06/3180   5 4 3 2 1 0

Preface

The conference on Semantics And digital Media Technologies (SAMT) covers a wide scope of subjects that contribute, from different perspectives, to narrow the large disparity between the low-level descriptors of multimedia content and the richness and subjectivity of semantics in user queries and human interpretations of content: The Semantic Gap. The second international conference SAMT 2007 took place in Italy, in the beautiful town of Genova, and continued to gather interest and high-quality papers from across Europe and beyond, bringing together forums, projects, institutions and individuals investigating the integration of knowledge, semantics and low-level multimedia processing, including new emerging media and application areas.

In response to the call for papers, 55 papers were submitted. After a thorough review process, only 16 contributions were accepted as full papers with oral presentation, in addition, 20 contributions were selected as short papers. We would like to thank all the members of the Technical Program Committee, the authors of submitted papers, and the additional reviewers for their efforts in setting the quality of this volume. The conference program also includes two invited keynote talks from Steffen Staab and Remco Veltkamp and we are very grateful to them for their insightful presentations. This volume includes a special section with three awarded short papers from the K-Space PhD Workshop that took place in October 2007 in Berlin.

We acknowledge and would like to thank the entire Organizing Committee for the excellent contribution to the coordination of the various events: the programme of SAMT 2007 included three workshops, one tutorial, three special sessions, project and demo session, the SAMT 2007 industry day, and two invited talks from European Commission representatives, Albert Gauthier of the INFSO.E2 Knowledge and Content Technologies, and Luis Rodríguez-Roselló of the INFSO.D2 Networked Media Systems.

The SAMT 2007 conference was organized by IMATI-CNR with the sponsorship of the European FP6 Networks of Excellence AIM@SHAPE and K-Space and of the Department of Information and Communication Technology of the CNR of Italy. Moreover, SAMT 2007 ran in cooperation with the European Commission and the SALERO project. Finally, we would like to thank Marinella Pescaglia and Sandra Burlando for their invaluable administrative support and the whole staff of the Shape Modelling Group of IMATI-CNR that contributed to making the conference happen.

December 2007

Bianca Falcidieno
Michela Spagnuolo
Yannis Avrithis
Ioannis Kompatsiaris
Paul Buitelaar

Organization

SAMT 2007 was organized by the department of Genova of the Institute of Applied Mathematics and Information Technology of the National Research Council of Italy (IMATI-CNR).

Conference Committees

General Chairs
Bianca Falcidieno
IMATI-CNR, Genova, Italy
Yannis Avrithis
National Technical Univ. of Athens, Greece

Program Chairs
Michela Spagnuolo
IMATI-CNR, Genova, Italy
Ioannis Kompatsiaris
Informatics and Telematics Institute, Greece
Paul Buitelaar
DFKI, Germany

Special Sessions
Werner Haas
Joanneum Research, Austria

Tutorials
Noel O'Connor
Dublin City University, Ireland

Workshops
Catherine Houstis
Informatics and Telematics Institute, Greece

Industry Day
Franca Giannini and Chiara Catalano
IMATI-CNR, Genova, Italy

Posters and Demos
Riccardo Albertoni and Michela Mortara
IMATI-CNR, Genova, Italy

Program Committee

Bruno Bachimont
Wolf-Tilo Balke
Jenny Benois-Pineau
Jesús Bescós
Nozha Boujemaa
Patrick Bouthemy
Pablo Castells
Andrea Cavallaro
Stavros Christodoulakis
Fabio Ciravegna
Thierry Declerck

Anastasios Delopoulos
Edward Delp
Martin Dzbor
Touradj Ebrahimi
Jerome Euzenat
Borko Furht
Moncef Gabbouj
Christophe García
William I. Grosky
Werner Haas
Christian Halaschek-Wiener

Siegfried Handschuh
Alan Hanjalic
Lynda Hardman
Andreas Hotho
Jane Hunter
Antoine Isaac
Ebroul Izquierdo
Franciska de Jong
Joemon Jose
Aggelos K. Katsaggelos
Hyoung Joong Kim
Joachim Koehler
Stefanos Kollias
Paul Lewis
Petros Maragos
Ferran Marques
José Martínez
Adrian Matellanes
Bernard Merialdo
Frank Nack
Jacco van Ossenbruggen
Jeff Pan
Thrasyvoulos Pappas

Ioannis Pitas
Dietrich Paulus
Eric Pauwels
Dennis Quan
Keith van Rijsbergen
Andrew Salway
Mark Sandler
Simone Santini
Guus Schreiber
Timothy Shih
Sergej Sizov
Alan Smeaton
John R. Smith
Giorgos Stamou
Fred S. Stentiford
Michael Strintzis
Rudi Studer
Vojtech Svatek
Murat Tekalp
Raphael Troncy
Paulo Villegas
Gerhard Widmer
Li-Qun Xu

Additional Reviewers

Samer Abdallah
Riccardo Albertoni
Alexia Briassouli
Stamatia Dasiopoulou
Alberto Machì

Joao Magalhaes
Vassilis Mezaris
Francesco Robbiano
Massimo Romanelli

Sponsoring Institutions

Institute of Applied Mathematics and Information Technology of the CNR, Italy
Department of Information and Communication Technology of the CNR, Italy
FP6 Network of Excellence AIM@SHAPE (http://www.aimatshape.net/)
FP6 Network of Excellence K-Space (http://www.k-space.eu/)

Table of Contents

Classification and Annotation of Multidimensional Content

Content Adaptation

MX: the IEEE Standard for Interactive Music

Semantic Multimedia Annotation II

Short Papers

K-Space Awarded PhD Papers

Improving the Accuracy of Global Feature Fusion Based Image Categorisation*

Ville Viitaniemi and Jorma Laaksonen

Laboratory of Computer and Information Science, Helsinki University of Technology,
P.O.Box 5400, FIN-02015 TKK, Finland
{ville.viitaniemi,jorma.laaksonen}@tkk.fi

Abstract. In this paper we consider the task of categorising images of the Corel collection into semantic classes. In our earlier work, we demonstrated that state-of-the-art accuracy of supervised categorising of these images could be improved significantly by fusion of a large number of global image features. In this work, we preserve the general framework, but improve the components of the system: we modify the set of image features to include interest point histogram features, perform elementary feature classification with support vector machines (SVM) instead of self-organising map (SOM) based classifiers, and fuse the classification results with either an additive, multiplicative or SVM-based technique. As the main result of this paper, we are able to achieve a significant improvement of image categorisation accuracy by applying these generic state-of-the-art image content analysis techniques.

1 Introduction

In this paper we consider categorisation of images into semantic classes. Image categorisation is a task closely related to the more general problem of image content understanding. The capabilities of image content analysis techniques can be demonstrated by applying them to the image categorisation task. In our earlier work [17] we have demonstrated image annotation and categorisation performance that compares favourable to methods presented in literature by using a system architecture that adaptively fuses a large set of global image features. One of the benchmark task concerns categorisation of images of the Corel collection, approached earlier by Andrews et al.[1], Chen and Zwang [3] and Qi and Han [16] with SVM-based multiple-instance learning methods.

In this work, we consider the same image categorisation task, taking our earlier PicSOM system as a baseline. We retain the overall feature fusion based system architecture since feature fusion has nowadays proven to be an effective approach to solving large-scale problems, such as analysis of the content

* Supported by the Academy of Finland in the projects *Neural methods in information retrieval based on automatic content analysis and relevance feedback* and *Finnish Centre of Excellence in Adaptive Informatics Research*. Special thanks to Xiaojun Qi and Yutao Han for helping with the experimental setup.

B. Falcidieno et al. (Eds.): SAMT 2007, LNCS 4816, pp. 1–14, 2007.

of large video collections. However, we refine the constituent components inside the system. We modify the set of visual features that are fused by including interest point SIFT feature [13] histograms. Such features have recently become popular and demonstrated a good performance in a variety of image content analysis tasks [20]. Based on the individual features in the feature set, the image categories are detected with support vector machine (SVM) classifiers in the improved system, instead of using self-organising map (SOM) based PicSOM classifiers. For fusing the detector outputs, we experiment with additive, multiplicative and SVM-based fusion mechanisms. With the improved system, we perform experiments that replicate the experimental setup of [16]. In addition to demonstrating a significant improvement of the overall image categorisation accuracy, we are able to experimentally compare the performance of different feature sets and fusion mechanisms.

The rest of the paper is organised as follows. Section 2 introduces the image categorisation problem and details the Corel categories benchmark task. In Sections 3 and 4 we describe the baseline system and the improvements, respectively. In Section 5 the performed experiments are described and their results reported. In Section 6 we draw conclusions from the results.

2   Image Categorisation

In the image categorisation task each image is assigned to exactly one of a list of possible categories on the basis of visual content of the image. We regard the task as a supervised learning problem, where a number of training images together with their ground truth categorisation is used to learn the connection between the images' visual properties and the category labels. The quality of the learned model is tested by predicting the categories of a previously unseen set of test images solely based on their visual contents and comparing the predictions with a manually specified ground truth. Another type of image content analysis problem—image annotation—is closely related to image categorisation. In that problem, however, an image may be assigned to any number of categories simultaneously. Sometimes the term annotation is used in literature also for categorisation problems.

As an experimental benchmark categorisation task we use the Corel categories task that was first defined by Chen and Wang [3] for 20 image categories. The image data consists of images from 20 Corel stock photograph CDs. Each of the CDs contains 100 images from a distinct topic and forms a target category for the task. A label is chosen to describe each of the categories. Table 1 shows the labels of the 20 categories. Some of the categories are very concrete, some are more abstract. The task was later extended to 60 categories by Qi and Han [16] who introduced 40 more Corel CDs. We use all the 60 categories in our experiments. The task also defines an ordering of the image categories and a corresponding succession of incremental subsets of categories: the set with M categories includes the first M categories of the ordering.

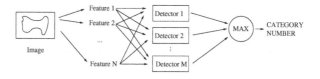

Fig. 1. The general architecture of our image categorisation system

As we want to compare our categorisation performance with that of the earlier efforts, we use the same metric for measuring the categorisation performance: the average accuracy. This is simply the fraction of correctly categorised test images over their total number. For measuring annotation quality, more sophisticated measures [18] are usually used, but the equal number of images in each category warrants the use of this simple average accuracy measure in this categorisation task.

Table 1. Labels of the first 20 categories of the Corel categories task ([3,16])

African people and villages	Beach	Historical buildings	Buses
Dinosaurs	Elephants	Flowers	Horses
Mountains and glaciers	Food	Dogs	Lizards
Fashion	Sunsets	Cars	Waterfalls
Antiques	Battle ships	Skiing	Desert

3 Baseline System

As a baseline system, we use our PicSOM image annotation system that was demonstrated to perform well in the Corel categories task in our earlier work [17]. Figure 1 shows the general architecture of the system. To categorise a test set image, a number of elementary visual features is extracted from it. The extracted set of feature vectors is fed parallel to several detectors, one for detecting each of the categories. The output of each detector is converted to the estimated posterior probability of the category. The predicted category of the test image is selected to be the category whose detector outputs the highest probability.

The rest of this section details the components of the baseline system: the used visual features (Sec. 3.1) and the detector modules (Sec. 3.2).

3.1 Visual Features

As a basis for the classification, a set of global visual features is extracted from the images. The extracted ten elementary features, listed in Table 2, are encoded as feature vectors with dimensionalities given in the rightmost column of the table. The first four rows correspond to features that more or less closely resemble the ColorLayout, DominantColor, EdgeHistogram and ScalableColor features of the MPEG-7 standard [8]. The column "Tiling" of the table shows that some of the features are calculated truly globally, such as the global colour histogram

Table 2. The elementary visual features extracted from the images

Feature	Tiling	Dim.
DCT coefficients of average colour in rectangular grid	global	12
CIE L*a*b* colour of two dominant colour clusters	global	6
MPEG-7 EdgeHistogram descriptor	4×4	80
Haar transform of quantised HSV colour histogram	global	256
Average CIE L*a*b* colour	5	15
Three central moments of CIE L*a*b* colour distribution	5	45
Co-occurence matrix of four Sobel edge directions	5	80
Magnitude of the 16×16 FFT of Sobel edge image	global	128
Histogram of four Sobel edge directions	5	20
Histogram of relative brightness of neighboring pixels	5	40

feature of the fourth row, others, such as the edge histogram feature of the third row, encode some spatial information by using a fixed image grid. The features calculated for five tiles employ a tiling mask where the image area is divided into four tiles by the two diagonals of the image, on top of which a circular center tile is overlaid.

3.2 Classifiers

The category detection is achieved by training a separate detector for each category in our PicSOM image content analysis framework. The framework classifies images by adaptively fusing information given by several different elementary low-level image features. The framework is readily described elsewhere (e.g. [12]).

In the PicSOM image content analysis framework, the input to the image classifier consists of three sets of images: training images annotated with a keyword (positive examples), training images not annotated with the keyword (negative examples), and test images. The task of the classifier is to associate a score to each test image so that the score reflects simultaneously the image's similarity to the positive examples and dissimilarity from the negative ones.

To obtain a detector for an image category, a set of visual features is first extracted from the example images. The PicSOM framework is then used to automatically generate representations of the features that are adaptive to both 1) the context of the totality of images in the image collection, and 2) the context of the present task, expressed in terms of sets of positive and negative example images. To this end, the components of the elementary feature vectors are divided into several overlapping subsets, feature spaces. As the number of all possible feature combinations is overwhelming, we must content ourselves with a rather arbitrarily chosen subset. In the baseline system, we consider all the 10 elementary feature vectors individually, almost all pairs of them, and some heuristically chosen triplets and quadruplets, resulting in 98 feature spaces in total.

Each of the feature spaces is quantised using a TS-SOM [11], a tree-structured variant of the self-organising map [10]. For the experiments reported here, we use TS-SOMs with three stacked levels, the bottom levels measuring 64×64 map units. For each two-dimensional quantised TS-SOM representation, positive and negative impulses are placed at the best-matching-unit (BMU) projections of the positive and negative example images on the TS-SOM surface. The impulses are then normalised and low-pass filtered. This associates a partial, feature-dependent classifier score to each map unit of the corresponding feature SOM. A total classifier score for a test set image is obtained by projecting the image to each of the SOM-quantised feature spaces and summing the partial classifier scores from the corresponding SOM's BMUs. In this procedure, the low-pass filtered summation of positive and negative examples effectively emphasises features that perform well in separating the positive and negative example images of that particular classification task.

To facilitate the comparison between outputs of detectors for different image categories, each output is converted to a probability estimate. We have observed that a simple logistic sigmoid model

$$p_i(w|s_w(i)) = \frac{1}{1 + e^{-\theta_1^w s_w(i) - \theta_0^w}} \tag{1}$$

suits this purpose well. Here $s_w(i)$ is the classifier score and and $p_i(w|s_w(i))$ the probability of image i belonging to category w. The model parameters $\theta_{\{0,1\}}^w$ are estimated separately for each category w by regarding all the training images as independent samples and maximising the likelihood

$$L_w = \prod_i \frac{I_w(i) + (1 - I_w(i)) e^{-\theta_1^w s_w(i) - \theta_0^w}}{1 + e^{-\theta_1^w s_w(i) - \theta_0^w}} \tag{2}$$

with respect to $\theta_{\{0,1\}}^w$. Here $I_w(i)$ is the binary indicator variable for the training image i to belong to the category w. The product is taken over all the training images. The maximisation is performed numerically using the classical Newton-Raphson method.

4   Improved System

The improvements of the image categorisation system are made to the components of baseline system, the overall system architecture of Figure 1 is not affected. Both the set of visual features (Sec. 4.1) and the category detectors (Sec. 4.2) are improved.

4.1   Improved Visual Features

In the improved system, the set of visual features is extended with histograms of interest point features. The interest points are detected using a Harris-Laplace detector [14]. The SIFT feature [13], based on local gradient orientation, is calculated for each interest point. Histograms of interest point features have proven to be efficient in image content analysis and are gaining popularity in various image analysis tasks [5,15].

Table 3. ROC AUC performance of interest feature histograms of different size in the VOC2006 image classification task. The different columns correspond to the ten classes of objects in the task. The rightmost column shows the average performance over all the ten classes. The bottom row shows the results for a SVM-based fusion of the classification results for all the listed histogram sizes.

	bicycle	bus	car	cat	cow	dog	horse	motorb.	person	sheep	sum
8 × 5	0.834	0.934	0.946	0.841	0.869	0.798	0.804	0.873	0.756	0.893	**0.855**
16 × 16	0.895	0.967	0.962	0.883	0.915	0.832	0.878	0.911	0.808	0.913	**0.896**
25 × 20	0.896	0.968	0.961	0.899	0.909	0.820	0.855	0.918	0.816	0.924	**0.897**
40 × 25	0.910	0.972	0.965	0.906	0.912	0.825	0.877	0.930	0.813	0.926	**0.904**
50 × 40	0.909	0.969	0.967	0.910	0.914	0.827	0.889	0.935	0.816	0.922	**0.906**
80 × 50	0.916	0.969	0.964	0.900	0.915	0.833	0.884	0.942	0.808	0.929	**0.906**
fused	0.917	0.974	0.968	0.916	0.925	0.839	0.897	0.942	0.829	0.934	**0.914**

We form histograms of the SIFT features according to codebook vectors selected using the self-organising map (SOM) algorithm [10]. Use of K-means algorithm for this purpose has also been reported (e.g. [20]), but in our preliminary experiments with Pascal VOC Challenge 2006 images and classification task [5], the SOM codebooks appeared to perform better than K-means codebooks with an equal number of codebook vectors.

We have experimented with the optimal size of the codebook in the VOC2006 task. SVM-based classification was used for these experiments. Table 3 shows Receiver Operating Curve (ROC) Area Under Curve (AUC) statistics used to quantify the classification performance in that task for ten target classes. There is some variation between classes and some differences are likely to be attributed to noise-like random variations. The general trend seems to be, however, that larger codebooks produce more accurate classification. Of course, the classification accuracy must start deteriorating at some point if the codebook size is continuously increased as the histogram comparison function we used in the experiments is a sum of bin-wise distances. However, with these experiments we had limited the size of the codebook to 80 × 50 for practical reasons. The turning point of the performance as function of codebook size appears to be beyond our maximum size. As the results on the codebook size were somewhat inconclusive, we decided to include codebooks of all the six considered sizes to our feature set. The last row of Table 3 shows that the SVM-based fusion of all the codebook sizes produces better results than any codebook alone. This points to the direction that the information in the differently-sized codebooks is not completely redundant.

4.2   Improved Detectors

In the improved system, the detection of categories was performed using support vector machines (SVM). A separate SVM was trained for each feature space. In the testing phase, each test image was first classified with each feature space SVM and then the SVM outputs were fused together. In principle, the forming of the feature spaces was accomplished similarly as in the baseline system by selecting combinations of the ten elementary and the six interest point features. However, to accommodate the increased computational cost of the improved detectors, pairs, triplets and quadruples of the features could not be utilised in such large a number as in the baseline system. Only a subset of twelve such combined features were included in the set of available features, in addition to the 16 elementary features. The feature combinations were chosen on basis of their individual performance they had shown earlier in the VOC2006 classification task performed with SOM classifiers.

We used a weighted version of the C-SVC variant of the SVM algorithm for individual features, implemented in the version 2.84 of the software package LIBSVM [2]. The outputs of the SVMs were normalised to probability estimates by using the method of [19]. As the kernel function g we used the χ^2-kernel

$$g_{\chi^2}(\mathbf{x}, \mathbf{x}') = \exp\left(-\gamma \sum_{i=1}^{d} \frac{(x_i - x_i')^2}{x_i + x_i'}\right) \qquad (3)$$

for the histogram features and the RBF kernel

$$g_{\mathrm{RBF}}(\mathbf{x}, \mathbf{x}') = \exp\left(-\gamma \|\mathbf{x} - \mathbf{x}'\|^2\right) \qquad (4)$$

for all other features.

Our own experiments with the VOC2006 data, and the results reported in [7] have revealed that careful selection of the free parameters of the SVM classifiers can markedly improve the classification results. In the experiments of this paper, the free parameters C of the C-SVC cost function and γ of the kernel function were chosen on basis of a search procedure that maximises the six-fold cross validation ROC AUC in the training set. The search procedure, inspired by [9,6], performs importance sampling in the parameter space by a suitably scaled increasing function of AUC,

$$t(\mathrm{AUC}) = (0.9b + 0.105 - \mathrm{AUC})^{-4}, \qquad (5)$$

and records the results in a kd-tree structure. Here b is the best AUC found during the course of the search. In our informal experiments, the procedure usually, but not always, found better parameter values with a smaller number of iterations than a grid search procedure. These results should still be regarded initial. The method should be further refined and the results confirmed more rigorously. We also tried averaging the final test set classifications among several of the best performing parameters. The additional computational cost of doing so is small compared to the cost of the SVM training and the cross-validation. The averaging did not, however, have a significant effect on the accuracy of classifications.

Weight parameter w compensates for the unbalanced class distribution in the training set in the cost function of C-SVC. For this parameter, we used a heuristically chosen value $w = 3$. A more rigorous alternative would have been to include the parameter w in the set of optimised parameters in the optimisation procedure described above. However, this was omitted in order to save computation time.

For the fusion of the detection results based on individual features we used three alternate mechanisms. The probability estimates were either summed, multiplied or combined with a SVM. In the SVM-based fusion the probability estimates were formed into a vector. Such feature vectors were used as input to a SVM with a RBF kernel function. The output of the SVM was once again converted to a probability estimate [19].

5   Experiments and Results

In the experiments we applied the image categorisation systems of Sections 3 and 4 to the Corel categories task of Section 2. We replicated the experimental setup of [16] as closely as possible. We used the same sets of categories with sizes 10 through 20, along with sets with 30, 40, 50 and 60 categories. In experiments with a certain set of categories, only the images that are labeled to one of the categories are considered. The categorisation was evaluated in a supervised manner by randomly partitioning the images of each category to training and test sets of equal size. To get an estimate of the accuracy of the results, all the experiments were repeated for five different random partitionings of the images into training and test sets. The same five partitionings are used throughout all the experiments.

In addition to reporting the average accuracies over the five trials, we have determined the 95% confidence intervals of the accuracies under the approximation of normal distribution, as in [16]. These confidence intervals apply to the comparison of the presented methods against literature methods. However, due to the use of the same partitionings, the results are statistically more reliable than the intervals would indicate for the comparison of the relative accuracies of the different configurations of our baseline and improved systems. A large fraction of the statistical fluctuation is due to the specific choice of image sets that applies identically to all our system configurations, and thus does not cause relative variability.

5.1   Overall Results

Figure 2 compares the overall image categorisation accuracy of the baseline method, the improved method and three methods reported in literature as a function of the number of image categories. The dashed lines indicate 95% confidence intervals. The abbreviation Qi refers to method proposed in [16]. DD-SVM method was introduced in [3] and the MI-SVM in [1]. Experimental results for DD-SVM and MI-SVM were not available for more than 20 categories. It can be observed that the improved method is clearly superior to the earlier PicSOM method. Both of our methods described in this paper in turn clearly outperform the three considered literature methods. The accuracies for the improved method are evaluated with the full feature set of Section 4.1. This choice of

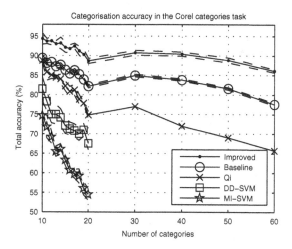

Fig. 2. Total image categorisation accuracies of different image categorisation systems in the Corel categories task

features is motivated by our prior belief that all the features we have extracted from the images are useful. The validity of this assumption will be examined in Section 5.3.

The results of Figure 2 use SVM for feature fusion up to 20 image categories. For 30 and more categories, accuracies are given for the multiplicative fusion technique. This practice will be justified in Sec. 5.2. The features used in the baseline method were chosen on basis of maximum performance in another image content analysis task we have performed earlier [18]. That set of features includes pairs, triplets and quadruples of the elementary image features of Section 3.1, but none of the individual features themselves.

Figure 3 compares the performance of the detector stage of the improved method with that of the baseline method with the same set of elementary visual features as input. This figure reveals the magnitude of difference in the detector accuracy of the two methods. By comparing with the previous figure, we see that the baseline method is partly able to compensate for the lower detector accuracy by the larger set of visual features it is able to use. This is not feasible in the improved method because of its larger computational cost.

The improvement in classification accuracy comes with increased computational cost. The improved method does not scale as well as the baseline PicSOM method when the number of categories and thus the number training images is increased. With the largest problem sizes, the relative difference in the required computation time may exceed two orders of magnitude.

5.2 Feature Fusion Mechanisms

Figures 4a and 4b show examples of comparing the relative accuracies of the three feature fusion mechanisms of Sec. 4.2. The figures show comparison for two

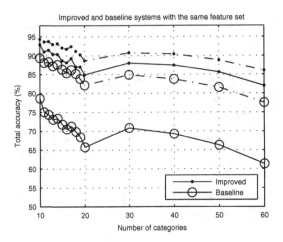

Fig. 3. Solid lines: accuracy of the baseline and improved categorisation systems with the elementary image features of the baseline system (Sec.3.1). Dashed lines: the performances of the two systems with the feature sets of Fig. 2.

sets of features, for other features we obtained similar results. First of all, the accuracies of the different fusion methods were usually not dramatically different. Nonetheless, some general trends could clearly be observed: the relative ordering of the fusion mechanisms changed as the number of categories was increased.

For a small number of categories, the SVM-based fusion mechanism was almost always the most accurate. For a large number of categories, multiplicative feature fusion produced best results, whereas SVM was often even the worst of the mechanisms, in particular with the largest sets of categories. The switchover point between these two mechanisms was usually between 20 and 30 categories.

This systematic change of ordering of fusion mechanisms as a function of the number of image categories was incorporated in the improved categorisation system architecture so that for up to 20 categories, the system uses SVM-based fusion mechanisms, and the multiplicative mechanism for 30 and more categories. This choice is reflected in Fig. 4 by setting the accuracy of this finally chosen scheme as zero-level.

5.3   Performance of Different Feature Subsets

In Sec. 5.1 we selected the set of visual features to include all the available ones for the improved method. This was based on our prior assumption that all the features we are able to use in the system are beneficial. The validity of this assumption is examined in Figure 5 by plotting the accuracies of several feature sets. The feature sets of subfigure (a) form a hierarchy of subsets of the feature set ALL, the full feature set of the improved system. ALL decomposes to feature sets FC and ELEM+IP. FC denotes the set of the twelve combined features of Sec. 4.1. ELEM+IP is the union of ELEM, the individual elementary elementary

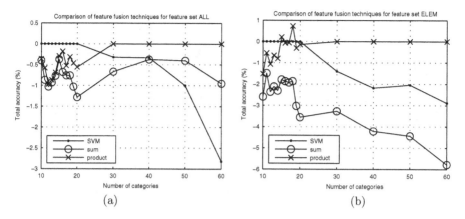

Fig. 4. The relative accuracy of different fusion mechanisms in the Corel categories task for two different feature sets. See Sec. 5.3 for a key to the feature set identifiers.

image features of Sec. 3.1, and IP, the set of six interest point feature histograms of Sec. 4.1. Subfigure (b) shows the performance of smaller feature sets composed of the four features SC (global colour histogram), IP80 (interest point histogram of size 80×50), SCIP80 (concatenation of SC and IP80), and COLM (colour moments). In this subfigure, the feature set ALL is included for reference.

With the larger feature sets of Figure 5a, we observe that the accuracies generally increase as more features are added to the set of features. In the light of this example, the fusion mechanism seems to be able to emphasise the important features and not get too distracted by the irrelevant ones. Looking at feature sets with a very few hand-picked features (Figure 5b) we see, however, that already these few features alone seem to contain almost all the information the system is able to extract from the full set of features. As it is not likely that the other features would be completely redundant in reality, it seems that the system actually is distracted by extraneous features.

Also on theoretical grounds, fusing badly-performing features with good ones usually leads to lower performance than the good features would produce alone. There is also empirical evidence to support this, both from our own experiments and from literature [4]. In this light, the observed increase of performance with increasing feature sets should be interpreted to mean that most of the features we used really convey useful information and are not completely redundant. This is due to the still relatively small number of features. If the feature set would be increased further, at some point the performance would probably saturate and even start to decrease.

The results with different feature sets suggest that feature selection could be beneficial also for this problem. How to do the feature selection partly depends on and partly determines the system architecture. In principle, at least some part of the feature selection process could be relegated to the feature fusion stage. The considered feature fusion mechanisms are probably not very good in assigning more weight to the essential ones and eliminating the redundant

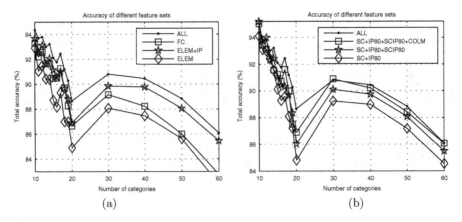

Fig. 5. The accuracy of two families of feature sets in the Corel categories task for SVM-based fusion

ones. However, something to this direction is already achieved by the proposed system architecture by converting the SVM outputs to probability estimates. For well-performing features, the values of the estimates are often close to the extremes and the interval of variation of the values is large. In contrast, for worse-performing features the values concentrate near the prior probability of a category. The amplitude of variation directly determines the weight of the corresponding feature in the implemented feature fusion schemes.

6   Conclusions and Discussion

In the experiments of this paper we have demonstrated a significant improvement in the state-of-the-art performance level in the Corel categories image categorisation task. This is achieved by applying general purpose image analysis techniques that are applicable also to other kinds of image analysis problems. Our experience [17] is that improvements in the present task translate also to other, more widely used benchmarks. The approach of fusing numerous different information sources is applicable and seems to be popular for even a larger variety of problems, such as concept detection from news videos [15]. Our success propones the view that it might be worthwhile to seek improvements in generic techniques that are simultaneously applicable to a large variety of problems, instead of concentrating only on specific problems.

We have experimentally investigated some technical details concerning the components of our improved system architecture. Our results on the codebook construction for histograms of SIFT-features were not conclusive in the sense that we would have been able to point out an optimal way to construct a codebook. However, we clearly saw that use of small codebooks is not advantageous.

We have also experimentally compared mechanisms for classifier fusion. The mechanisms we studied were relatively elementary, yet they have often been used in data fusion problems. For example, the SVM-based data fusion method has been popular among the participants of the TRECVID 2006 video retrieval evaluation campaign. The small performance difference between multiplicative and SVM-based data fusion mechanisms was somewhat surprising, as was the fact that the multiplicative combination actually works better for a large number of image categories. Without further experiments, the underlying mechanisms can only be speculated upon. One explanation could be that the multiplicative fusion enjoys from the shunting effect of estimated probabilities near zero: even one of the multiple detectors based on different features could effectively inhibit the other detectors of indicating an image as belonging to a category. This unsymmetric right to "veto" a decision could be beneficial for a large number of categories. SVM-based fusion is more symmetric in terms of different features. However, this is only speculation, and needs to be confirmed in further experiments.

In our improved image categorisation system, the feature fusion mechanisms performed satisfactorily in the sense that the best performance levels were achieved by fusing together all the 28 features of the improved system. It seems that even with the current fusion mechanism and the quality of our visual features, we are still below the saturation point where the inclusion of additional features would no longer improve the system performance. This gives motivation to develop the detector stage so that even a larger number of features can be used, for example all the features of the baseline system. Of course, the balance point is dependent both on the quality and redundancy of the feature set and on the fusion mechanism. Improving either of these has still performance potential. In case of the fusion mechanism this can be inferred from the observation that very small hand-picked feature sets almost achieved the performance of the full set of features.

The performed experiments confirmed our earlier hypothesis [17] that our baseline image categorisation system used features somewhat inefficiently. Even with a subset of the earlier features, our improved system produced significantly more accurate image categorisations. We have thus partly addressed the problem, but with a price: the new system needs considerably more computation time. This is emphasised with the rigorous five-fold statistical testing procedure of the experiments. Furthermore, the computational cost rapidly grows with the number of image categories. Our future plans thus include addressing this issue of computational cost, as we plan to apply our improved system architecture to large vocabulary image annotation problems with vocabulary sizes of hundreds of keywords and extend the set of visual features we use.

References

1. Andrews, S., Tsochantaridis, I., Hoffman, T.: Support vector machines for multiple-instance learning. In: NIPS 15, pp. 561–568. MIT Press, Cambridge (2003)
2. Chang, C.-C., Lin, C.-J.: LIBSVM: a library for support vector machines, Software (2001), available at http://www.csie.ntu.edu.tw/~cjlin/libsvm

3. Chen, Y., Zwang, J.Z.: Image categorization by learning and reasoning with regions. Journal of Machine Learning Research 5, 913–939 (2004)
4. Snoek, C.G.M., et al.: The MediaMill TRECVID 2006 semantic video search engine. In: TRECVID. TRECVID Online Proceedings (November 2006), http://www-nlpir.nist.gov/projects/tvpubs/tv.pubs.org.html
5. Everingham, M., Zisserman, A., Williams, C.K.I., Van Gool, L.: The PASCAL Visual Object Classes Challenge (VOC2006) Results (2006), http://www.pascal-network.org/challenges/VOC/voc2006/results.pdf
6. Hämäläinen, P., Aila, T., Takala, T., Alander, J.: Mutated kd-tree importance sampling. In: SCAI 2006. Proceedings of the The Ninth Scandinavian Conference on Artificial Intelligence, Espoo, Finland, October 2006, pp. 39–45 (2006)
7. Hauptmann, A.G., Chen, M.-Y., Christel, M., Lin, W.-H., Yan, R., Yang, J.: Multilingual broadcast news retrieval. In: TRECVID. TRECVID Online Proceedings (November 2006), http://www-nlpir.nist.gov/projects/tvpubs/tv.pubs.org.html
8. ISO/IEC: Information technology - Multimedia content description interface - Part 3: Visual, 15938-3:2002(E) (2002)
9. Kajiya, J.T.: The rendering equation. In: SIGGRAPH 1986, pp. 143–150 (1986)
10. Kohonen, T.: Self-Organizing Maps, 3rd edn. Springer Series in Information Sciences, vol. 30. Springer, Berlin (2001)
11. Koikkalainen, P., Oja, E.: Self-organizing hierarchical feature maps. In: Proceedings of International Joint Conference on Neural Networks, San Diego, CA, USA, vol. II, pp. 279–284 (1990)
12. Laaksonen, J., Koskela, M., Oja, E.: PicSOM—Self-organizing image retrieval with MPEG-7 content descriptions. IEEE Transactions on Neural Networks 13(4), 841–853 (2002)
13. Lowe, D.G.: Distinctive image features from scale-invariant keypoints. International Journal of Computer Vision 60(2), 91–110 (2004)
14. Mikolajcyk, K., Schmid, C.: Scale and affine point invariant interest point detectors. International Journal of Computer Vision 60(1), 68–86 (2004)
15. Over, P., Ianeva, T., Kraaij, W., Smeaton, A.F.: TRECVID 2006 - an introduction. In: TRECVID. TRECVID Online Proceedings (November 2006), http://www-nlpir.nist.gov/projects/tvpubs/tv.pubs.org.html
16. Qi, X., Han, Y.: Incorporating multiple SVMs for automatic image annotation. Pattern Recognition 40, 728–741 (2007)
17. Viitaniemi, V., Laaksonen, J.: Empirical investigations on benchmark tasks for automatic image annotation. In: VISUAL 2007. LNCS, vol. 4781, pp. 93–104. Springer, Heidelberg (2007)
18. Viitaniemi, V., Laaksonen, J.: Evaluating the performance in automatic image annotation: example case by adaptive fusion of global image features. Signal Processing: Image Communications 22(6), 557–568 (2007)
19. Wu, T.-F., Lin, C.-J., Weng, R.C.: Probability estimates for multi-class classification by pairwise coupling. J. of Machine Learning Research 5, 975–1005 (2005)
20. Zhang, J., Marszałek, M., Lazebnik, S., Schmid, C.: Local features and kernels for classification of texture and object categories: a comprehensive study. International Journal of Computer Vision 73(2), 213–238 (2007)

Stopping Region-Based Image Segmentation at Meaningful Partitions

Tomasz Adamek and Noel E. O'Connor

Centre for Digital Video Processing, Dublin City University, Dublin 9, Ireland
{adamekt,oconnorn}@eeng.dcu.ie

Abstract. This paper proposes a new stopping criterion for automatic image segmentation based on region merging. The criterion is dependent on image content itself and when combined with the recently proposed approaches to *syntactic* segmentation can produce results aligned with the most salient semantic regions/objects present in the scene across heterogeneous image collections. The method identifies a single iteration from the merging process as the stopping point, based on the evolution of an accumulated merging cost during the complete merging process. The approach is compared to three commonly used stopping criteria: (i) required number of regions, (ii) value of the least link cost, and (iii) Peak Signal to Noise Ratio (PSNR). For comparison, the stopping criterion is also evaluated for a segmentation approach that does not use syntactic extensions. All experiments use a manually generated segmentation ground truth and spatial accuracy measures. Results show that the proposed stopping criterion improves segmentation performance towards reflecting real-world scene content when integrated into a syntactic segmentation framework.

1 Introduction

The problem of partitioning an image into a set of homogenous regions or semantic entities is a fundamental enabling technology for understanding scene structure and identifying relevant objects. A large number of approaches to image segmentation have been proposed in the past [1]. This paper focuses on automatic *visual feature-based* segmentation of images that does not require developing models of individual objects, thereby making the approach generic and broadly applicable.

A popular class of automatic segmentation techniques are the *Graph Theoretic* approaches in which images are represented as weighted graphs, where nodes correspond to pixels or regions and the edges' weights encode the information about the segmentation, such as pairwise homogeneity or edge strength [2,3,4,5,6,7]. The segmentation is obtained by partitioning (cutting) the graph so that an appropriate criterion is optimized. Several approaches from this category utilize *Region Adjacency Graphs* (RAGs) where nodes represent regions while edges contain region pairwise dissimilarities [2,8,9]. The pairwise dissimilarities are

B. Falcidieno et al. (Eds.): SAMT 2007, LNCS 4816, pp. 15–27, 2007.

typically computed using a distance between regions' colour features. Such dissimilarities are often referred to as *region homogeneity criteria*. RAGs can be simplified by successive mergings of neighboring regions. A fast and yet effective (and therefore popular) approach from this category is the *Recursive Shortest Spanning Tree* (RSST) algorithm [2,3,9]. Another example of an approach from this category is the *Normalized Cut* [5] approach that has attracted considerable attention within the content-based information retrieval (CBIR) community in recent years due to its state of the art performance [10].

Unfortunately however, since such approaches are based purely on low-level image features, the resulting segmentation results do not necessarily reflect the real-world content of the scene, thereby limiting their usefulness both for feature extraction for CBIR or as a pre-processing step prior to automatic or semi-automatic semantic object segmentation. This is manifested by either under-segmentation, where semantic objects or regions[1] are merged in the final segmentation mask, or over-segmentation, where single semantic objects or regions are composed of many small irrelevant regions. Thus, in an effort to make region-based segmentation more useful in applications that require a semantic interpretation of the scene, recently several researchers have proposed to improve the nature and quality of the segmentation produced by region merging by utilizing additional cues. These additional cues, whilst not semantic in themselves, reflect the semantic nature of the scene by encapsulating its geometrical and spatial configuration properties. In this way, the resulting approach is still generic but potentially provides a much more stable basis upon which to perform semantic knowledge extraction. *Ferran and Cassas* used the term *syntactic features* to refer to such quasi-semantic features [11]. In practice, integrating such features is achieved by using the geometrical properties and the spatial configuration of regions as merging criteria in one of the Graph Theoretic approaches referred to above [12,13,11,14,7]. In fact, using such a merging framework, some approaches have attempted to divide the merging process into stages and use different homogeneity criterion in each stage [11,14]. Whilst these approaches are promising, they are still hampered by the key difficulty of knowing when to stop the merging process in order to obtain the best possible segmentation result, particularly in the case of hierarchical segmentation.

The region merging (or splitting) approaches referred to above are particularly attractive in constructing hierarchical representations of images, e.g. all merges (or cuts) performed during the region merging (splitting) process can be stored in a *Binary Partition Tree* (BPT). The usefulness of such hierarchical representations, and BPT in particular, has been advocated by many researchers as an important pre-processing step in applications such as region-based compression, region-based feature extraction in the context of MPEG-7 description of content, and semi-automatic segmentation [4,15,7]. However, at the moment

---

[1] Note that we use the term 'semantic regions' to refer to image regions that would be defined by a human annotator of the scene. Whilst they may not constitute full semantic objects as required by any given application/user they are parts thereof that reflect real physical structure of an object.

many applications, including CBIR systems, can utilize only a single partition of the scene. Moreover, even in scenarios where the segmentation is used to produce a hierarchical representation of the image it is often necessary, e.g. due to reasons of efficiency, to identify a single partition within such a representation most likely to contain a meaningful segmentation (or the most representative impression of the scene) – see for example [15].

In the case of segmentation via region merging, a single partition is obtained simply by defining a criterion for stopping the merging process. In other words, the stopping criterion is needed to identify those elements within the hierarchical structure which are most likely to be relevant in a given application. To date, only very simple stopping criteria have been used, e.g. the required number of regions [3] or the minimum value of *Peak Signal to Noise Ratio* (PSNR) between the original and the segmented image reconstructed using mean region colour [15]. Although intuitive, in cases of heterogeneous image collections such criteria often fail to produce partitions containing the most salient objects present in the scene.

This paper proposes a new stopping criteria for a syntactic region-based segmentation approach that facilitates the generation of single partitions that contain the most salient objects present in the scene (or partitions corresponding to the most representative impression of the scene) and that works across heterogeneous image collections. The method identifies a single iteration from the merging process corresponding to the most salient partition based on the evolution of the accumulated merging cost during the overall merging process. The proposed approach is compared to three different commonly used stopping criteria: (i) required number of regions, (ii) value of the least link cost, and (iii) *Peak Signal to Noise Ratio* (PSRN) between the original and segmented image (reconstructed using mean region colour) [15]. The different stopping criteria are evaluated within a merging framework that includes two different region homogeneity criteria corresponding to the commonly used RSST [3] that uses only low-level features, and a recently proposed improvement of this [7,16] that uses syntactic features.

It should be noted that recently a stopping criterion also based on the analysis of the accumulated cost was proposed in [17]. However, this alternative criterion was aiming mainly at producing the initial partition used in the BPT, i.e. a partition correctly characterizing all objects in the scene with a "reasonable" amount of regions. In other words, the stopping criterion from [17] was proposed to detect anchor points for object detection algorithms able to further merge regions from the initial partition and improve the BPT representation. Also, the merging criteria used in [17] are quite different to those used in this work. Merging criteria used in [17] aimed at maximizing PSNR while at the same time minimizing the variance of region sizes produced at each iteration. However, typical scenes contain objects of various sizes and therefore encouraging creation of regions similar in size may not produce the most intuitive results in partitions containing only a small number of the most relevant regions. In the work presented here the balance between colour distance and the size dependent scaling factor changes as the merging progresses in order to provide balance appropriate for each stage.

The remainder of this paper is organized as follows: the next section presents an overview of the region merging framework used. Then, section 3 discusses three different commonly used stopping criteria and section 4 introduces our new approach. The results of exhaustive evaluation with two different merging criteria and an image collection with ground-truth segmentation of semantic regions are presented and discussed in section 5. Section 6 concludes the paper.

2   Region Merging Framework

The proposed stopping criteria are evaluated using two region merging approaches. The first is the commonly used *Recursive Shortest Spanning Tree* (RSST) algorithm [2,3,9]. The second one is a new extension to the original RSST, proposed recently in [7], that uses so-called syntactic features. However, theoretically the criterion could be integrated into any approach based on region merging or splitting.

2.1   Original RSST

The original RSST algorithm starts by mapping the input image into a weighted graph [2], where the regions (initially pixels) form the nodes of the graph and the links between neighboring regions represent the merging cost, computed according a selected homogeneity criterion. Merging is performed iteratively. At each iteration two regions connected by the least cost link are merged. Merging two regions involves creating a joint representation for the new region (typically its colour is represented by average colour of all its pixels [3]) and updating its links with its neighbors. The process continues until a certain stopping condition is fulfilled, e.g. the desired number of regions is reached or the value of the least link cost exceeds a predefined threshold.

The merging order is exclusively controlled by the function used to compute the merging cost. Let r_i and r_j be two neighboring regions. The merging cost is based solely on a simple colour homogeneity criterion defined as [18,19]:

$$C_{\text{orig}}(i,j) = \|\mathbf{c}_i - \mathbf{c}_j\|_2^2 \cdot \frac{1}{a_{\text{img}}} \cdot \frac{a_i a_j}{a_i + a_j} \qquad (1)$$

where $\mathbf{c}_i$ and $\mathbf{c}_j$ are the mean colours of r_i and r_j respectively, a_i and a_j denote region sizes, and $\| \cdot \|_2$ denotes the $\mathcal{L}_2$ norm. a_{img} is the size of the entire image. Such normalization by a_{img} does not affect the merging order allowing to use the value of C_{orig} as a stopping criterion with collections containing images of different sizes. Alternative merging criteria based on colour can be found in [2,15,11].

2.2 Enhanced RSST

In the region merging approach proposed in [7] additional evidence for merging is provided by the syntactic visual features advocated by Ferran and Casas [11], representing geometric properties of regions and their spatial configuration, e.g. homogeneity, compactness, regularity, inclusion or symmetry. It has been shown that these features can be used in bottom-up segmentation approaches as a way of partitioning images into more meaningful entities without assuming any application dependent semantic models.

The segmentation process is divided into two stages. The initial partition is obtained by the RSST algorithm with the original region homogeneity criterion implemented as in [3] since it is capable of producing good results when regions are uniform and small. This stage ensures the low computational cost of the overall algorithm and also avoids analysis of geometrical properties of small regions with meaningless shape. This initial stage is forced to stop when a predefined number of regions is reached (100 for all experiments presented in this paper). Then, in the second stage, homogeneity criteria are re-defined based on colour and syntactic visual features and the merging process continues until a certain stopping criterion is fulfilled. Regions' colour is represented using a fine and compact representation motivated by the *Adaptive Distribution of Colour Shades* (ADCS) [20] whereby each region contains a list of pairs of colour/population (where population refers to the ratio between the number of pixels with this colour and the total size of the region) that represents its complex colour variations more precisely than the mean value. Two region geometric properties, adjacency and changes in global shape complexity, are included as syntactic features. The new merging order is based on evidence provided by different features (colour and geometric properties) fused using an integration framework based on *Dempster-Shafer* (DS) theory [21] which takes into account the reliability of different sources of information as well as the fact that certain measurements may not be precise (doubtful) or even "unknown" in some cases. Full details can be found in [7].

3 Existing Stopping Criteria

This section describes and briefly discusses three simple but commonly used stopping criteria that are used to evaluate our approach. They are: (i) required number of regions, (ii) maximum value of the least link cost, and (iii) minimum value of *Peak Signal to Noise Ratio* (PSRN) between the original and segmented image (reconstructed using mean region colour) [15].

3.1 Number of Regions

In this very simple method, the merging process continues until a desired number of regions is reached independent of the image content. Clearly, application of this criterion is limited to cases where the number of required regions/object is known.

3.2   Merging Cost

In this case, the merging process is stopped when the merging cost exceeds a predefined threshold. The threshold is typically chosen in an ad-hoc manner to suit a particular application. Although the method is intuitive, one should note that for heterogeneous collections it is often impossible to choose a single threshold resulting in segmentation corresponding to the most salient regions present in the scene. Looking at Fig. 1, for example, a threshold high enough to segment the person's torso in Fig. 1(c) as one region would result in the "jet" being merged into the backgroundd in Fig. 1(a).

3.3   PSNR Value

In this case, a single partitioning of the image is achieved based on value of PSNR between the original and segmented image (reconstructed using mean region colour) – see for example [15]. Typically, all merges performed during the region merging process are stored in a *Binary Partition Tree* (BPT) together with values of PSNR at each merging iteration. A single partition is then obtained from the BPT by deactivating nodes following the merging sequence until the value of PSNR falls below a pre-defined threshold. In our work, only the luminance information is used in order to limit the computational complexity. Formally PSNR for an $M \times N$ image is computed as:

$$PSNR = 20 \log_{10} \frac{MAX_L}{\sqrt{\frac{1}{MN} \sum_{i=0}^{M-1} \sum_{j=0}^{N-1} \left(L(i,j) - K(i,j) \right)^2}} \tag{2}$$

where L is the original image and K is the segmented image reconstructed using mean region intensities and MAX_L is the maximum luminance value. Again, it may be impossible to choose a single threshold resulting in segmentation corresponding to the most salient regions present in the scene for heterogeneous collections of images.

4   Proposed Stopping Criterion

This section discusses a new stopping criteria to obtaining a single partition that reflects meaningful image content. The approach is not based on a pre-defined threshold, but rather evaluates the 'goodness' of segmentation at a single iteration with a final decision made based on the evolution of the merging cost accumulated during the overall merging process. As in the PSNR-based approach, the selected partition is obtained by deactivating nodes from the BPT built during the merging process.

First, let us define an accumulated merging cost measure C_{cum} which measures the total cost of all mergings performed to produce t regions as:

$$C_{\mathrm{cum}}(t) = \begin{cases} \sum_{n=t}^{N_I-1} C_{\mathrm{mrg}}(n) & \text{if } 1 \leq t < N_I \\ 0.0 & \text{otherwise} \end{cases} \tag{3}$$

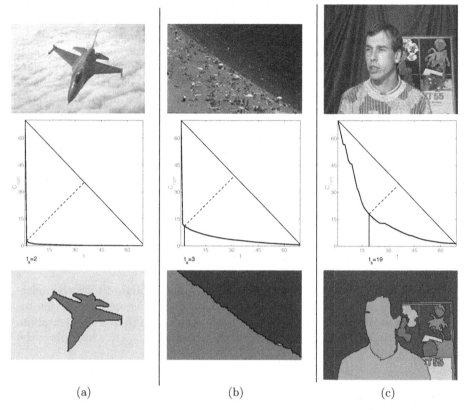

Fig. 1. Stopping criterion based on accumulated merging cost (C_{cum}) for $T_{\mathrm{cum}} = 70$. The merging order is computed using the extended colour representation together with syntactic visual features. Values of C_{cum} are re-scaled to the range $[0, T_{\mathrm{cum}}]$ for visualization purposes.

where $C_{\mathrm{mrg}}(n)$ denotes cost of merging of a single pair of regions reducing the number of regions from $n + 1$ to n computed using a given merging criterion. N_{I} denotes the number of regions in the initial partition, i.e. produced in the first merging stage. In the case of the enhanced RSST, only the costs of merging performed during the second stage contribute to the value of C_{cum} (i.e. $N_{\mathrm{I}} = 100$). Also, N_{I} was set to 100 in the case of the original RSST. It should be stressed that the value of N_{I} has a minimal effect on the final result since C_{cum} changes very slowly for the initial merges.

Figure 1 shows measure C_{cum} plotted for each iteration t of the merging process for three images, each presenting a different segmentation challenge. The basic idea behind the stopping criterion is to find the number of regions t_{s} which partitions the curve $C_{\mathrm{cum}}(t)$ into two segments in such a way so that with decreasing values of t, values of $C_{\mathrm{cum}}(t)$ within segment $[1, t_{\mathrm{s}}]$ increase significantly faster than for the segment $[t_{\mathrm{s}}, N_{\mathrm{I}}]$.

In the proposed approach t_s is found using the method proposed in [22] for bi-level thresholding designed to cope with uni-modal distributions (histogram based thresholding). The algorithm is based on the assumption that the main peak of the uni-modal distribution has a detectable corner at its base which corresponds to a suitable threshold point. The approach has been found suitable for various problems requiring thresholding such as edge detection, image difference, optic flow, texture difference images, polygonal approximations of curves and even parameter selection for the split and merge image segmentation algorithm itself [23].

Here, the approach is used to determine the stopping criterion t_s based on the accumulated merging cost measure C_{cum}. Let us assume a hypothetical reference accumulated merging cost measure C_{ref} having a form of a line passing through points $(1, C_{cum}(1))$ and $(T_{cum}, C_{cum}(T_{cum}))$, where T_{cum} is a parameter whose role is explained later. The threshold point t_s is selected to maximize the perpendicular distance between the reference line and the point $(t_s, C_{cum}(t_s))$. It can be shown that this step is equivalent to an application of a single step of the standard recursive subdivision method for determining the polygonal approximation of a curve [24].

Parameter T_{cum} can be used to bias the stopping criterion towards under- or over-segmentation, i.e. smaller values of T_{cum} bias the stopping criterion towards under-segmentation while larger values tend to lead to over-segmentation. However, it should be stressed that it is the shape of $C_{cum}(t)$ which plays the dominant role in selecting t_s.

5   Experiments

5.1   Image Collection

The collection used consists of 100 images from the Corel gallery and 20 images from various sources such as keyframes from well known MPEG-4 test sequences and a private collection of photos. Ground-truth segmentation masks of semantic regions in the scene were created manually using a tool developed specifically for this task [7]. Using this tool creation of a new reference region involves the annotator choosing a label for the new region, drawing its boundary on the original image and then filling its interior. To ensure high accuracy of the masks, the area around the cursor is automatically zoomed. Furthermore, the mask can be edited in a *Boundary Mode* in which only region boundaries superimposed on the original image are displayed allowing accurate localization of the borders on the original image and fine corrections. The typical time required to manually partition an image was between 5-10 minutes, depending on user drawing skills and complexity of the scene. Although this time seems acceptable, in fact manual segmentation of several images proved extremely tiresome.

Taking into account the relatively small number of images available the proposed criteria are evaluated by two-fold cross validation. The dataset is divided into two subsets, each containing 60 images, i.e. 50 images from the Corel dataset and 10 from the other sources. Each time one of them is used for parameter tuning, the other subset is used as the test set. The final result is computed as the average error from the two trials. All images showing selected segmentation results are generated during the cross-validation.

5.2   Evaluation Criterion

The evaluation measure proposed in [25] assesses the quality of partitions in terms of spatial accuracy error. The method allows evaluation of segmentation in cases where both the evaluated segmentation and the ground-truth mask might contain several regions. Special consideration is devoted not only to the accuracy of boundary localization but also to over- and under-segmentation. To tackle such issues, the evaluation starts by establishing exclusive correspondence between regions in the reference and evaluated masks. Then, three different types of errors are taken into account: (i) accuracy errors for the associated pairs of regions from both reference and evaluated masks, (ii) errors due to under-segmentation computed based on non-associated regions from the reference mask, and finally (iii) errors due to over-segmentation computed from non-associated regions from the evaluated mask. Although the method is quite simple, a set of convincing evaluation results was provided in [25] indicating that the method correlates well with subjective evaluation.

In this work, a simplified fast version of the above evaluation method, implemented by the author, is used [7]. Let $\mathcal{S} = s_1, s_2, \ldots, s_K$ be the segmentation mask to be evaluated comprising K regions s_k, and $\mathcal{R} = r_1, r_2, \ldots, r_Q$ be the reference mask comprising Q reference regions. Let also a region be defined as a set of pixels ($|r|$ denotes number of pixels in region r). In our implementation the spatial error is computed based directly on the number of misclassified pixels. Prior to calculation of segmentation errors, the correspondence between reference and evaluated regions is established. The pairing strategy involves exclusive associations of regions s_k from the evaluated mask with regions r_q from the reference mask on the basis of region overlapping. In particular, maximization of $s_k \bigcap r_q$ was suggested in [25]. In the author's implementation an additional pairing requirement $(s_k \bigcap r_q)/(s_k \bigcup r_q) > 0.5$ is imposed. It should be noted that due to over and under-segmentation of certain parts of the image not all regions from the evaluated and the reference mask are associated. Once correspondences are established, the spatial errors can be computed. Let $\mathcal{A} = (s_k, r_q)$ denote the set of region pairs identified using the pairing procedure, and let $\mathcal{N}_S$, $\mathcal{N}_R$ denote the sets of non-associated regions of masks $\mathcal{S}$ and $\mathcal{R}$ respectively. The segmentation error is computed as the sum of accuracy errors for all associated pairs and all non-associated regions from both evaluated and reference masks:

$$E = \frac{1}{a_{\text{img}}} \left[\sum_{\forall (s_k, r_q) \in \mathcal{A}} \left[|s_k - s_k \cap r_q| + |r_q - s_k \cap r_q| \right] + \right.$$

$$\left. + \alpha \sum_{\forall s_k \in \mathcal{N}_S} |s_k| + \sum_{\forall r_q \in \mathcal{N}_R} |r_q| \right] \tag{4}$$

Table 1. Results of cross-validation

STOPPING CRITERION	Original colour homogeneity criterion [3]	Homogeneity based on ADCS and the Syntactic Visual Features [7]
"Manual"	0.99	0.50
Number of regions	0.93	0.72
Merging cost	0.77	0.69
PSNR	0.84	0.82
Accum. merging cost	0.81	**0.63**

where, α is a scaling factor allowing different penalty for over-segmentation and, following the suggestion in [25], is set to 2.0 in all experiments. To enable comparison of results for images with different sizes, the segmentation accuracy measure is normalized by the size of the image (a_{img}).

5.3   Results

All parameters (thresholds) were found during the training phase by an optimization process. Table 1 shows the average spatial accuracy error for all evaluated stopping criteria. Additionally, the first column contains results of a "manual" criterion where the merging process is stopped based on the known number of regions in the ground-truth.

The first observation to be made is that utilization of the merging cost in the enhanced RSST always results in a lower value of the average spatial error, irrespective of the stopping criterion used. Secondly, stopping criteria based on the merging cost usually perform better that the PSNR stopping criterion. Finally, the proposed stopping criterion significantly outperforms all other criteria when used with the merging cost based on ADCS and the syntactic visual features. It should be noted that this combination of merging cost together with the new stopping criterion leads to significantly lower average spatial segmentation error than the original RSST approach even with the "manual" stopping criterion based on the number of regions in the ground-truth mask.

Figure 2 shows selected segmentation results obtained by the original RSST algorithm with the PSNR based stopping criterion and segmentations produced by the merging criterion integrating the extended colour representation combined with the syntactic visual features, together with the new stopping criterion. The first five rows show examples where the new approach improves the results compared to the original RSST and the last five rows show examples where the proposed approach obtained higher spatial segmentation error than the original RSST approach. However, it should be stressed that the latter cases are very rare (31 from 120 images) and in fact even in such cases the partitions appear somewhat intuitive.

For comparison, the last column from Fig. 2 shows the results of the Blobworld segmentation algorithm[2]. The Blobworld algorithm has been extensively tested and used as a segmentation processing stage in the literature [26].

---

[2] Source code obtained from http://elib.cs.berkeley.edu/src/blobworld/

Fig. 2. Selected segmentation results. Column: (a) Original image; (b) Reference mask created by manual annotation; (c) Original RSST [3] with PSNR Stoping Criterion; **(d) RSST extended with syntactic features [7] with the proposed stopping criterion**; (e) Blobworld.

6   Conclusions

This paper presented an automatic stopping criterion for image segmentation algorithms based on iterative region merging. The approach is particularly applicable to segmentation approaches that integrate syntactic visual features. Whilst such approaches potentially offer the possibility of segmentation results that reflect real-world structure, they still need an appropriate stopping criterion to determine when such a result has been reached. Unlike other approaches, the proposed approach is not based on a single pre-defined threshold, but rather evaluates the 'goodness' of a segmentation result at a single iteration and makes a final decision based on the merging cost accumulated during the overall process. This makes the overall segmentation approach broadly applicable and experimental results show strong performance against a manually generated ground truth for a heterogeneous image collection.

Although, in many cases the proposed stopping criterion does not produce a perfect segmentation, i.e. ideally aligned with the manual segmentation of semantic regions, it successfully identifies the most visually salient regions in almost all evaluated images. This presents a tremendous opportunity for utilizing such salient regions in CBIR systems. It should be stressed that the method performs well on images presenting very different challenges with fixed value of the parameter T_{cum} and unlike other stopping criteria, e.g. based on the value of merging cost or PSNR, very erroneous segmentations are extremely rare.

Acknowledgments

The research leading to this paper was supported by the European Commission under contract FP6- 027026 (K-Space) and by Science Foundation Ireland under grant 03/IN.3/I361.

References

1. Cheng, H.D., Jiang, X.H., Sun, Y.: Color image segmentation: Advances & prospects. Pattern Recognition 34(12), 2259–2281 (2001)
2. Morris, O.J., Lee, M.J., Constantinides, A.G.: Graph theory for image analysis: an approach based on the shortest spanning tree. IEE Proceedings 133, 146–152 (1986)
3. Alatan, A.A., Onural, L., Wollborn, M., Mech, R., Tuncel, E., Sikora, T.: Image sequence analysis for emerging interactive multimedia services - the European COST 211 Framework. IEEE Trans. CSVT 8(7), 802–813 (1998)
4. Salembier, P., Garrido, L.: Binary partition tree as an efficient representation for image processing, segmentation, and information retrieval. IEEE Trans. on Image Processing 9(4), 561–576 (2000)
5. Shi, J., Malik, J.: Normalized cuts and image segmentation. IEEE Trans. Pattern Anal. and Machine Intell. 22(8), 888–905 (2000)
6. Nock, R., Nielsen, F.: Statistical region merging. IEEE Trans. Pattern Anal. and Machine Intell. 26(11), 1452–1458 (2004)
7. Adamek, T.: Using Contour Information and Segmentation for Object Registration, Modeling and Retrieval, Ph.D. thesis, School of Electronic Engineering, Dublin City University (June 2006)

8. Salembier, P., Garrido, L.: Binary partition tree as an efficient representation for filtering, segmentation, and information retrieval. In: ICIP 1998. Proc. IEEE Int'l Conf. on Image Processing, Chicago (IL), USA (October 1998)
9. Kwok, S.H., Constantinides, A.G., Siu, W.-C.: An efficient recursive shortest spanning tree algorithm using linking properties. IEEE Trans. Circuits Syst. Video Technol. 14(6), 852–863 (2004)
10. Barnard, K., Duygulu, P., Guru, R., Gabbur, P., Forsyth, D.: The effects of segmentation and feature choice in a translation model of object recognition. In: CVPR 2003. Proc. IEEE Conf. On Computer Vision and Pattern Recognition (2003)
11. Ferran Bennstrom, C., Casas, J.R.: Binary-partition-tree creation using a quasi-inclusion criterion. In: IV 2004. Proc. 8th Int'l Conf. on Information Visualization, London, UK (2004)
12. Vasseur, P., Pégard, C., Mouaddib, E.M., Delahoche, L.: Perceptual organization approach based on dempster-shafer theory. Pattern Recognition 32(8), 1449–1462 (1999)
13. Zlatoff, N., Tellez, B., Baskurt, A.: Region-based perceptual grouping: a cooperative approach based on dempster-shafer theory. In: Proc. of the SPIE, vol. 6064, pp. 244–254 (2006)
14. Adamek, T., O'Connor, N.E., Murphy, N.: Region-based segmentation of images using syntactic visual features. In: WIAMIS 2005. Proc. 6th Int'l Workshop on Image Analysis for Multimedia Interactive Services, Montreux, Switzerland (April 2005)
15. Salerno, O., Pardas, M., Vilaplana, V., Marqués, F.: Object recognition based on binary partition trees. In: ICIP 2004. Proc. Int'l Conf. on Image Processing, vol. 2, pp. 929–932 (October 2004)
16. Adamek, T., O'Connor, N.: Using dempster-shafer theory to fuse multiple information sources in region-based segmentation. In: ICIP 2007. Proc. of the 14th IEEE Int'l Conf. on Image Processing, San Antonio, Texas, USA (2007)
17. Vilaplana, V., Marques, F.: Region-based hierarchical representation for object detection. In: CBMI 2007. Proc. 5th Int'l Workshop on Content-Based Multimedia Indexing, pp. 157–164 (2007)
18. Ward, J.H.: Hierarchical grouping to optimize an objective function. American Stat. Assoc. 58, 236–245 (1963)
19. Cooray, S., O'Connor, N.E., Marlow, S., Murphy, N., Curran, T.: Semi-automatic video object segmentation using recursive shortest spanning tree and binary partition tree. In: WIAMIS 2001. Proc. 3rd Int'l Workshop on Image Analysis for Multimedia Interactive Services, Tampere, Finland (2001)
20. Fauqueur, J., Boujemaa, N.: Region-based image retrieval: Fast coarse segmentation and fine color description. Journal of Visual Languages and Computing, special issue on Visual Information Systems 15, 69–95 (2004)
21. Smets, P., Mamdami, E.H., Dubois, D., Prade, H.: Non-Standard Logics for Automated Reasoning. Academic Press, Harcourt Brace Jovanovich Publisher (1988), ISBN 0126495203
22. Rosin, P.L.: Unimodal thresholding. Pattern Recognition 34(11), 2083–2096 (2001)
23. Horowitz, S., Pavlidis, T.: Picture segmentation by a tree traversal algorithm. J. Assoc. Compt. Math. 23(2), 368–388 (1976)
24. Ramer, U.: An iterative procedure for the polygonal approximation of plane curves. Computer, Graphics and Image Processing 1, 244–256 (1972)
25. Mezaris, V., Kompatsiaris, I., Strintzis, M.G.: Still image objective segmentation evaluation using ground truth. In: Proc. 5th COST 276 Workshop, Berlin, pp. 9–14 (2003)
26. Carson, C., Belongie, S., Greenspan, H., Malik, J.: Blobworld: Color- and texture-based image segmentation using EM and its application to image querying and classification. IEEE Trans. Pattern Anal. and Machine Intell. 24(8), 1026–1037 (2002)

Hierarchical Long-Term Learning for Automatic Image Annotation

Donn Morrison, Stéphane Marchand-Maillet, and Eric Bruno

Centre Universitaire d'Informatique
University of Geneva, Geneva, Switzerland
{donn.morrison,marchand,eric.bruno}@cui.unige.ch
http://viper.unige.ch/

Abstract. This paper introduces a hierarchical process for propagating image annotations throughout a partially labelled database. Long-term learning, where users' query and browsing patterns are retained over multiple sessions, is used to guide the propagation of keywords onto image regions based on low-level feature distances. We demonstrate how singular value decomposition (SVD), normally used with latent semantic analysis (LSA), can be used to reconstruct a noisy image-session matrix and associate images with query concepts. These associations facilitate hierarchical filtering where image regions are matched based on shared parent concepts. A simple distance-based ranking algorithm is then used to determine keywords associated with regions.

1 Introduction

The semantic gap, recognised as the major hurdle in image retrieval, can be narrowed by tracking patterns of user interaction during query [1,2,3,4,5]. Previous research tends to focus on fully automatic methods using low-level features such as colour, texture, and shape [6,7,8,9,10] or structured augmentation using ontologies [11]. As with patterns in web traffic analysis, users of image retrieval systems exhibit useful information via their browsing and searching habits [12,5].

The inherent limitations of using only low-level features in image retrieval become apparent after a brief appraisal of the available literature. Retrieval systems cannot reliably glean high-level semantics from low-level features due to a lack of image understanding in computer vision. There are many facets to semantic meaning and images can be described in many ways [13]. Subjectivity and intent in photography as well as in retrieval play a critical role. Therefore, we feel it is necessary to place more focus on user interaction in image retrieval and annotation. To ignore this information can be likened to marketing goods or services without some knowledge of consumer purchase patterns. In this paper, our goal is to semantically describe the images users are searching for, thus facilitating subsequent queries. This involves the propagation of keywords across partially annotated databases using a mixture of long-term learning and low-level image features.

In a previous paper, we demonstrated the use of singular value decomposition for the reconstruction of missing values in a session-image matrix, where each session represents a query concept [14]. The advantage of this method was that it relied only

B. Falcidieno et al. (Eds.): SAMT 2007, LNCS 4816, pp. 28–40, 2007.

on long-term learning via relevance feedback on a partially annotated image database. However, a fundamental limit was found during the annotation process where new annotations were selected based on the most popular concept keywords. The result was a quantised annotation where each image belonging to a concept was annotated with similar keywords. In this paper, we improve this by dividing each image into regions which can be represented by specific keywords. The two feature types have very different meaning on the semantic level, and therefore must be hierarchically fused.

The article is structured as follows: Section 2 gives a lengthy review of related work, ranging from fully automatic annotation methods to semi-supervised methods that utilise long-term learning for annotation propagation. We have omitted works dealing with annotation by ontologies, except some studies which use WordNet. Section 3 introduces our method for automatic annotation using regions of low-level features and long-term learning. Next, Section 4 details the image database we use and the experiments followed. Section 5 reviews the experimental results and Section 6 closes with a conclusion and some proposed improvements.

2   Related Work

Automatic image annotation can be approached with a variety of machine learning methods, from supervised classification to probabilistic to clustering. It is common to borrow latent and generative models from text retrieval such as latent semantic analysis (LSA) [15] and it's probabilistic cousin, PLSA [16]. These two latent-space models are compared in [17]. The authors pose the question of whether annotation by propagation is better than annotation by inference. LSA is shown to outperform PLSA. However, they explain that some of the reasons for this may be that LSA is better at annotating images from uniformly annotated databases.

In a later paper, the authors introduce an improved probabilistic latent model, called *PLSA-words*, which models a set of documents using dual cooperative PLSA models. The intention is to increase the relevance of the captions in the latent space. The process is divided into two stages: parameter learning, where the latent models are trained, and annotation inference, where annotations are projected onto unseen images using the generated models. In the first stage, the first PLSA model is trained on a set of captions and a new latent model is trained on the visual features of the corresponding images. In the second stage, the standard PLSA technique projects a latent variable onto the new image, and annotations of an aspect are assigned if the probability exceeds a threshold [18].

Extensions of PLSA have been described, for example *latent Dirichlet allocation (LDA)*, introduced in [19], which models documents as probabilistic mixtures of topics which are comprised of sets of words [20]. This model was applied to image annotation in a slightly modified version called *correspondence latent Dirichlet allocation (Corr-LDA)* [21]. In this study, the authors compare the algorithm with two standard hierarchical probabilistic mixture models. Three tasks are identified: modelling the joint distribution of the image and it's caption, determining the conditional distribution of words in an image, and determining the conditional distribution of words in an image region. The Corr-LDA model first generates region descriptions from the image using an LDA model. Then corresponding caption words and image regions are selected, based on how the image region was selected.

In addition to low-level image features, a semantic modality can be introduced to harness the knowledge generated by users or groups of users interacting with an image database, whether it be browsing or performing longer queries (including but not limited to relevance feedback). By observing these interactions and the associations made between relevant and non-relevant images during a query, semantic themes can start to become apparent. These themes need not be named entities such as words describing objects or concepts, but can simply be relationships between images indicating some level of semantic similarity.

This type of learning is dubbed *inter-query learning* due to the feature space spanning multiple (or all) query sessions. The converse is the traditional *intra-query learning*: the utilisation of relevance feedback examples during the current query only (after the session has ended the weights are discarded). Inter-query learning takes an approach similar to collaborative filtering; interaction (in the form of queries with relevance feedback) is required to increase density in the feature space. It is in this way that a collection can be incrementally annotated. The more interaction and querying, the more accurate the annotations become.

The Viper group produced one of the first studies which looked at inter-query learning [1]. The authors analysed the logs of queries using the *GIFT (GNU Image Finding Tool)* demonstration system over a long period of time and used this information to update the *tf-idf* feature weightings. Images were paired based on two rules: images sharing similar features and also marked relevant have a high weight while images sharing similar features but marked both relevant and irrelevant should have a low weight (indicating a semantic disagreement). Two factors were introduced to manage the relevance feedback information. The first being a measure of the difference between the positively and negatively rated marks for each feature and the second re-weighting the positively and negatively marked features differently such that the ratio is scaled non-linearly.

Later, in [22], the authors focus more formally on annotation. A general framework is described which annotates the images in a collection using relevance feedback instances. As a user browses an image database using a CBIR system, providing relevance feedback as the query progresses, the system automatically annotates images using the relationships described by the user.

Taking a direction toward the fusion of the two modalities, [3] combine inter-query learning with traditional low-level image features to build semantic similarities between images for use in later retrieval sessions. The similarity model between the request and target images are refined during a standard relevance feedback process for the current session. This refinement and fusion is facilitated by a *barycenter*. The paper also discusses the problems with asymmetrical learning, where the irrelevant images are marked irrelevant by the user for a variety of reasons, whereas relevant images are marked relevant only because they relate semantically to the query. Therefore, the authors reduce the relevance of irrelevant images during the fusion of feedback stages. Similarly, in [23], a statistical correlation model is built to create semantic relationships between images based on the co-occurrence frequency that images are rated relevant to a query. These relationships are fused with low-level features (256 colour histogram, colour moments, 64 colour coherence, Tamura coarseness and directionality) to propagate the annotations onto unseen images.

In [2], inter-query learning is used to improve the accuracy of a retrieval system and latent semantic indexing (LSI) is used in a way such that the interactions are the documents and the images correspond to the term vocabulary of the system. The authors perform a validation experiment on image databases consisting of both texture and segmentation data from the MIT and UCI repositories. Random queries were created and two sessions of relevance feedback were conducted to generate the historical information to be processed by LSI. From experiments on different levels of data, they conclude that LSI is robust to a lack of data quality but is highly dependent on the sparsity of interaction data.

This method of exchanging RF instances and images for the documents and term vocabulary was also used in a later study where the authors use long-term learning in the PicSOM retrieval system [4]. PicSOM is based on multiple parallel tree-structured *self-organising maps (SOMs)* and uses MPEG7 content descriptors for features. The authors claim that by the use of SOMs the system automatically picks the most relevant features. They note that the relevance feedback information provided by the users is similar to hidden annotations. Using Corel images with a ground truth set of 6 classes, MPEG7 features scalable colour, dominant colour, colour structure, colour layout, edge histogram, homogeneous texture, and region shape, the authors reported a significant increase in performance.

In [24], a system is proposed which shifts the document retrieval paradigm from content-based features to document similarity based on user interaction with a retrieval system. The system is built using principles from collaborative filtering (CF) which completely replace the traditional content-based technique. CF data was obtained by having users group similar images in a test environment. The CF data collected comprised 5010 similarity records 4010 of which were used as training data, and the remaining 1000 as testing data. The result of the classification experiment showed an increase in performance over a feature vector based on histograms. They concluded by stating that there exists "good inter-subject transferability of interpretation."

In [5] long term user interaction with a relevance feedback system is used to make better semantic judgements on unlabelled images for the purpose of image annotation. Relationships between images which are created during relevance feedback can denote similar or dissimilar concepts. The authors also try to improve the learning of semantic features by "a moving of the feature vectors" around a group of concept points, without specifically computing the concept points. The idea is to cluster the vectors around the concept centres.

3 Proposed Annotation Model

The following proposed hierarchical annotation model works by selecting a general (parent) concept based on relevance feedback over past query sessions. This concept comprises a subset of images from the database, each of which may have associated keywords, depending on the amount of initial annotation. With the concept selected for a particular image, each region is matched with similar regions in the concept space

Table 1. Concept-category relationships

Concept	Category
landscape	beach
	sunrise/sunset
sky	beach
	cloud
	sunrise/sunset
animals	bird
	insect
	leopard
	lizard
plants	flower
	mushroom
man-made	architecture

based on low-level image features. A ranked list of the closest matches is used to annotate each region in the unannotated image, hence a propagation of annotation.

In this paper, similar to what has been done previously [5], we created a set of artificial queries based on concepts each of which comprise semantically similar image classes (i.e., the concept "animals" contains image categories "birds," "insects," "leopards," and "lizards"). The query data is used to compose an image-session matrix, in which the rows contain the images in the database and the columns represent the relevance feedback values for each query session [2, 4]. The cells of the session columns can have the following values: -1, meaning the image is irrelevant to the query; 0, where no judgement is given; or +1, where the image is considered relevant to the query by the user.

We define a session to be a query where a typical user has performed a search using relevance feedback to locate an image belonging to a particular concept (in our study these concepts are manually defined in Table 1) based on the image database. This knowledge can be used in a hierarchical manner to filter available keywords for propagation. Figure 1 shows a flow diagram for the annotation of an unseen image i.

Consider an unannotated image, i_u^C, with regions R_i, which as been found through long-term learning to belong to a concept class C. Then, the subset of annotated images within C, denoted as I_a^C, are used solely for the nearest matches on low-level features. Next, for each of the n regions r_j belonging to the image i_u^C, the top k matching histograms from the relevant concept are ranked and the most common keyword is propagated to that region. This approach assumes images share similar concepts with respect to the regions. For example, an image of an insect, in the context of our collection, has a high likelihood of also being accompanied by regions depicting leaves or plants; an image of a sunset has a high likelihood of having regions depicting water or cloud.

The ranking algorithm is simply a ranked list of k Euclidean distances between the unannotated image region and all other regions sharing the same parent concept. The keyword with the highest vote in this ranked list is used as the new annotation. As we will see in the following section, it is possible to have no associated concept with some

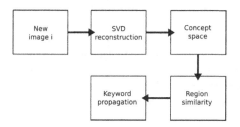

Fig. 1. Flow diagram for hierarchical annotation of an unannotated image i. The image is first added to the image-session matrix. SVD is used to decompose, filter, and reconstruct the matrix. Concepts then become apparent and are used to filter the keyword space. The nearest matching regions are ranked and the associated keywords are propagated onto the new unannotated image.

images due to the sparsity of the image-session matrix. In this case, we simply fall back to using the low-level feature distances to propagate region annotations. The only drawback with this is that the probability of matching irrelevant regions is increased.

The long-term learning works by storing all previous queries in a matrix A. After each query involving relevance feedback, this matrix is updated and SVD is used to associate images with concepts. In this way the annotations are never completely fixed, but can evolve with use of the retrieval system. An example artificial matrix is show in Figure 3 (a). It is highly redundant because of the large number of RF sessions generated and the low number of concepts.

In this experiment, only the positive examples ($A(i,j) == +1$) are used in order to simplify the propagation stage (we will ignore irrelevant concepts for the moment). Next, to simulate missing relevance feedback data, the values of A are randomly dropped (set to 0) to form a new noisy matrix, A_n. Singular value decomposition (SVD) is applied to this matrix to yield:

$$A_n = U \Sigma V^T. \tag{1}$$

The diagonal matrix Σ contains the singular values. We retain only $k = 5$ concepts as Σ' to filter out unimportant concepts and reconstruct A as A_r.

$$A_r = U \Sigma' V^T \tag{2}$$

With A_n reconstructed as A_r, we now apply a thresholding measure to allow diffusion of relevance feedback examples into cells with missing data. As a result of SVD, cells previously zero will now be non-zero. These values are normalised into the same space as A_n and then empirically thresholded at 0.7, giving:

$$A_r(i,j) = \begin{cases} 1 & \text{where} \quad A_r(i,j) > 0.7 \\ 0 & \text{otherwise} \end{cases} \tag{3}$$

The result is the reconstructed matrix with values that should minimise the difference from A_n:

$$D_{nr} = \sum_{i,j} |A_n(i,j) - A_r(i,j)| \tag{4}$$

We intend to annotate the unlabelled images in the database to allow for keyword-based queries. Image similarities are specified by the user by way of relevance feedback. This alone could be sufficient for labelling, but normally the feature space is very sparse, and some diffusion is needed to propagate image labels throughout the collection.

During the matching of low-level features, we use a 64 bin histogram for each of the RGB channels segmented by normalised cuts. The Euclidean distance metric was used to find the closest matching regions using these histograms:

$$D = \sqrt{\sum_{i=1}^{n} (x_i - y_i)^2}. \tag{5}$$

Other measures such as histogram intersection could easily be used here, but for this initial experiment the Euclidean measure is sufficient.

Due to the fact that the regions are dynamically sized by normalised graph cuts, each region is normalised with respect to its area. Normalised graph cuts requires the number of regions to be specified as a parameter. In our experiments we set this parameter to 4. This inflexibility can cause problems during the segmentation process because if there exist three very obvious regions, a fourth will be created by dividing one of the three. This could be alleviated with some preprocessing of the image (or manual specification) to determine an optimal number of regions. To simplify our approach we left the number of regions static.

4   Experiments

For the purposes of an initial investigation, a small, uniformly distributed subset of images was taken from the Corel collection based on 10 predefined semantic categories

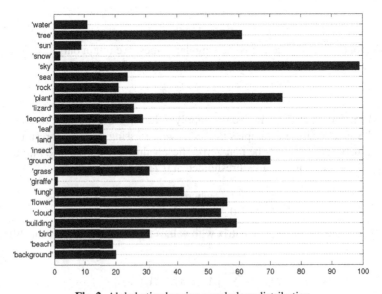

Fig. 2. Alphabetised region vocabulary distribution

(recall Table 1). Twenty images from each category were taken at random so as to reduce a bias towards low-level similarity. Each image was segmented into four regions using normalised graph cuts [25]. Next, each region was manually annotated with a keyword which best described the majority of the region. For example, if a region exists containing a small bird on a large sky, the word 'sky' would be used as the annotation. In total, 200 images were collected. The final vocabulary comprised 23 words. Figure 2 shows the distribution of words in the vocabulary.

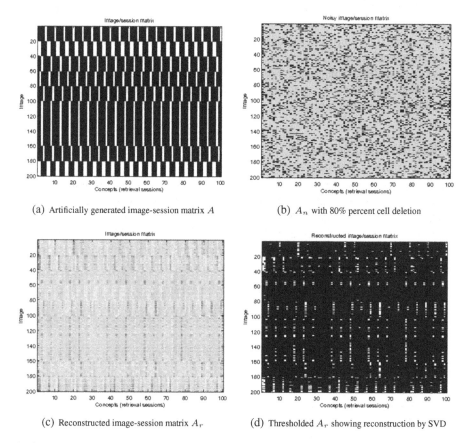

(a) Artificially generated image-session matrix A

(b) A_n with 80% percent cell deletion

(c) Reconstructed image-session matrix A_r

(d) Thresholded A_r showing reconstruction by SVD

Fig. 3. The various stages of the image-session matrix during singular value decomposition: (a) shows the original matrix, (b) shows the matrix A_n with entries removed simulating sparsity, (c) shows the matrix A_r reconstructed with $K = 5$ concepts, and (d) shows A_r after thresholding

A pool of 100 artificial relevance feedback sessions was created by setting all images under a concept as relevant to that query. In essence, the matrix created is a ground truth matrix where all concepts are related to the categories through artificial sessions. This data simulates query sessions where users would have a concept image in mind (for example, images depicting animals), and would construct the query by selecting a number of positive and negative examples.

Figure 3 (b) shows A_n, which results from the random cell deletion on A at 80%. In our experiments, the percentage of cell deletion was varied to see how SVD handles incremental missing values. Finally, Figures 3 (c) and (d) show the reconstructed image/session matrix A_r, before and after thresholding, respectively. It can be seen that one category of images (Figure 3 (d), images 61-80) suffers more corruption after reconstruction than the rest, with almost no associated concepts. This is because the category in question, "cloud", belongs to only one concept ("sky"), while the other members of that concept belong to two concepts ("landscape" and "sky"). This causes the "cloud" category to have less influence, and thus, SVD tries to map the "cloud" concept onto these images.

To simulate a partially annotated database, we use hold-one-out cross validation to pick an unannotated image and use the remaining for distance matching.

As can be realised from Figure 3 (d), there will be missing values in the matrix that can cause some images to be unassociated with any particular concept. In this case, our algorithm falls back to simply matching the low-level feature regions.

5 Results and Discussion

Figure 4 shows the prediction accuracy versus k top ranked low-level feature matches. The distribution of the vocabulary plays a part here because keywords with high distribution will eventually dilute the rankings provided their histograms are relatively close to that of the unannotated region. The accuracy (%) is calculated by strictly counting the number of predicted region keywords that match the ground truth region keywords.

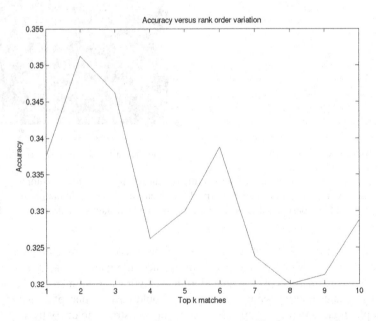

Fig. 4. Strict by-region prediction accuracy varied by k nearest regions

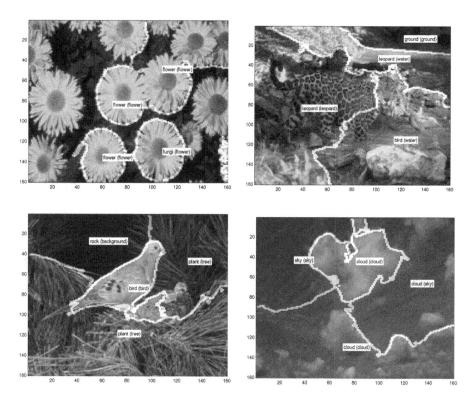

Fig. 5. Examples of well labelled regions after 80% cell deletion in the image-session matrix. Predicted labels precede ground truth labels (in parentheses).

If this restriction is relaxed so that keywords are just associated with the image, as is the case with the Corel data set, the accuracy is significantly improved.

Accuracy reaches a peak, just above 35%, when there are $k = 2$ top results, and declines with local maxima for $k > 2$. However, with $k = 2$ there is no majority vote, with a keyword being picked at random from the ranking if there are two suggestions. A more stable value is $k = 3$, where a majority can be found in more cases.

Some example results are pictured below in Figures 5 and 6, depicting favourable and less favourable results, respectively. It was observed that images with simple histograms were annotated more accurately (see the flower and sky images in Figure 5). Images in the animals concept were often given wrong labels for the main subject, for example, commonly mis-annotating lizards for leopards. This is partially due to the fact that the generated regions do not always directly surround semantic objects, so the colour histograms will be diluted with other areas of the image.

Further improvement could be gained by utilising WordNet to find words in similar semantic branches. In the case of the third image (bird) in Figure 5, the labels are not actually very far from the ground truth. According to WordNet, "tree" falls under the category of "plant", which in this case is the predicted annotation.

Fig. 6. Examples of poorly labelled regions after 80% cell deletion in the image-session matrix. Predicted labels precede ground truth labels (in parentheses).

In Figure 6, we have an example of a lizard being mistaken for a leopard. In cases of animals, especially those which exhibit patterns (scales on lizards, dots on leopards), a texture-based feature could be useful for further discrimination. The distribution of annotations in the ground truth vocabulary also has an affect on the predicted annotations. In the example of the sunset, the sun has been mis-annotated as sky. Looking back at Figure 2, we can see that the word "sun" is only associated with roughly 10 regions in the database whereas "sky" is the most common word in the vocabulary. This will have a direct effect on the predictions because the difference in the distribution is so great.

In our previous study, the reliance on only relevance feedback information demonstrated the need to segment the images into regions which could more closely model specific keywords with low-level features [14]. These experiments show that image regioning provides a much finer grained approach after concept selection from relevance feedback.

Figure 6 shows examples where incorrect keywords were propagated due to the over-simplistic nature of the colour histograms used in the distance measure. Improvement could be found by adding more discriminant features such as texture and shape, although shape features would require regions to be better suited to object shape.

Due to the redundancy in the artificial data, we expect to see a large drop in performance when performing the same experiments on natural data. The natural data will normally have a much sparser image-session matrix, and thus many more images will not have category information.

6   Conclusion

From the foundation of an earlier study, this paper has demonstrated a hierarchical annotation system that combines relevance feedback and low-level colour-based features. The relevance feedback is crucial for determining the concept to which an image belongs and provides a narrowing of the secondary distance-based feature space. Because of the semantic gap, user interaction – which can be seen as a sparse approximation of image semantics, is very important for automatic image annotation. In this study, the two sets of features are complimentary. The low-level features are used to find similar image regions within the same concept space as specified by the relevance feedback information, thus allowing a much more accurate propagation.

In the longer term, we hope to add more low-level features to the distance measure, compare the distance measure with a classification approach, and use a larger image database to verify these initial findings. We also expect to begin gathering real-world data to use in place of the artificially generated relevance feedback instances.

Acknowledgements

This study was funded by the Swiss NCCR Interactive Multi-modal Information Management (IM2).

References

1. Mueller, H., Mueller, W., Squire, D.M., Marchand-Maillet, S., Pun, T.: Long-term learning from user behavior in content-based image retrieval. Technical report, Université de Genève (2000)
2. Heisterkamp, D.: Building a latent-semantic index of an image database from patterns of relevance feedback (2002)
3. Fournier, J., Cord, M.: Long-term similarity learning in content-based image retrieval (2002)
4. Koskela, M., Laaksonen, J.: Using long-term learning to improve efficiency of content-based image retrieval (2003)
5. Cord, M., Gosselin, P.H.: Image retrieval using long-term semantic learning. In: IEEE International Conference on Image Processing (2006)
6. Wang, J.Z., Li, J.: Learning-based linguistic indexing of pictures with 2-d mhmms. In: MULTIMEDIA 2002. Proceedings of the tenth ACM international conference on Multimedia, pp. 436–445. ACM Press, New York, NY, USA (2002)
7. Kosinov, S., Marchand-Maillet, S.: Multimedia autoannotation via hierarchical semantic ensembles. In: LAVS 2004. Proceedings of the Int. Workshop on Learning for Adaptable Visual Systems, Cambridge, UK (2004)

8. Kosinov, S., Marchand-Maillet, S.: Hierarchical ensemble learning for multimedia categorization and autoannotation. In: MLSP 2004. Proceedings of the 2004 IEEE Signal Processing Society Workshop, São Luís, Brazil, pp. 645–654 (2004)

9. Goh, K.S., Chang, E.Y., Li, B.: Using one-class and two-class svms for multiclass image annotation. IEEE Transactions on Knowledge and Data Engineering 17(10), 1333–1346 (2005)

10. Tang, J., Hare, J.S., Lewis, P.H.: Image auto-annotation using a statistical model with salient regions. In: ICME. Proceedings of IEEE International Conference on Multimedia & Expo, Hilton Toronto, Toronto, Ontario, Canada (2006)

11. Srikanth, M., Varner, J., Bowden, M., Moldovan, D.: Exploiting ontologies for automatic image annotation. In: SIGIR 2005. Proceedings of the 28th annual international ACM SIGIR conference on Research and development in information retrieval, pp. 552–558. ACM Press, New York, NY, USA (2005)

12. Baldi, P., Frasconi, P., Smyth, P.: Modeling the Internet and the Web: Probabilistic Methods and Algorithms. John Wiley & Sons, West Sussex, England (2003)

13. Smeulders, A.W.M., Worring, M., Santini, S., Gupta, A., Jain, R.: Content-based image retrieval at the end of the early years. IEEE Trans. Pattern Anal. Mach. Intell. 22(12), 1349–1380 (2000)

14. Morrison, D., Marchand-Maillet, S., Bruno, E.: Automatic image annotation with relevance feedback and latent semantic analysis. In: Proceedings 5th International Workshop on Adaptive Multimedia Retrieval (2007)

15. Deerwester, S.C., Dumais, S.T., Landauer, T.K., Furnas, G.W., Harshman, R.A.: Indexing by latent semantic analysis. Journal of the American Society of Information Science 41(6), 391–407 (1990)

16. Hofmann, T.: Unsupervised learning by probabilistic latent semantic analysis. IEEE Trans. on PAMI 25 (2000)

17. Monay, F., Gatica-Perez, D.: On image auto-annotation with latent space models. In: Proc. ACM Int. Conf. on Multimedia (ACM MM), Berkeley (2003)

18. Monay, F., Gatica-Perez, D.: Plsa-based image auto-annotation: constraining the latent space. In: MULTIMEDIA 2004. Proceedings of the 12th annual ACM international conference on Multimedia, pp. 348–351. ACM Press, New York, NY, USA (2004)

19. Blei, D.M., Ng, A.Y., Jordan, M.I.: Latent dirichlet allocation. Journal of Machine Learning Research 3, 993–1022 (2003)

20. Barnard, K., Duygulu, P., Forsyth, D., de Freitas, N., Blei, D., Jordan, M.: Matching words and pictures. Machine Learning Research 3, 1107–1135 (2003)

21. Blei, D.M., Jordan, M.I.: Modeling annotated data. In: SIGIR 2003. Proceedings of the 26th annual international ACM SIGIR conference on Research and development in informaion retrieval, pp. 127–134. ACM Press, New York, NY, USA (2003)

22. Wenyin, L., Dumais, S., Sun, Y., Zhang, H., Czerwinski, M., Field, B.: Semi-automatic image annotation (2001)

23. Li, M., Chen, Z., Zhang, H.: Statistical correlation analysis in image retrieval (2002)

24. Kanade, T., Uchihashi, S.: User-powered content-free approach to image retrieval. In: DLKC 2004. Proceedings of International Symposium on Digital Libraries and Knowledge Communities in Networked Information Society 2004, pp. 24–32 (2004)

25. Shi, J., Malik, J.: Normalized cuts and image segmentation. IEEE Transactions on Pattern Analysis and Machine Intelligence 22(8), 888–905 (2000)

LSA-Based Automatic Acquisition of Semantic Image Descriptions

Roberto Basili[1], Riccardo Petitti[2], and Dario Saracino[2]

[1] University of Rome "Tor Vergata",
Department of Computer Science, Systems and Production, Roma, Italy
[2] Exprivia S.p.A, Via Cristoforo Colombo 456, 00145, Roma, Italy

Abstract. Web multimedia documents are characterized by visual and linguistic information expressed by structured pages of images and texts. The suitable combinations able to generalize semantic aspects of the overall multimedia information clearly depend on applications. In this paper, an unsupervised image classification technique combining features from different media levels is proposed. In particular linguistic descriptions derived through Information Extraction from Web pages are here integrated with visual features by means of Latent Semantic Analysis. Although the higher expressivity increases the complexity of the learning process, the dimensionality reduction implied by LSA makes it largely applicable. The evaluation over an image classification task confirms that the proposed model outperforms other methods acting on the individual levels. The resulting method is cost-effective and can be easily applied to semi-automatic image semantic labeling tasks as foreseen in collaborative annotation scenarios.

1 Introduction

Retrieval processes acting over multimedia material are now facing new problems for the size and the heterogeneity of the available data and annotations on the Web. They ask for flexible, efficient classification and retrieval tools. As opposed to usual search engines, image search is supported either via keyword-based SQL interfaces or image queries in query-by-example settings. This strict separation is unsatisfactory. More efficient methods for retrieving semantic information from multimedia content are required suitable for search, classification or knowledge extraction from visual data, as it is traditionally carried out over texts. Synergies of multimedia and knowledge technologies [9], [5] are expected to enhance services by the creation of machine-understandable content. The automation of these processes is then the only way to a conclusive switch to a true Web of semantic objects.

The critical problem here has ben recently defined as the "semantic gap" between descriptions obtained from different media levels ([9]). The gap is the "*lack of coincidence between the information that one can extract from the visual data and the interpretation that the same data have for a user in a given situation*" [9]. The role of language in this process is apparent. The simplicity of the available visual features is far weaker than the inherent power of language to express

B. Falcidieno et al. (Eds.): SAMT 2007, LNCS 4816, pp. 41–55, 2007.
© Springer-Verlag Berlin Heidelberg 2007

to the richness of the user semantics. Advancements in content-based retrieval is critically tight to the filling of this *semantic gap*, as the meaning of an image is rarely self-evident. The major methodological position behind these conclusions is that much of the interesting work in attempting to bridge the semantic gap automatically is tackling the gap between low-level descriptors and linguistic labels as effective surrogates of the full visual semantics. This is even strengthened by the fact that queries to multimedia data are typically formulated with respect to their semantics, via more or less complex linguistic descriptions, i.e. descriptors.

The linking between the data visual properties and their semantics is strictly connected with an *annotation* task, i.e. the selection of proper linguistic symbolic labels able to express most of the semantics underlying visual properties. However, labeling introduces here a subtle problem: as noticed by [5], the discrete nature of symbolic labels that are rigidly assigned to images, provide a too strict semantic position. Annotating by a given label, i.e. deciding the appropriateness of a label, is in fact a discrimination problem much prone to fallacies. *Hard* annotation techniques make explicit use of (sets of) labels by which an image is fully characterized and errors are much possible. The so-called *soft* approach has been recently introduced as a solution to these problems [5]. It makes a better use of the full power of geometrical modeling of the image properties. A vector space is introduced where visual properties, i.e. low level descriptors, *live with* textual properties, i.e. terms or keywords. This *duality* property characterizes latent semantic spaces [3].

In this view, annotations are no longer rigid descriptors but *clouds* of terms "*close*" enough to some of the corresponding visual properties. Vice versa, some visual properties can be seen as expressions of some concepts whenever they are named by keywords that are "*close*" enough. Notice that set of terms are more expressive than single keywords: the inherent linguistic ambiguities characterizing a single word in fact disappear in the larger context of an entire term cluster.

Quantitative measures of similarity in this vector space capture flexible properties and imply a smoothed notion of annotation. This strength is just an example of the very general power that geometrical models show in dealing with the vague and eluding notion of meaning [10].

The focus of this work is a specific problem: how can we classify and index multimedia information with a semi-automatic annotation of the contents, in an accurate and cost-effective way? In other words, how can we drastically reduce the huge costs of manual annotation. We emphasize here the subtle difference between annotation and classification. By adopting a soft annotation model, we cope with clusters of features (e.g. linguistic labels). This provides a natural notion of *class* as widely studied in automatic text classification [2]. As in [5], we produce vector representations for images including both visual and textual (i.e. symbolic) properties. However, in this paper the evaluation of the selective power of this representation is carried out with respect to a classification task.

Clusters of features are first automatically built and then used to characterize image categories and train a supervised classifier.

This paper presents a geometric model for semi-automatic image labeling and its evaluation over a classification problem: it represents a semi-supervised machine learning approach for automating a realistic image classification. Two major differences exist with previous approaches. First, it exploits a combination of low-level and symbolic descriptors in the derivation of a space where an expressive notion of semantic similarity can be modeled. Second, it applies to unlabeled raw data directly extracted from the Web. The structural properties of usual HTML pages are here used to derive candidate labels for images. We define as potential labels the words as they are found in text paragraphs adjacent to images according to the Web layout.

Section 2 describes how multimediality is managed in the model, through the combination of evidence derived from visual and textual properties. In Section 3, the methods employed to learn from the extracted heterogenous features are presented, resulting in a general architecture for Image Classification. In Section 4, the evaluation of the experimental results is presented.

2 Feature Engineeing over Multimedia Data

Documents containing texts and images are widely distributed in the Web and represent a relevant body of evidence for extracting possible correlations. As our objective is to manage the image domain as it is currently possible to do with texts, the most difficult task is the representation of the visual information coming from images. This representation must contain a valid description of the information to be indexed, thus reducing the unavoidable semantic loss. A possible solution is to learn the mapping and relationships between low level image features and words (or concepts) observable in texts *as an emerging property of available data*, even when the latter are unlabeled. In this work the semantic information brought by the observable language facts about images is thus exploited to improve the design efficiency and the scalability of the image annotation task. Multimediality is addressed here by the combination of two abstraction levels, the visual and textual one. It aims to enable both the precision of the semantic annotations the recognition of correlations between the visual and the textual descriptors. We propose an analysis based on an initial separation of the visual and textual domains: reference to the originating context (e.g. a Web page) is preserved. Images and texts are thus processed according to different modalities (see Section 3) acting over the same similarity model, i.e. Latent Semantic Analysis [4]. The resulting combined representation is a comprehensive model supporting a unified and expressive similarity model.

The technology applied to the textual corpus is Deep Knowledge (DK) that proceeds by a set of application specific queries to Web search engines[1]. It carries out the following tasks over the retrieved content:

---

[1] *Deep Knowledge* is a text mining platform produced and commercialized by Exprivia S.p.A, URL:www.exprivia.it

- *Web pages indexing*, that generates unique references to the source images and pages
- *morphological recognition* and *terminology extraction* that produces the seed term vocabulary
- *segmentation*, that split pages into units in which the coexistence of one (or more) image(s) and some text is represented

Moreover, image processing is also applied for the derivation of the visual properties characterizing the Web data. Images are analysed through LTI-lib, [7], an object-oriented open source library with dedicated algorithms and data structures for image processing. It has been developed at the Aachen University of Technology and is frequently used in image processing and computer vision tasks. These activities generate the multimedia corpus that is used in all the experimental phases described in the next sections.

2.1   The Extracted Features

Visual and textual properties are the observable evidences used as features of the target individual (Web) units.

Textual features include *POS tagged atomic terms* (e.g. grass, fields, airport), *terminological expressions* (e.g. *civil airport, air traffic*) and proper nouns (e.g. *Rome Fiumicino, Golden Gate Bridge*). Traditional weighting schemes (as adopted in vector space models [8]), are assigned to individual terms, depending on their local and global distribution, in documents and in the collection, respectively. Terms with less than 5 occurrences in the corpus are excluded from the vocabulary. No other feature selection filter is applied.

Visual Features refer to three major classes: *chromatic, textures* and *transformation features*. Chromatic features express the color properties of the image. Textures emphasize the background properties and their composition. Transformations are also modeled in order to take into account possible shapes that are invariant according to some geometrical transformations. These features should allow to compute similarities of the depicted discrete objects independently from their position in the images. The adopted platform is LTI-LIB [7] that offered the different libraries to extract weights for these features.

Feature relevance is expressed through real valued numbers, i.e. the weight. A geometrical space can be thus derived by juxtaposing the dimensions generated by individual (textual and visual) features. However, this is problematic in several senses. First, weights for different features belongs to different domains so that a sampling problems arise. Feature types are in fact not homogeneous, and different sampling and models are required. Second the distributions are also very different as their variance change strongly across individual domains.

The aim is to derive a rich space where distinguishing features can trigger useful generalizations across the visual and textual dimensions: for example, terms like *grass* or *fields* should be represented "*close*" enough to high levels of the *green* band for the RGB chromatic representation. We aim thus to capture correlations among the subdomains of visual and textual properties, as they can be found with a significant regularity in the data collected from the Web.

3   A Geometrical Model for Combining Visual and Textual Features

As textual and visual features vary in different domains, the aims is to fully capture their correlations: inner correlations (e.g. correlations among pairs/set of words) and external correlations (e.g. correlations between visual properties, like color dominance, and words). Higher order correlations can be captured in vector space models (e.g. VSM [8]) via Latent Semantic Analysis methods [4]. LSA attempts to capture the term semantic dependencies using the purely automatic matrix decomposition process, called *Singular Value Decomposition (SVD)*. The original term-by-document matrix M describing the traditional term-based document space is transformed in the product of three new matrices: T, S, and D such that their product $TSD^T = M$. They capture the same statistical information than M in a new k-dimensional space where each dimension represents one of the derived LSA features (or concepts). These may be thought of as artificial concepts and represent emerging meaning components from many different words and documents [4]. LSA maps documents from a vector space representation to new space with a lower number of dimensions. Terms are mapped to vectors into this lower dimensional space. Following the usual vector space metrics, e.g., calculating cosine similarities in the new space, we can evaluate term-by-term and document-by-document similarity. Moreover, we are also able to compute term-by-document similarities, a property that has been often referred to as the *duality principle*. When multimedia properties are targeted, the ability to decompose the original matrix (with mixed visual and textual features) and map it into an LSA space enable the detection of specific correlations between multimedia features and Web images. Closeness in the resulting space should be able to suggest latent semantic relationships between terms (like *green, fields* or *vegetation*) and colors dominating some images (e.g. green band in an RGB chromatic representation). This emerges automatically from an LSA-like transformation of the original space. However, correlations are able to emerge if a proper alignment of individual domains is applied, as their significantly different distributions may affect the quality of the direct application of LSA.

Accordingly, LSA is applied in a structured fashion. First, LSA is run separately over the vectorial representations of visual and textual features with an SVD decomposition process applied to each visual feature (chromatic, texture and transformation).

The combination of visual features is then applied as a second phase. In this step two possible approaches are possible:

In *Method 1*, a $3k$-dimensional space is obtained by creating three comprehensive T, S and D^T matrices by juxtaposing accordingly the matrices derived in the previous step for each graphical feature: the T_i, S_i and D_i^T matrices (derived from each visual feature class i) are here combined. Then, we sort the diagonal of the new S matrix, and, accordingly, swap the columns of the remaining two matrices T and D^T. Finally, the three matrices are reduced to the first k dimensions. With this approach we obtain a k-dimensional representation simulating a global LSA process: all features are combined before any SVD transformation over all the three views on image graphics is applied. The aim is to avoid that the emergence of too many dimensions for a given graphical feature, by hiding the effects of other relevant features. As local SVD transformations are carried out separately, features are represented in a comparable space and ranking can realize a more precise feature selection. The above process is represented in Fig. 1

In *Method 2*, a $3k$-dimensional space is derived by first juxtaposing the three k-dimensional vector of the same objects and then applying another SVD step to gather the final, unified, k-dimensional representation. The new latent semantic space should integrate all the original graphical features in the target k-dimensions.

In either of the above visual feature processing modes, a unified k-dimensional space representing the visual domain is obtained.

Two independent k-dimensional LSA spaces are thus made available for texts and visual features expressing correlations lying in the individual domains, i.e. the visual and linguistic domain. In a third phase a $2k$-dimensional space is obtained by juxtaposing the two k-dimensional vectors of the same objects. On the resulting LSA space, a last SVD step is applied to gather the unified k-dimensional representation. The final latent semantic space should integrate in

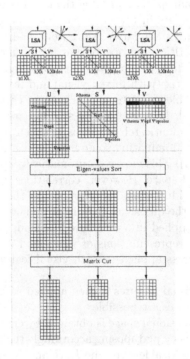

Fig. 1. Approach 1: The combination model of visual features

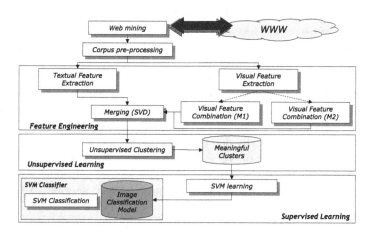

Fig. 2. The general architecture of the Image classification system

k-dimensions all the original multimedia properties and should make the useful correlations explicit: these emerge from the comparable k-dimensional vectors derived through the SVD chain. Cosine similarity in the resulting space is a distance metrics useful for capturing the suitable notion of similarity helpful to clustering objects and features. Notice how individual vectors in the final space represent groups of visual or textual features in the original space: clustering over this space can finally exploit a large body of evidence derived from the preceding independent analysis. The workflow sketched above gives rise to a general architecture for unsupervised image classification, as discussed in the next section.

3.1   Multimedia Semantic Classification

In an unsupervised approach to image classification we aim to exploit the inherent redundancy of features that we can observe in Web pages as well as the combination of different learning methods this including Latent Semantic Analysis, unsupervised clustering and automatic image classification, i.e. Support Vector Machine (SVM) learning. This requires the definition of a complex architecture for image classification. The overall data flow is shown in Figure 2 where the four major phases are sketched: *Web mining, Feature Engineering, Unsupervised Learning* and *Supervised Image Classification*.

The processing chain works as follows

– First, Web is searched for pages related to the types of images of interest. During this phase images and their surrounding text in the page layout is derived through the DK engine. In this phase the major extraction process relates the detection of interesting HTML material, the pruning of irrelevant mark-up and the feeding of an internal DB with all the potentially relevant Web data.

- Then a *Corpus pre-processing* phase is applied to build the vector representations of the target images including visual as well textual descriptions. In this phase the role of DK is to segment individual pages into significant parts. This is done in the following steps:
  - First, segments are derived according to the DK technology able to express meaningful information units that include texts and images.
  - Textual information is then processed by DK to derive the proper vocabulary and map segments into an internal vector representation. Complex terms are automatically detected so that jargon or technical expressions (e.g. *remote sensing* or *Very High Resolution Radiometer*) are given independent and specific weights. This provides a traditional vector space model enriched with linguistic information very specific to the target domain.
  - Corpus preprocessing proceeds then by applying the dimensionality reduction process discussed in Section 3. Two k-dimensional pseudo vectors are obtained by the SVD transformations representing one data item (a segment) according to two independent views: text and visual pseudo vectors are separate but they refer to the same Web segment[2].
  - Finally, *merging* the two k-dimensional representations in the LSA spaces is carried out, over the individual $2k$-dimensional vectors obtained by juxtaposing the original ones. As discussed in section 3 a second LSA stage is applied and the result is a k-dimensional vector for each unit of information.
- *Unsupervised Learning* is applied by clustering the derived vectorial data with all the available multimedia properties. In this phase efficient agglomerative clustering algorithm (i.e. k-mean [1]) is applied. The results are groups of images (and segments) that are similar according to their textual or visual properties (or both).
A cluster $C = \{s_i | s_i$ describe "similar" segments $\}$ provides interesting information:
  - C is characterized by a *centroid*, c, given by:
$$c = \frac{\sum_{s_i \in C} s_i}{|C|}$$
  - The properties that characterize the centroid are *terminological expressions* and visual features that are typical for most of its members. They justify C. The terminological expressions are helpful to describe the cluster's semantics
  - Clusters can be automatically characterized by their internal coherence: more coherent clusters are a more promising source of information (i.e. examples to manually label) and should be preferred for the training of a supervised classifier. A score capturing this notion is the following:
$$\sigma(C) = \sigma(c) = \mu_i(||c - s_i||) \cdot max_i(||c - s_i||)$$

---

[2] Multiple images in the same segment are treated by replicating the (unique) textual properties for all the images. This may possibly introduce noise in the resulting data sets.

where $||c - s_i||$ is the distance between the centroid and the cluster members, μ and max are the *average* and *maximal* values across the members, respectively.

As we will see in Section 4, the combination of textual and visual features obtains usually the best result.

- *Supervised Classification.* Given the quality of clusters derived automatically, a supervised image classification approach can be flexibly applied. In fact, the different clusters can be more easily annotated by looking only at the best cluster instances and to their terminological descriptions. As labeling one cluster provides, as a side effect, the availability of many examples, this increases the overall scalability of a supervised approach. Every member in a cluster is an individual example for a supervised image classification model but several members are labeled, by receiving the single label decided at cluster level. In the last phase, clusters are thus used as training seeds for a supervised classifier. As in all our experiments, we applied Support Vector Machines ([11], [2]) as a learning paradigm: we thus call these stages *SVM training* and *SVM classification*, respectively. During training an SVM develops (according to the risk minimization principle, [11]) the best image classification model able to separate negative from positive examples of one class. Different SVMs are trained for the different classes relying on separate models. The ensemble of these models is called *Image Classification Model* in Fig. 2. During classification, the different SVMs run on a new example in a binary fashion: they can accept or reject the example as valid member of each class. A final combination of the individual decisions selects the best global class[3]. The integration of the *Image Classification Model* with the combined set of SVMs realizes the overall SVM classifier shown in Fig. 2.

Notice how the improvement over an image classification task is just one of the outcome of the proposed methodology. In fact, clusters with their relevant (centroid) properties define prototypical ways of referring and describing a target category: the best features (e.g. most important terms characterizing the cluster centroid) can be thus seen as typical ways of querying a class of visual objects. This supports the learning of linguistic ways of referring to categories useful for searching images by NL queries.

4 Evaluation

The objectives of the experiments were to validate the idea of enabling fast and accurate image classification via the combination of heterogeneous media properties and the adoption of clustering as a semisupervised learning process. In this perspective, at least two dimensions of the proposed approach should be explored: the *type of required annotation* and the *different training evidence* made available (media levels).

---

[3] Binary classifiers can be combined through the method known as *one vs. all* classification ([2]).

- *Type of Annotation.* The architecture proposed in Section 3.1 supports two types of training. The simplest and cheapest form of preparation of training data consists in labeling an entire cluster C (rather than an image) with a single label denoting a target concept (e.g. all members of C are *bridges*). In this case every image, inside a cluster labeled as *label(C)*, is considered a positive example for the training of the *label(C)* class SVM categorizer. Although the cluster may include also some wrong (i.e. misplaced) images, the advantage is a more productive annotation that scale faster to larger sets of labeled images (training sets). We refer to this annotation as a *cluster-based. Image-based annotation* corresponds to the labeling of individual images, irrespectively from their membership in clusters.
- *Training Evidence.* Images in Web pages convey different information at the visual as well as at the textual level. We will refer to *textual features* as the terms and terminological expressions derived from the Web pages texts, close enough to the target image. *Visual features* are divided into chromatic properties (e.g. colors, ...) and texture oriented features. Specific combinations of these types of features are also allowed although all tests have been run by using all the visual features we call chromatic.

4.1   Experimental Set-Up

The corpus adopted for the experiment has been obtained by Web spidering and it includes Web pages potentially well related to two target concepts *bridges* and *airplanes*. Although dominated by true instances of these concepts, the corpus includes also a large portions of errors (like banners or other advertising pictures). Each page is segmented into parts that include one picture and a corresponding text, from which all the features are derived. Overlapping are also possible: the same text can be attached to different images if they are enough close in the page. This allows a bit of redundancy at the price of possible noise in the data. Table 1 describes the main aspects of the corpus.

Due to the presence of significant noise in the original data, the experiments in automatic image classification have been targeted to three classes, i.e. *bridges, airplanes* and *other*, the latter being used to recognize pictures irrelevant for the task. The experiments have been all run by coupling the clustering process over the corpus (as a fully unsupervised step) with a supervised classifier based on SVM[4]. The Latent Semantic Analysis has been applied first separately over the textual and visual properties and then after the merging of the first decomposition process. The order k of the decomposition in all cases was 100.

Experiments have been designed to measure the effectiveness of the semi-supervised learning approach on the image categorization task. We compared here the use of unsupervised clustering to train an SVM classifier with a traditional (i.e. fully supervised) use of the SVM. In the first case, the annotations of automatically derived clusters are used to gather positive and negative class

---

[4] The SVM implementation called SMO made available by the WEKA machine learning platform [12] has been applied in all the experiments.

Table 1. Description of the Web corpus used in the experiments

Corpus property	Value
Number of target categories	2 + 1 (*Others*)
Number of HTML pages	23,520
Number of sections derived	1,527
Number of images	3,959
Number of selected terms	53,251
Number of words (types)	121,545
Number of Annotated images	867
Number of images in the Test Set	432
Number of images in the Training Set	3,527

instances (classified images plus their text) and then train the SVM. The reference supervised model makes no use of clusters and it is trained directly on hand-annotated images. The evaluation has been done in all cases by computing the *accuracy* measure, that is the percentage of correct classification decisions in the different tests.

4.2   Evaluating the Semi-supervised Image Classifier

These experiments aim to measure the effectiveness of the semi-supervised learning approach on the image categorization task by comparing it with a fully supervised training process. We call here *semi-supervised*, any process that makes use of a small set of labeled images to collect available training evidence and automatically tag new images. It aims to exploit the information gained during the clustering phase: a cluster built around few annotated images is generalized and the most likely label in the class is extended to all images of the cluster. The presence of noise inside clusters is very likely: for example, it is possible that a cluster gathering mainly images of *bridges* includes also banners (i.e. class=*other*). This is particularly true, when banners are highly related to Web pages dealing with bridges. Evaluation here aims to measure the impact of this potential noise sources on the training of the SVM classifier. Obviously, the supervised learning scheme of the SVM (that is trained only on labeled images as examples, or counterexamples, of a class) constitutes an upper-bound for the accuracy of the semi-supervised setting.

1. Given the entire training set, we labeled some images so that each of the three classes is equally represented (e.g. 20 aircrafts, 20 bridges and 20 other). Images are randomly chosen although they are selected so that each class is equally represented (about 289 images for each class). The derived test set is balanced and the baseline is thus around 33%. Some of the labeled images are used as *seeds* of the semisupervised training process.
2. We apply the clustering algorithm on the entire training set obtaining a set of clusters, some of them including annotated images.
3. A majority policy is used to assign a unique class to the entire cluster: the information provided by the seed labeled images in the cluster is employed.

For example, if the labeled examples of a cluster amount to 55% of *aircrafts*, 20% *bridges* and 25% of *other*, all the images belonging to that cluster will be labeled as *aircrafts*. At the end of this phase, some clusters are assigned to one of the three classes. However clusters not including any seed image are not labeled and will not be used for training.

4. Using the extended set of labeled images (i.e. the training data obtained in an semi-supervised way), we trained an SVM and evaluate its performance over the test images, that have been kept out of the seeding and labeling process.

As the number of seed (i.e. labeled) images is critical to the quality of the semisupervised learning, we compared the performance of different SVMs trained over growing sizes of the seeding image set.

Hand annotation has been applied by two small teams of three experts with very high annotation agreement scores (over 95%). An annotation was accepted when it was given by at least two annotators. The annotators working on a cluster basis (i.e. by annotating all images of a cluster with the single decision over the entire cluster) were concluding their job 12 times faster than those labeling images individually. A training-test split of about 90% vs. 10% was applied for the SVM training. This process wasn't totally random: the test images have been chosen to balance the representativeness of the three classes. However, when the number of test images for each class was fixed, a random selection process was applied. Using the same approach we selected small subsets of seed labeled images and triggered the clustering algorithm over all the remaining images in the training data set.

In Tables 2 and 3 we report the results of the automatic semisupervised labeling process in which training images (Colum 4) are derived from sets of seed images of growing sizes, according to different feature sets. In the first table we use only the textual feature, while in the second one we show results from the combination of both kinds of features. This tables shows the *Number of seed images*, the *labeled clusters*, i.e. the number of automatically tagged clusters and *the number of labeled images* that is the number of automatically labeled images added to the SVM training set. Columns 5 reports the *annotation rate*, i.e. the ratio between the resulting labeled images and the seeding images.

Accuracy measures of experiment 2 are reported in Figure 4. At a first glance is to be noticed that the performance is very high. Good accuracy values ($> 70\%$) are obtained even when very small seeding sets are employed (about 1% of the

Table 2. Labeling rates over textual features only

Percentage of Seeding images	Number of seed images	Number of labeled clusters	Number of labeled images	Annotation rate
1.22%	43	9	1262	29.21
3.67%	130	23	962	7.42
6.12%	216	27	1321	6.11
8.57%	302	36	1446	4.78

Table 3. Labeling rates via combined textual and visual features

Percentage of Seeding images	Number of seed images	Number of labeled clusters	Number of labeled images	Annotation rate
1.22%	43	23	1840	42.59
3.67%	130	39	2140	16.51
6.12%	216	47	2324	10.76
8.57%	302	58	2447	8.09

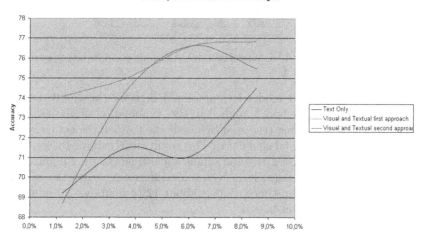

Fig. 3. Learning curves of different classifiers in Experiment 2

entire training set, and about 5% of the annotated images). This suggests that the employed feature spaces are very effective. It is to be also noticed that, although the task (i.e. only three target classes) is not very difficult, the source data that fed the unsupervised clustering phase are realistic and thus complex and very noisy.

The plots in figure 3 shows that the classification accuracy grows according to increasing sizes of the seeding set, even if possibly noisy semi-supervised image annotation is applied. There are some particular exceptions, for example in the *text only* case there is a little loss of performance, however the gap is only of 1% and this is probably due to the random process of seed image selection. Notice how better performance levels are provided by the combination of textual and visual properties as these give a positive contribution to the overall accuracy reachable. Method 1 described in section 3 is particularly impressive as it achieves the best result (76.86%) and it provides a relevant accuracy with about 1% of the training evidence.

Results in Figure 4 show also the performance of a supervised SVM trained over the seeding set of images, i.e. about the 8.6% of the training set, or over the entire set of 867 annotated images, respectively (last two bars in the Figure).

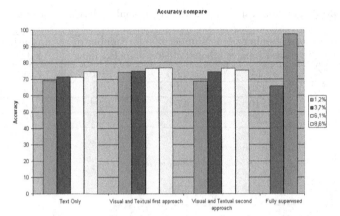

Fig. 4. Semi-supervised vs. supervised image classification in Experiment 2

The best semi-supervised model is represented here by the eighth bar. This model corresponds to Method 1 applied to the combination of all features and to a semisupervised training triggered by about the 8.6% of the training images (about 50% of the seeding labeled data). This result clearly shows the superiority of the semi-supervised model over the simple supervised classifier applied to the same amount of seeding data. It is even more interesting when the estimated speed-up in the annotation times, that is about 12 (4 hours vs. 20 minutes for about 1,000 images), is considered.

5   Conclusions

In this paper, a robust and accurate semi-supervised approach to image categorization has been proposed. It exploits LSA for developing efficient vector representations of Web multimedia documents. The aim here is to reduce the unavoidable loss in capturing the semantics of available contents. We showed that the application of SVD, an automatic matrix decomposition process, constitutes a very promising track to fill the semantic gap in multimedia analysis. Finally, our approach also supports image search via keyword-based queries, thanks to the integration of texts and visual features.

The approach presented here is particularly close to that in [5], although this latter is a pure image annotation task. The major difference is the adoption of a fully automatic approach to data collection, that in our case correspond to downloaded pages from the Web. However, the good accuracy resulting from our experiments is in line with the previous results. As in [6], LSA seems a very effective way to capitalize the available heterogenous features.

The future research will include a study on the full adoption of a co-training approach to SVM training, over the LSA representation. Incremental learning and boosting have been shown very effective in improving the learning rate in statistical automatic classifiers and their application to the image annotation task is very interesting. Moreover, a study of the generalization power of our method in search tasks is foreseen, as linguistic labels for describing image clusters are here provided. Finally, larger corpora and different domains should be studied to assess the generality and effectiveness of the presented approach.

References

1. Alsabti, K., Ranka, S., Singh, V.: An efficient k-means clustering algorithm. In: First Workshop High Performance Data Mining (1998)
2. Basili, R., Moschitti, A.: Automatic Text Categorization: from Information Retrieval to Support Vector Learning. Aracne (2005)
3. Berry, M.W., Dumais, S.T., O'Brien, G.W.: Using linear algebra for intelligent information retrieval. SIAM Review 37(4), 573–595 (1995)
4. Deerwester, S., Dumais, S., Furnas, G., Harshman, R., Landauer, T.: Indexing by latent semantic analysis. Journal of the American Society for Information Science 41(6), 391–407 (1990)
5. Hare, J.S., Lewis, P.H., Enser, P.G.B., Sandom, C.J.: Mind the gap: Another look at the problem of the semantic gap in image retrieval. In: Proceedings of Multimedia Content Analysis, Management and Retrieval 2006 SPIE (2006)
6. Monay, F., Gatica-Perez, D.: On image auto-annotation with latent space models. In: Proceedings of the 11th annual ACM international conference on Multimedia (2003)
7. RWTH: Lti-lib - computer vision library. Website, University of Aachen (September 2006)
8. Salton, G.: Automatic Text Processing–The Transformation, Analysis, and Retrieval of Information by Computer. Addison-Wesley, Reading, Massachusetts (1989)
9. Smeulders, A.W.M., Worring, M., Santini, S., Gupta, A., Jain, R.: Content-based image retrieval at the end of the early years. IEEE Transactions on Pattern Analysis and Machine Intelligence 12(22), 1349–1380 (2000)
10. van Rijsbergen, C.J.: The Geometry of Information Retrieval. Cambridge University Press, Cambridge (2004)
11. Vapnik, V.: The Nature of Statistical Learning Theory. Springer, New York (1995)
12. Witten, I.H., Frank, E.: Data Mining: Practical machine learning tools and techniques. Morgan Kaufmann, San Francisco (2005)

Ontology-Driven Semantic Video Analysis Using Visual Information Objects

Georgios Th. Papadopoulos[1,2], Vasileios Mezaris[2], Ioannis Kompatsiaris[2], and Michael G. Strintzis[1,2]

[1] Information Processing Laboratory
Electrical and Computer Engineering Department
Aristotle University of Thessaloniki, Greece
[2] Informatics and Telematics Institute/CERTH
1st Km Thermi-Panorama Road
Thessaloniki, GR-57001 Greece
{papad,bmezaris,ikom}@iti.gr, strintzi@eng.auth.gr

Abstract. In this paper, an ontology-driven approach for the semantic analysis of video is proposed. This approach builds on an ontology infrastructure and in particular a multimedia ontology that is based on the notions of Visual Information Object (VIO) and Multimedia Information Object (MMIO). The latter constitute extensions of the Information Object (IO) design pattern, previously proposed for refining and extending the DOLCE core ontology. This multimedia ontology, along with the more domain-specific parts of the developed knowledge infrastructure, supports the analysis of video material, models the content layer of video, and defines generic as well as domain-specific concepts whose detection is important for the analysis and description of video of the specified domain. The signal-level video processing that is necessary for linking the developed ontology infrastructure with the signal domain includes the combined use of a temporal and a spatial segmentation algorithm, a layered structure of Support Vector Machines (SVMs)-based classifiers and a classifier fusion mechanism. A Genetic Algorithm (GA) is introduced for optimizing the performed information fusion step. These processing methods support the decomposition of visual information, as specified by the multimedia ontology, and the detection of the defined domain-specific concepts that each piece of video signal, treated as a VIO, is related to. Experimental results in the domain of disaster news video demonstrate the efficiency of the proposed approach.

1 Introduction

Over the past decades, access to multimedia content has become the cornerstone of several everyday activities, as well as a key enabling factor at professional level. However, due to the fact that literally vast amounts of multimedia data are generated, stored and distributed from multiple information sources, new needs arise regarding their effective and efficient manipulation. This has triggered intense research efforts towards the development of intelligent systems capable of automatically locating, organizing, accessing and presenting such huge and heterogeneous

B. Falcidieno et al. (Eds.): SAMT 2007, LNCS 4816, pp. 56–69, 2007.

amounts of multimedia information in an intuitive way, while attempting to understand the underlying semantics of the multimedia content [1].

Among the proposed solutions for the problem of semantic analysis of multimedia content, i.e. bridging the so called *semantic gap* between the low-level numerical audio-visual data and the higher-level human perceivable concepts and entities [2], the exploitation of *a priori* knowledge emerges as a very promising one. Approaches belonging to this category require the specification of appropriate knowledge structures for defining a representation of the prior knowledge necessary for analyzing multimedia content and providing support for learning possible links between low-level audiovisual information and semantically meaningful concepts [3].

Regarding the possible domain knowledge representation formalisms, ontologies have been particularly favored due to the significant advantages they present. In particular, they achieve to exhibit a coherent domain knowledge representation model, provide machine-processable semantics definitions and allow automatic analysis and further processing of the extracted semantic descriptions [4]. Concerning the process of semantic video analysis, ontologies have been broadly used in a wide range of approaches. In [5], an ontology framework is proposed for detecting events in video sequences, based on the notion that complex events are constructed from simpler ones by operations such as sequencing, iteration and alternation. A large-scale concept ontology for multimedia (LSCOM) is designed in [6] to simultaneously cover a large semantic space and increase observability in diverse broadcast news video data sets. Additionally, in [7], a pictorially enriched ontology is used both to directly assign multimedia objects to concepts and to extend the initial knowledge for the soccer video domain.

In this paper, an ontology-driven approach for the semantic analysis of video is proposed. The approach builds on an ontology infrastructure and principally a multimedia ontology, whose design is based on the notion of the MMIO. The developed infrastructure is accompanied with signal-level video processing techniques, that are necessary for associating the developed ontology infrastructure with the signal domain. The proposed system supports the decomposition of the visual information and the detection of the defined ontological concepts, thus resulting in a higher-level semantic representation of the video content. Experimental results in the domain of disaster news video demonstrate the efficiency of the proposed approach. The remainder of the paper is organized as follows: Section 2 presents the overall system architecture. Sections 3 and 4 describe the employed high-level knowledge and the low-level information processing, respectively. Sections 5 and 6 detail individual system components. Experimental results are presented in Section 7 and conclusions are drawn in Section 8.

2   System Overview

The first step to the development of the proposed ontology-driven semantic video analysis architecture is the definition of an appropriate knowledge infrastructure that will model the knowledge components that need to be explicitly defined

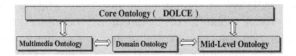

Fig. 1. Knowledge Infrastructure

for the analysis process. For that purpose, ontologies were used, due to the advantageous characteristics that they present and were discussed in the previous section. The developed knowledge architecture, which is depicted in Fig. 1, consists of four individual modules: the Core Ontology (DOLCE), the Mid-Level Ontology, the Domain Ontology and the Multimedia Ontology.

The Core Ontology, which is based on the DOLCE core ontology [8], contains specifications of domain independent concepts and relations based on formal principles derived from philosophy, mathematics, linguistics and psychology. In the proposed framework, it is introduced in order to facilitate the integration and alignment of the individual ontological modules. The Mid-Level Ontology aims to include additional concepts that are generic and not included in the core ontology, thus attempting to ease the alignment of the abstract philosophy of the Core Ontology and the concrete philosophy of the Domain Ontology. Moreover, the Domain Ontology provides a conceptualization of the domain of interest by defining a taxonomy of domain concepts, which are in turn separated into global and local ones. The latter can be used to further characterize parts of a video signal that can be associated with a global one. Furthermore, the Multimedia Ontology, which models the content of multimedia data, serves as an intermediate layer between the Domain Ontology and the audiovisual features, through which the associations of the domain concepts are realized, and includes algorithms for processing the content.

The design of the Multimedia Ontology is based on the notion of the MMIO, and in particular the VIO. The latter constitute extensions and adaptations of the IO design pattern, previously proposed for refining and extending the DOLCE core ontology. Each VIO represents a piece of the video signal to be analyzed, defines its relations and interactions with other VIOs and encompasses the means and methods for its semantic interpretation. The aforementioned ontological modules are suitably aligned and are used to drive the semantic video analysis process. Regarding the particular domain of experimentation, the disaster news video domain was selected and an appropriate Domain ontology was developed.

At the signal level, the video processing procedure, that is necessary for associating the developed ontology infrastructure with the visual domain, is initiated with the application of a temporal segmentation algorithm for segmenting the video sequence into shots and is followed by a keyframe extraction step. More specifically, for every shot a single keyframe is extracted. Subsequently, low-level global frame descriptors are estimated for every keyframe and form a *frame feature vector*. This is utilized for associating the keyframe with the global concepts defined in the domain ontology based on global-level descriptors, serving as

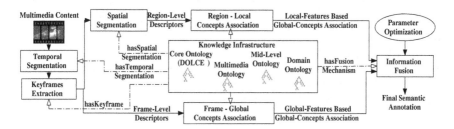

Fig. 2. System Architecture: The dashed arcs denote properties of the developed ontologies that correspond to specific multimedia content processing algorithms

input to a set of SVMs, where each SVM has been trained to detect instances of a particular concept. Every SVM returns a numerical value which denotes the degree of confidence to which the corresponding frame is assigned to the ontology global concept associated with the particular SVM.

In parallel to this process, spatial segmentation is performed for every keyframe and low-level descriptions are estimated for every resulting segment. These are employed for generating hypotheses regarding the region's association to an ontology concept. This is realized by evaluating the low-level *region feature vector* and using a second set of SVMs, where each SVM is trained this time to identify instances of a local concept defined in the domain ontology. SVMs were selected for the aforementioned tasks due to their reported generalization ability [9]. The computed region-level hypothesis sets are subsequently introduced to a *decision function* that is defined in the Multimedia Ontology and which realizes keyframe-global concept association based on local-level information.

Then, a fusion mechanism is introduced, which implements the fusion of the computed keyframe-global concept association based on global- and local-features, in order to make a final keyframe semantic annotation decision. A GA is employed for optimizing the parameters of the fusion mechanism. The choice of a GA for this task is based on its extensive use in a wide variety of global optimization problems [10], where they have been shown to outperform other traditional methods.

Since the final semantic annotation decision is made for every keyframe, it is in turn used to indicate the respective video shot semantic interpretation. Thus, the output of the proposed semantic video analysis framework is a set of shots, to which the input video sequence is decomposed to, and a global concept, defined in the domain ontology, associated with each shot. The overall architecture of the proposed system is illustrated in Fig. 2.

3   Multimedia Ontology

As was described in the previous section, the Multimedia Ontology generally models the content of the multimedia data, serves as an intermediate layer between the Domain Ontology and the audiovisual features, through which the associations of the domain concepts are realized, and includes algorithms for processing the content. Because of its crucial role in the overall semantic video analysis approach, it is described in detail in this section.

Under the proposed approach, the role of the multimedia ontology is to provide the adequate amount of knowledge so that the semantic video analysis procedure is tailored to the specific requirements of a particular application case. More specifically, the multimedia ontology aims to suitably model the content layer of video, define a mapping between low-level audio-visual features or video processing techniques and high-level domain concepts, and generally drive the overall semantic video analysis procedure.

Since the multimedia ontology objective is to guide the semantic video analysis process, its structure should be designed in a way so that both the multimedia properties for specific domain concepts can be described in an arbitrary way and the actual multimedia material is appropriately modeled. For that purpose, the IO design pattern, previously proposed for refining and extending the DOLCE core ontology [11], was adapted and suitably extended. In particular, the DOLCE IO was enriched with two additional properties, namely the 'hasDecomposition' and the 'refersTo' properties, and the resulting information object is denoted with the term MMIO [12]. The MMIO model combines the DOLCE IO pattern with the MPEG-7 standard for the representation of media content and multimedia features [13][14].

Regarding the IO extensions, the 'hasDecomposition' property, the range of which is 'Decomposition', is introduced for describing the decomposition of multimedia objects. Every piece of multimedia information is considered as a multimedia object. 'Decomposition' will provide the MMIO with the needed concepts for the structural description of the multimedia content, in accordance to the respective MPEG-7 description scheme. For that purpose, a variation of the MPEG-7 subpart of the SWIntO [15] ontology, which is in turn an MPEG-7 based ontology for semantic annotation of multimedia content, was adopted. This part of the SWIntO ontology focusses on the MPEG-7 Content Description and Content Management Description Scheme that suffice to model concepts describing storage features (e.g. format, encoding), spatial, temporal and spatio-temporal components (e.g. scene cuts, region segmentation, motion tracking), and low-level features (e.g. color, shape, texture, timbre, melody) of multimedia content. Additionally, the 'refersTo' property is introduced for realizing the connection between MMIOs that are expressed in different modalities but refer to the same content unit. Thus, objects that derive through decomposition of the multimedia content can constitute independent MMIOs with the capability of one referring to another.

The remaining MMIO properties that were inherited from the prototypical DOLCE IO are: the 'about' property, which is used for refering to domain ontology classes or properties; the 'realizedBy' property, which is used as the link to the 'MultimediaFile' class, which in turn describes the physical realization of the MMIO; the 'orderedBy' property, which denotes the format that expresses the MMIO (it can represent a multimedia standards' format, e.g. JPG, UTF-8,

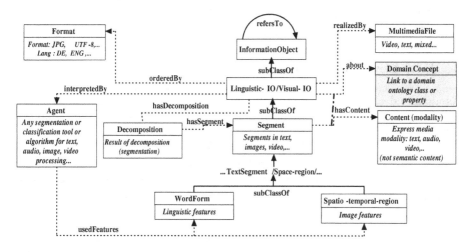

Fig. 3. Multimedia Information Object (MMIO) Design Pattern

etc., or a language, e.g. DE, EN, etc., the latter for the case of textual multimedia information); the 'interpretedBy' property, which is used to represent any segmentation or classification tool or multimedia algorithm that can be used for processing the MMIO. A schematic description of a MMIO is depicted in Fig. 3.

The Multimedia Ontology was also enriched with the 'hasMMAnalysisContextProperty' property, which was defined in order to provide the appropriate information for reinforcing the semantically driven video analysis process. More specifically, this property associates individual domain ontology concepts in terms of multimedia evidence. Thus, it covers the analysis context of a particular concept in terms of its relationship with other concepts defined in the domain ontology. According to its usage, it covers every modality that multimedia includes (image, text, sound), while it can be further analyzed in a list of properties in order to represent more specific contextual information. For example, the 'hasMMAnalysisContextProperty' property can be particularized to the property 'isLocalConceptOf' for denoting the relation of concept *debris* to the concept *earthquake* in a possible domain ontology, i.e. for denoting the relation of a local to a global concept defined in the domain ontology, as already mentioned. Another example is the 'hasFrequencyOfAppearance' property, which is introduced for indicating the degree of association of a specific local concept to a particular global concept of the domain ontology, i.e. a quantitative interpretation of the aforementioned 'isLocalConceptOf' property.

In the developed framework, the MMIO can be sub-divided into three sub-classes, namely the Visual Information Object (VIO), the Linguistic Information Object (LIO) and the Audio Information Object (AIO), that each bears all the information about the corresponding distinct modality (Visual, Textual and Audio). In the presented work, only the visual medium is considered, i.e. only the VIO notion is utilized in the semantic video analysis process. A VIO, as being sub-class of the MMIO, uses the same model of object relations that connect it to other concepts and data-type relations, which add to the visual information it carries.

4   Low-Level Visual Information Processing

As already described in Section 2, the video processing procedure, that is necessary for associating the developed ontology infrastructure with the visual domain, is initiated with the application of a temporal segmentation algorithm for segmenting the video sequence into shots and followed by a keyframe extraction step, as denoted by the 'hasDecomposition' property of the Multimedia ontology. The segmentation algorithm proposed in [16] is adopted for that purpose, due to its low computational complexity. Additionally, for every shot a single keyframe is extracted and specifically the median frame is selected.

The association of every extracted keyframe with global concepts of the domain ontology based on global-level information, as will be described in detail in the sequel, requires that appropriate low-level descriptions are extracted at the frame level and form a *frame feature vector*. The frame feature vector employed in this work comprises three different descriptors of the MPEG-7 standard, namely the *Scalable Color*, *Homogeneous Texture* and *Edge Histogram* descriptors. Following their extraction, the frame feature vector is produced by stacking all extracted descriptors in a single vector. This vector constitutes the input to the SVMs structure which realizes the association of every keyframe with global concepts of the domain ontology using global-level information, as described in Section 5.1.

Moreover, in order to perform the association of frame regions with local concepts of the domain ontology, every keyframe has to be spatially segmented into regions, as denoted again by the 'hasDecomposition' property of the Multimedia ontology, and suitable low-level descriptions have to be extracted for every resulting segment. In the current implementation, a modified K-Means-with-connectivity-constraint pixel classification algorithm has been used for segmenting the keyframes [17]. Output of this segmentation algorithm is a segmentation mask S, $S = \{s_i , i = 1, ..., N\}$, where s_i, $i = 1, ...N$ are the created spatial regions. For every generated frame segment, the following MPEG-7 descriptors are extracted: *Scalable Color*, *Homogeneous Texture*, *Region Shape* and *Edge Histogram*. The above descriptors are then combined to form a single *region feature vector*. This vector constitutes the input to the SVMs structure which computes the hypothesis sets regarding the association of every frame region with the local concepts of the domain ontology (Section 5.2).

5   Keyframe-Concept Association

5.1   Keyframe-Concept Association Using Global Features

In order to perform the association of every extracted keyframe with the global concepts defined in the domain ontology using global level descriptions, a *frame feature vector* is initially formed, as described in Section 4. Then, a SVMs structure is utilized to associate each keyframe with the appropriate global concept. This comprises R SVMs, one for every defined global concept C_r^G, each trained under the '*one-against-all*' approach. For the purpose of training the SVMs, a set of keyframes belonging to the domain of interest is assembled, Q_{tr}, as described in Section 7, and is used as training set. The aforementioned frame feature vector constitutes the input to each SVM, which at the evaluation stage returns for each keyframe of unknown global concept association a numerical value in the range $[0, 1]$ denoting the degree of confidence to which the corresponding frame is assigned to the global concept associated with the particular SVM. The metric adopted is defined as follows: For every input feature vector the distance z_r from the corresponding SVM's separating hyperplane is initially calculated. This distance is positive in case of correct classification and negative otherwise. Then, a sigmoid function is employed to compute the respective degree of confidence, h_r^G, as follows:

$$h_r^G = \frac{1}{1 + e^{-t \cdot z_r}} \quad , \tag{1}$$

where the slope parameter t is experimentally set. For each keyframe, the maximum of the R calculated degrees of association indicates its global concept assignment based on global-level information, whereas all degrees of confidence, h_r^G, constitute its respective global concept hypotheses set H^G, where $H^G = \{h_r^G, \ r = 1, ...R\}$.

5.2 Keyframe-Concept Association Using Local Features

As already described in Section 2, the SVMs structure, used in the previous section for global concept assignment using global features, is also utilized to compute the association of every keyframe region with local concepts of the domain ontology. Similarly to the global case, an individual SVM is introduced for every local concept C_j^L, to detect the corresponding association.

For that purpose, a training process similar to the one performed for the global concepts is followed. The differences are that now the region feature vector, as defined in Section 4, is utilized and that each SVM returns a numerical value in the range $[0, 1]$ which in this case denotes the degree of confidence to which the corresponding segment is assigned to the local concept associated with the particular SVM. The respective metric adopted for expressing this degree is defined as follows: Let $h_{ij}^L = I_M(g_{ij})$ denote the degree to which the visual descriptors extracted for segment s_i match the ones of local concept C_j^L, where g_{ij} represents the particular assignment of C_j^L to s_i. Then, $I_M(g_{ij})$ is defined as

$$I_M(g_{ij}) = \frac{1}{1 + e^{-t \cdot z_{ij}}} \quad , \tag{2}$$

where z_{ij} is the distance from the corresponding SVM's separating hyperplane for the input feature vector used for evaluating the g_{ij} assignment. The pairs of all supported local concepts of the domain ontology and their respective degree of confidence h_{ij}^L computed for segment s_i comprise the segment's local concept hypotheses set H_i^L, where $H_i^L = \{h_{ij}^L, \ j = 1, ... J\}$.

After the local concept hypotheses sets, H_i^L, are generated for every keyframe region s_i, a decision function, which is defined in the multimedia ontology, is introduced for realizing the global concept association based on local features, i.e. estimating the global concept assignment for every keyframe on the basis of the local concept hypotheses sets of its constituent regions:

$$d(C_r^G) = \sum_{s_i, \ where \ C_j^L \subset C_r^G} I_M(g_{ij}) \cdot (a_r \cdot freq(C_j^L, C_r^G) + (1 - a_r) \cdot area(s_i)) \ (3)$$

where $\subset$ denotes the 'isLocalConceptOf' property (already defined in the multimedia ontology), $area(s_i)$ is the percentage of the frame area captured by region s_i and $freq(C_j^L, C_r^G)$ is the frequency of appearance of local concept C_j^L with respect to the global concept C_r^G of the domain ontology. The latter is denoted by the 'hasFrequencyOfAppearance' property of the Multimedia ontology, as already described in Section 3. Regarding the computation of its value, the keyframe set, Q_{tr}, assembled as described in Section 7, is utilized. The reported frequency of appearance of each local concept C_j^L with respect to the global concept C_r^G, $freq(C_j^L, C_r^G)$, is defined as the percentage of the keyframes associated with the global concept C_r^G where the local concept C_j^L appears. The computed values are stored in the multimedia ontology. Parameters a_r are introduced for adjusting the importance of the aforementioned frequencies against the regions' areas for every defined global concept. Their values are estimated according to the procedure described in Section 6.

5.3   Information Fusion for Final Keyframe-Concept Association

After global concept association has been performed using global-, h_r^G, and local-level, $d(C_r^G)$, information, a fusion mechanism is introduced for deciding upon the final global concept association for every keyframe. This has the form of a weighted summation, based on the following equation:

$$D(C_r^G) = \mu_r \cdot d(C_r^G) + (1 - \mu_r) \cdot h_r^G \tag{4}$$

where μ_r, $r = 1, ..., R$ are global-concept-specific normalization parameters, which adjust the magnitude of the global features against the local ones upon the final outcome and their values are estimated according to the procedure described in Section 6. The global concept with the highest $D(C_r^G)$ value constitutes the final global concept association for every keyframe and consequently the semantic annotation of the respective video shot.

6   Optimizing Information Fusion

In Sections 5.2 and 5.3, variables a_r and μ_r are introduced for adjusting the importance of the frequency of appearance against the frame region's area and the global- against the local-level information on the final global concept association decision, respectively. A GA is employed for estimating their values (Section 2).

Initially, the keyframe set $\mathcal{Q}_{tr}$, that was assembled as described in Section 7, is divided into two equal in terms of amount subsets, namely a sub-training $\mathcal{Q}_{tr}^2$ and a validation $\mathcal{Q}_v^2$ set. $\mathcal{Q}_{tr}^2$ is used for training the employed SVMs framework and $\mathcal{Q}_v^2$ for validating the overall system global concept association performance during the optimization process.

Subject to the problem of concern is to compute the values of parameters a_r and μ_r that lead to the highest correct global concept association rate. For that purpose, *Global Concept Association Accuracy*, *GCAA*, is used as a quantitative performance measure and is defined as the fraction of the number of the keyframes that are associated with the correct global concept to the total number of keyframes to be examined.

Under the proposed approach, each chromosome F represents a possible solution, i.e. a candidate set of values for parameters a_r and μ_r. In the current implementation, the number of genes of each chromosome is predefined and set equal to $2 \cdot r \cdot 2 = 4 \cdot r$. The genes represent the decimal coded values of parameters a_r and μ_r assigned to the respective chromosome, according to the following equation:

$$F \equiv [\, f_1 \ f_2 \ ...f_{4 \cdot r} \,] = [\mu_1^1 \ \mu_1^2...\mu_r^1 \ \mu_r^2 a_1^1 \ a_1^2...a_r^1 \ a_r^2] \tag{5}$$

where $f_k \in \{0, 1, ...9\}$ represents the value of gene k and μ_p^q, a_p^q represent the qth decimal digit of variable μ_p, a_p, respectively. The genetic algorithm is provided with an appropriate *fitness function*, which denotes the suitability of each solution. More specifically, the fitness function $W(F)$ is defined as equal to the *GCAA* metric already defined, $W(F) \equiv GCAA(F)$, where $GCAA(F)$ is calculated over all keyframes that comprise the validation set $\mathcal{Q}_v^2$, after applying the fusion mechanism (Section 5.3) with parameter values for a_r and μ_r denoted by the genes of chromosome F.

Regarding the GA's implementation details, an initial population of 50 randomly generated chromosomes is employed. New generations are iteratively produced until the optimal solution is reached. Each generation results from the current one through the application of the following operators:

- Selection: a pair of chromosomes from the current generation are selected to serve as parents for the next generation. In the proposed framework, the Tournament Selection Operator [10], with replacement, is used.
- Crossover: two selected chromosomes serve as parents for the computation of two new offsprings. Uniform crossover with probability of 0.2 is used.
- Mutation: every gene of the processed offspring chromosome is likely to be mutated with probability of 0.4.

To ensure that chromosomes with high fitness will contribute to the next generation, the overlapping populations approach was adopted. More specifically, assuming a population of m chromosomes, m_s chromosomes are selected according to the employed selection method, and by application of the crossover and mutation operators, m_s new chromosomes are produced. Upon the resulting $m + m_s$ chromosomes, the selection operator is applied once again in order to select the m chromosomes that will comprise the new generation. After experimentation, it was shown that choosing $m_s = 0.4m$ resulted in higher performance and faster convergence. The above iterative procedure continues until the diversity of the current generation is equal to/less than 0.001 or the number of generations exceeds 30. The final outcome of this optimization procedure are the optimal values of parameters a_r and μ_r, used in Eq. 3 and 4, and which are stored in the Multimedia ontology.

7   Experimental Results

In this section, experimental results of the application of the proposed approach to videos belonging to the disaster news domain are presented. The first step to the application of the presented approach for semantic video analysis is the development of the appropriate knowledge infrastructure for representing the knowledge components that need to be explicitly defined, as described in detail in Section 3. Regarding the particular domain of experimentation, an individual domain ontology was developed. This defines the domain concepts of concern, their separation into global and local ones, and the relations among them. The taxonomy of these concepts for the selected domain is depicted in Fig. 4.

Regarding the tasks of SVMs training, parameter optimization and evaluation of the proposed system performance, a number of keyframe sets needs to be formed. More specifically, a set of 400 keyframes, Q, that were extracted from respective disaster news videos according to the procedure described in Section 4 and include global concepts of the developed domain ontology, was assembled. Each keyframe was manually annotated (i.e. assigned to a global concept and, after segmentation is applied, each of the resulting frame regions associated with a local concept of the domain ontology). This set was divided into two equal in terms of amount sub-sets, Q_{tr} and Q_{te}. Q_{tr} was used for training purposes, while Q_{te} served as a test set for the evaluation of the proposed system performance.

According to the SVMs training process (Section 5), a polynomial function was used as a kernel function by each SVM for both global and local concept association cases. The respective low-level *frame* and *region feature vector* are composed of 405 and 445 values respectively, normalized in the interval $[-1, 1]$.

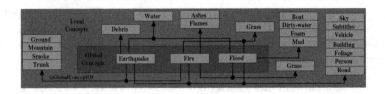

Fig. 4. Taxonomy of Domain Concepts: The arcs denote the 'isGlobalConceptOf' property, which connects the domain concepts in terms of multimedia evidence

Extracted Keyframe			
Keyframe-Concept Association Using Global-Level Information	Earthquake :**0.84** Fire :0.22 Flood :0.43	Earthquake :0.21 Fire :**0.87** Flood :0.11	Earthquake :**0.54** Fire :0.21 Flood :0.31
Keyframe-Concept Association Using Local-Level Information	Earthquake :0.51 Fire :0.18 Flood :**0.52**	Earthquake :0.24 Fire :**0.69** Flood :0.31	Earthquake :0.22 Fire :0.14 Flood :**0.62**
Keyframe-Concept Association Using Information Fusion	**Earthquake**	**Fire**	**Flood**

Fig. 5. Indicative Keyframe-Concept Association Results

The disaster news videos to be analyzed, were initially temporally segmented and corresponding keyframes were extracted, following the procedure described in Section 4. Then, based on the trained SVMs structure, keyframe-concept association based on global level features is performed, as described in Section 5.1. In parallel, after spatial segmentation is applied to the extracted keyframes, local concept hypotheses are generated for each frame segment and a decision function realizes keyframe-concept association based on local features (Section 5.2). Afterwards, the approach described in Section 5.3 is employed for implementing the fusion of the global and the local features based keyframe-concept association information and computing the final keyframe-concept assignment for every keyframe, which in turn constitutes the semantic interpretation of the respective video shot. The values of the fusion mechanism parameters are estimated according to a GA-based optimizer (Section 6).

In Fig. 5 indicative keyframe-concept association results are presented, showing the extracted keyframe, the keyframe-concept association using only global (row 2) and only local (row 3) information and the final keyframe-concept assignment after the implementation of the fusion mechanism. Additionally, in Fig. 6 exemplary region-concept association results are presented, showing the extracted keyframe (row 1) and, after spatial segmentation is applied, the association of the local concepts of the domain ontology (row 2). Furthermore, in Table 1, the respective quantitative performance measures for every individual algorithm are given in terms of accuracy for each global concept and overall. Accuracy is defined as the percentage of the keyframes that are associated with the correct global concept. From the results presented in Table 1, it can be verified

Table 1. Keyframe-Concept Association Accuracy

Method	Accuracy			
	Earthquake	Fire	Flood	Overall
Keyframe-Concept Association Using Global-Level Information	**93.75%**	**98.08%**	72.13%	86.96%
Keyframe-Concept Association Using Local-Level Information	83.33%	75.00%	59.02%	71.43%
Keyframe-Concept Association Using Information Fusion	**93.75%**	94.23%	**91.80%**	**93.17%**

Fig. 6. Indicative Region-Concept Association Results

that the keyframe-concept association based only on global information generally outperforms the respective association based only on local information. Furthermore, it must be noted that the proposed global and local features information fusion approach leads to a significant performance improvement, compared to the keyframe-concept association based solely on global or local features.

8 Conclusions

In this paper, an ontology-driven approach to semantic video analysis that is based on the notion of the Visual Information Object was presented. The proposed framework can easily be extended or applied to additional domains, provided that the employed knowledge infrastructure is appropriately modified and that the utilized training set is enriched with suitable training samples. Future plans include the introduction of audio signal processing tools and text analysis algorithms, so that the entire capabilities of the developed framework can be fully exploited and multi-modal semantic multimedia analysis is realized.

Acknowledgements

The work presented in this paper was supported by the European Commission under contracts FP6-001765 aceMedia, FP6-027685 MESH and FP6-027026 K-Space, and by the GSRT under project DELTIO.

References

1. Al-Khatib, W., Day, Y.F., Ghafoor, A., Berra, P.B.: Semantic modeling and knowledge representation in multimedia databases. IEEE Trans. on Knowledge and Data Engineering 11, 64–80 (1999)
2. Smeulders, A.W.M., Worring, M., Santini, S., Gupta, A., Jain, R.: Content-based image retrieval at the end of the early years. IEEE Trans. on Pattern Analysis and Machine Intelligence 22(12), 1349–1380 (2000)

3. Zlatoff, N., Tellez, B., Baskurt, A.: Image understanding and scene models: a generic framework integrating domain knowledge and Gestalt theory. In: ICIP 2004. Int. Conf. on Image Processing, vol. 4, pp. 2355–2358 (2004)
4. Staab, S., Studer, R.: Handbook on ontologies. In: Int. Handbooks on Information Systems, Springer, Berlin (2004)
5. Francois, A.R.J., et al.: VERL: an ontology framework for representing and annotating video events. IEEE Multimedia 12(4), 76–86 (2005)
6. Naphade, M., et al.: Large-scale concept ontology for multimedia. IEEE Multimedia 13(3), 86–91 (2006)
7. Bertini, M., Cucchiara, R., del Bimbo, A., Torniai, C.: Video Annotation with Pictorially Enriched Ontologies. In: Proc. of ICME, pp. 1428–1431 (July 2005)
8. Laboratory for Applied Ontology, `http://www.loa-cnr.it/DOLCE.html`
9. Kim, K.I., Jung, K., Park, S.H., Kim, H.J.: Support vector machines for texture classification. IEEE Trans. on Pattern Analysis and Machine Intelligence 24, 1542–1550 (2002)
10. Mitchell, M.: An introduction to genetic algorithms. MIT Press, Cambridge (1995)
11. Gangemi, A., et al.: Sweetening ontologies with DOLCE. In: Gómez-Pérez, A., Benjamins, V.R. (eds.) EKAW 2002. LNCS (LNAI), vol. 2473, Springer, Heidelberg (2002)
12. SmartWeb project: `http://www.smartweb-projekt.de/`
13. Buitelaar, P., Sintek, M., Kiesel, M.: A Lexicon Model for Multilingual/Multimedia Ontologies. In: Sure, Y., Domingue, J. (eds.) ESWC 2006. LNCS, vol. 4011, Springer, Heidelberg (2006)
14. Romanelli, M., Buitelaar, P., Sintek, M.: Modeling Linguistic Facets of Multimedia Content for Semantic Annotation. In: SAMT 2007. LNCS, vol. 4816, pp. 240–251. Springer, Heidelberg (2007)
15. SWIntO: SmartWeb Integrated Ontology, `http://smartweb.dfki.de/ontology_en.html`
16. Kobla, V., Doermann, D., Lin, K.: Archiving, indexing, and retrieval of video in the compressed domain. In: Proc. SPIE Conf. on Multimedia Storage and Archiving Systems, vol. 2916, pp. 78–89 (1996)
17. Mezaris, V., Kompatsiaris, I., Strintzis, M.G.: Still Image Segmentation Tools for Object-based Multimedia Applications. Int. Journal of Pattern Recognition and Artificial Intelligence 18(4), 701–725 (2004)

On the Selection of MPEG-7 Visual Descriptors and Their Level of Detail for Nature Disaster Video Sequences Classification

Javier Molina[1], Evaggelos Spyrou[2], Natasa Sofou[2], and José M. Martínez[1]

[1] Grupo de Tratamiento de Imágenes, Universidad Autónoma de Madrid, Spain
[2] Image, Video and Multimedia Laboratory, National Technical University of Athens, Greece
`Javier.Molina@uam.es, espyrou@image.ece.ntua.gr,`
`natasa@image.ntua.gr, JoseM.Martinez@uam.es`

Abstract. In this paper, we present a study on the discrimination capabilities of colour, texture and shape MPEG-7 [1] visual descriptors, within the context of video sequences. The target is to facilitate the recognition of certain visual cues which would then allow the classification of natural disaster-related concepts. Low-level visual features are extracted using the MPEG-7 "eXperimentation Module" (XM) [2]. The extraction times associated to the levels of detail of the descriptors are measured. The pattern sets obtained as combination of significant levels of detail of different descriptors are the input to a Support Vector Machine (SVM), resulting on the classification accuracies. Preliminary results indicate that this approach could be useful for the implementation of real-time spatial regions classifiers.

Keywords: Image Classification, Semantic Retrieval, Visual Descriptors.

1 Introduction

Due to the huge amount of multimedia contents, the automatic extraction of visual information from videos is widely required. Time performance evaluation is gaining more and more importance in visual descriptor based applications, although not very much work has been done on this line [3]. Studies mainly focus on evaluation of images retrieval accuracies as in [4]. We propose a methodology that by operating on image regions, studies the relation between the classification accuracy and the computational cost of the required extractions of descriptors.

2 Visual Descriptors Profiles and Their Levels of Detail Selection

In this paper the term *Profile* refers to a combination of visual descriptors and the term *Levels of detail* refer to the different combinations of *elements* of a specific descriptor. Since this work gives focus to spatial regions of still images, it appears that some visual descriptors do not make sense in the presented application field. First of all, motion descriptors are discarded, because currently we are working at the

B. Falcidieno et al. (Eds.): SAMT 2007, LNCS 4816, pp. 70–73, 2007.
© Springer-Verlag Berlin Heidelberg 2007

frame-by-frame level. The *Group-of-Frame/Group-of-Picture Descriptor* is discarded since it is used for joint representation of a group of frames or pictures. Moreover, the *3-D Shape Descriptor* is directly ruled out from this study, simply because we are working with 2-D projections of the 3-D real world. The *Edge Histogram Descriptor* is only applicable to rectangular images, and not to arbitrary shape regions, thus, its use makes sense for global extraction from an image. Towards the goal of obtaining useful information for real-time implementations, it makes sense to rule out the *Texture Browsing Descriptor*, since its extraction (with the XM) is about 20 times slower than the *Homogeneous Texture Descriptor* as documented in [5]. Table 1 presents the extraction times for different levels of detail of each descriptor, grouping them (one label for each group) in the case they present low variation (Groups of levels of detail of a descriptor: mean extraction time $\pm$ percentage variation). As a result, each group of levels of detail gets represented by the most detailed descriptor.

Table 1. Levels of detail Extraction times.VP(Variances Present); BN (Bins Number); SC (Spatial Coherency); NBPD (Number of Bit Planes Discarded); QR (Quantification Resolution). Extraction times measured while executing the XM version 6.1 [2] on an Intel(R) Core(TM) 2 Duo CPU T7200 @ 2Ghz with 1GB of RAM.

Label	Descriptors and levels of detail[1]	Extraction time(msec)
DCD1	DCD: VP (1,0), BN (256,128,64,32,16,8,4,2), **SC (1)**	$1832\pm0.50\%$
DCD2	DCD: VP (1,0), BN (256,128,64,32,16,8,4,2), **SC (0)**	$1577\pm1.10\%$
SCD	SCD: NBPD(0,1,2,3,4,5,6,8), NC(256,128,64,32,16)	$196\pm0.05\%$
CSD	CSD: QR(256,128,64,32)	$191\pm0.15\%$
CLD	CLD	65
HTD	HTD: layer 1 or layer 0	$2652\pm0.05\%$
rSD	region-based SD	1933
cSD	contour-SD	315

Using the equation (1) with values: $n_{DCD} = 2+1$, $n_{SCD} = 1+1$, $n_{CSD} = 1+1$, $n_{CLD} = 1+1$, $n_{HTD} = 1+1$, $n_{region-basedSD} = 1+1$ and $n_{contour-SD} = 1+1$, the obtained number of patterns´ sets (Ns) is 96. The '$+1$' is added in order to consider the non usage of descriptors.

$$Ns = n_{DCD} \times n_{SCD} \times n_{CSD} \times n_{CLD} \times n_{HTD} \times n_{\text{region-based SD}} \times n_{\text{contour-SD}} \tag{1}$$

Each pattern set is analyzed with a Support Vector Machine using the LIBSVM implementation [6] and applying 10-fold cross-validation. The relations of computational cost and classification accuracy are presented in section 0.

3   Classification Results

The dataset is composed of a subset of MESH repository, images from the *Labelme* dataset [7] and various images collected from the *world wide web*. The goal was the

---

[1] DCD (*Dominant Color Descriptor*); SCD (*Scalable Color* Descriptor) ; CSD (*Color Structure Descri ptor*); HTD (*Homogeneous Texture Descriptor*); CSD (*Color Layout Descriptor*); region-based SD (*Region-Based Shape Descriptor*); contour-SD (*Contour-Shape Descriptor*).

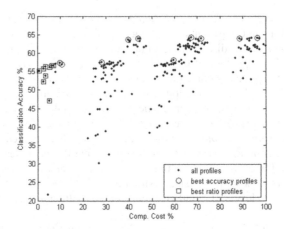

Fig. 1. Relation between computational cost and classification accuracy for the combinations of descriptors and their levels of detail. The computational cost is expressed as the percentage of the maximum value of computational cost (7184 msec.) which corresponds to the profile [DCD1 SCD CSD CLD HTD rSD cSD].

detection of certain visual cues (*flames, smoke, vegetation, buildings, water, snow* etc) which, when combined, would assist to the detection of semantic concepts (*forest fires,floods, volcanic eruptions* etc). Around 100 spatial regions of each visual cue have been manually obtained and annotated.

Table 2. Profiles with best classification accuracies of computational cost segments

Profile	Cost (%)	Acc. (%)	Acc./Cost
[SCD CSD cSD]	(0,10]	57.41	5.8762
[SCD CSD CLD cSD]	(20,30]	56.98	5.3352
[DC2 SCD CSD CLD]	(30,40]	57.56	2.0382
[SCD HTD]	(30,40]	63.66	1.606
[SCD HTD cSD]	(40,50]	63.95	1.4524
[DC2 SCD CSD CLD HTD]	(50,60]	61.63	0.94583
[SCD CLD HTD rSD]	(60,70]	64.24	0.95227
[SCD CLD HTD rSD cSD]	(70,80]	63.95	0.89017
[DC2 SCD HTD rSD]	(80,90]	63.95	0.7226
[DC1 SCD HTD rSD cSD]	(90,100]	64.1	0.66466

Table 3. Profiles with best relation between accuracy and computational cost

Profile	Cost (%)	Acc. (%)	Acc./Cost
[CLD]	0.9	55.09	61.211
[SCD]	2.73	55.81	20.443
[CSD]	2.66	52.18	19.617
[SCD CLD]	3.63	56.4	15.537
[CSD CLD]	3.56	53.78	15.107
[SCD CSD]	5.39	56.25	10.436
[SCD CSD CLD]	6.29	56.69	9.0127
[CLD cSD]	5.29	47.09	8.9017
[SCD cSD]	7.11	56.69	7.9733
[CSD cSD]	7.04	51.89	7.3707

It can be inferred from the compiled data in Table 1 that in some combinations the use of a concrete descriptor is counterproductive, worsening the classification results. It makes sense to highlight the profiles which show the highest accuracy/computational cost ratios (Table 2). This gives us a hint of which MPEG-7 visual descriptors are most recommendable for real time contexts.

4 Conclusions and Future Work

In this work, a method for the estimation of the interdependency between classification accuracy of spatial regions and computational cost of the required descriptor extraction has been proposed. We can conclude that certain descriptors seem to be more recommendable for real time applications, since they improve the classification accuracy with a very small increase of the computational cost. Future work includes study on temporal dependant descriptors, such as motion or shape evolution descriptors.

Acknowledgments. The work presented in this paper was partially supported by the European Commision under contractFP6-027685 MESH, and by the Spanish Government under project TIN2004-07860-C02 (MEDUSA) and by the Comunidad de Madrid under project S-0505/ TIC-0223 (ProMultiDis-CM). Evaggelos Spyrou is funded by PENED 2003 – Project Ontomedia 03ED475.

References

1. Manjunath, B.S., Salembier, P., Sikora, T.: Introduction to MPEG-7, 1st edn. John Wiley & Sons, Ltd., West Sussex, England
2. MPEG-7: Visual experimentation model (xm) version 10.0. ISO/IEC/JTC1/SC29/WG11, Doc. N4062 (2001)
3. Mikolajczyk, K., Schmid, C.: A performance Evaluation of Local Descriptors. IEEE Transactions on Pattern Analysis and Machine Intelligence 27(10), 1615–1630 (2007)
4. Ojala, T., Aittola, M., Matinmikko, E.: Empirical Evaluation of MPEG-7 XM Color Descriptors in Content-Based Retrieval of Semantic Image Categories. In: ICPR 2002. 16th International Conference on Pattern Recognition, vol. 2, p. 21021.
5. Ojala, T., Mäenpää, T., Viertola, J., Kyllönen, J., Pietikäinen, M.: Empirical evaluation of MPEG-7 texture descriptors with a large-scale experiment. In: Proc. 2nd International Workshop on Texture Analysis and Synthesis, Copenhagen, Denmark, pp. 99–102 (2002)
6. Chang, C.-C., Lin, C.-J.: LIBSVM : a library for support vector machines (2001), software available at http://www.csie.ntu.edu.tw/ cjlin/libsvm
7. Russell, B.C., Torralba, A., Murphy, K.P., Freeman, W.T.: LabelMe: a database and web-based tool for image annotation, MIT AI Lab Memo AIM-2005-025 (September 2005)

A Region Thesaurus Approach for High-Level Concept Detection in the Natural Disaster Domain

Evaggelos Spyrou and Yannis Avrithis

Image, Video and Multimedia Systems Laboratory,
School of Electrical and Computer Engineering
National Technical University of Athens
9 Iroon Polytechniou Str., 157 80 Athens, Greece
espyrou@image.ece.ntua.gr
http://www.image.ece.ntua.gr/~espyrou/

Abstract. This paper presents an approach on high-level feature detection using a region thesaurus. MPEG-7 features are locally extracted from segmented regions and for a large set of images. A hierarchical clustering approach is applied and a relatively small number of region types is selected. This set of region types defines the region thesaurus. Using this thesaurus, low-level features are mapped to high-level concepts as model vectors. This representation is then used to train support vector machine-based feature detectors. As a next step, latent semantic analysis is applied on the model vectors, to further improve the analysis performance. High-level concepts detected derive from the natural disaster domain.

1 Introduction

High-level concept detection in both image and video documents remains still an unsolved problem. However, due to the continuously growing volume of audiovisual content, this problem attracts a lot of interest within the multimedia research community. Its two main and most interesting aspects appear the selection of the low-level features that will be extracted and the approach that will be used for assigning these low-level descriptions to high-level concepts, a problem commonly referred to as the "Semantic Gap".

There exist plenty of works towards the solution of this problem. In [1] a multimedia analysis and retrieval system using multi-modal machine learning techniques in order to model semantic concepts in video is presented. Moreover, in [2], a region-based approach in content retrieval that uses Latent Semantic Indexing (LSI) techniques is presented. In [3] the features are extracted by segmented regions of an image. Also, in [4], a region-based approach is presented, that uses knowledge encoded in the form of an ontology. Moreover, a hybrid thesaurus approach for semantic object recognition and identification within video news archives is presented in [5]. Finally, a lexicon, used in an approach for an interactive video retrieval system is presented in [6].

B. Falcidieno et al. (Eds.): SAMT 2007, LNCS 4816, pp. 74–77, 2007.

Fig. 1. An input image and its coarse segmentation

2   Low-Level Feature Extraction

For the representation of the low-level features, descriptors from the MPEG-7 standard [7] were used and more specifically, the *Color Layout Descriptor*, the *Scalable Color Descriptor*, the *Color Structure Descriptor* and the *Homogeneous Texture Descriptor*. For the extraction of the aforementioned descriptors, the MPEG-7 eXperimentation Model (XM)[8] was used.

Instead of extracting descriptors globally, the color and texture descriptors are extracted from image regions. A multiresolution variation of the RSST color segmentation algorithm [9] is first applied, tuned to produce a coarse segmentation. This way, one can intuitively describe the image with respect to the image segments. To explain this, an input image along with its coarse segmentation is depicted in figure 1. In this example, one could easily describe the input image as a set of regions. Here a user could see "a light blue region" (sky), "two green regions" (vegetation), "an orange region" (fire) etc. Then, the MPEG-7 descriptors are extracted locally, from each image region.

3   Region Thesaurus Construction

Given the entire set of images and their extracted low-level features as described in section 2, it is rather obvious that regions belonging to similar semantic concepts, also have similar low-level descriptions. Also, images that contain the same semantic concepts include similar regions. To exploit these observations, we try to formalize an image description in terms of the regions it is consisted of.

A hierarchical clustering algorithm [10] is applied on the low-level descriptions of all the regions that occur from segmenting all images from the training set. After this clustering, the number of clusters to keep is selected experimentally. This way, by keeping the centroids of those clusters, we select the more often encountered regions. These regions will be referred to as "region types" and form the region thesaurus. Its purpose is to formalize a conceptualization between the low and the high-level features and facilitate their association.

Each region type is represented as a feature vector that contains all the extracted low-level information for it. As it is obvious, a low-level descriptor does not carry any semantic information. On the other hand, a high-level concept carries only semantic information. A region type lies in-between and contains the necessary information to formally describe the color and texture features, but can also be described with a "lower" description than the high-level concepts. I.e., one can describe a region type as "a green region with a coarse texture".

After this clustering procedure, we can easily observe that each cluster may or may not contain regions from the same high-level feature and regions from the same high-level feature may be encountered in more than one clusters. For example, the high-level concept *vegetation* can have more than one instances differing in i.e. the color of the leaves of trees. Moreover, in a cluster that contains instances from the semantic entity i.e. *sea*, these instances could be mixed up with parts from i.e. *sky*.

4   Image Analysis

Having calculated the distance of each region of the image to all the words of the constructed thesaurus we construct a model vector to semantically describe the visual content of the image, in terms of the region thesaurus. This vector is formed by keeping the smaller distance for each region type among all image regions. The distances are calculated using the MPEG-7 XM [8] and linearly combined. Let: $d_i^1, d_i^2, ..., d_i^j, i = 1, 2, 3, 4$ and $j = N_C$, where N_C is the number of region types and d_i^j the distance of the i-th region of the clustered image to the j-th region type. Then, the model vector D_m is described by equation 1.

$$D_m = [min\{d_i^1\}, min\{d_i^2\}, ..., min\{d_i^{N_C}\}], i = 1, 2, 3, 4 \qquad (1)$$

Using these model vectors to describe low-level image features, we train one Support Vector Machine for each high-level concept. We also perform experiments using a Latent Semantic Analysis [11](LSA) approach. This way, we try to exploit the relationships between the high-level concepts and the region types they contain more often. Images correspond to documents and region types to terms. This way, the co-occurrence matrix is formed. Then, matrices Σ and U are determined using the training set of images. The value of k is selected experimentally. Finally, each input model vector is driven to the concept space. This way, the model vectors are transformed and used to train the concept detectors.

5   Experimental Results

For the evaluation of the presented framework a dataset[1] from various images collected from the world wide web. This set consists of approximately 600 images from the following semantic classes: *fire, rocks, smoke, snow, trees, water*. A separate detector was trained for each concept. Results are shown in table 1, before and after the application of Latent Semantic Analysis. The presented approach was also tested on TRECVID 2007 [12] development data for the same concepts. Considering the complexity of the TRECVID data, the results appear promising.

---

[1] Special thanks to Javier Molina for sharing his collection.

Table 1. Accuracy for all 6 concepts. Set 1 contains images collected from the web, set 2 from TRECVID 2007 development data.

Concept	Set 1 Without LSA	Set 1 With LSA	Set 2 Without LSA	Set 2 With LSA
Fire	76.0%	84.0%	20.0%	46.0%
Rocks	69.5%	73.9%	21.0%	63.0%
Smoke	60.0%	64.0%	-	-
Snow	80.9%	90.5%	46.0%	72.0%
Vegetation	94.7%	89.4%	31.0%	47.0%
Sea	65.3 %	72.0%	-	-
Sky	72.0 %	75.0%	41.0%	47.0%

Acknowledgements

The work presented in this paper was partially supported by the European Commission under contracts FP6-027026 K-Space and FP6-027685 MESH. Evaggelos Spyrou is funded by PENED 2003 Project Ontomedia 03ED475.

References

1. IBM: MARVEL Multimedia Analysis and Retrieval System. IBM Research White paper (2005)
2. Souvannavong, F., Mérialdo, B., Huet, B.: Region-based video content indexing and retrieval. In: CBMI 2005. Fourth International Workshop on Content-Based Multimedia Indexing, June 21-23, 2005, Riga, Latvia (2005)
3. Saux, B., Amato, G.: Image classifiers for scene analysis. In: International Conference on Computer Vision and Graphics (2004)
4. Voisine, N., Dasiopoulou, S., Mezaris, V., Spyrou, E., Athanasiadis, T., Kompatsiaris, I., Avrithis, Y., Strintzis, M.G.: Knowledge-assisted video analysis using a genetic algorithm. In: WIAMIS 2005. 6th International Workshop on Image Analysis for Multimedia Interactive Services (April 13-15, 2005)
5. Boujemaa, N., Fleuret, F., Gouet, V., Sahbi, H.: Visual content extraction for automatic semantic annotation of video news. In: IS&T/SPIE Conference on Storage and Retrieval Methods and Applications for Multimedia, part of Electronic Imaging symposium (2004)
6. Snoek, C.G., Worring, M., Koelma, D.C., Smeulders, A.W.: Learned lexicon-driven interactive video retrieval. In: Sundaram, H., Naphade, M., Smith, J.R., Rui, Y. (eds.) CIVR 2006. LNCS, vol. 4071, Springer, Heidelberg (2006)
7. Chang, S.F., Sikora, T., Puri, A.: Overview of the mpeg-7 standard. IEEE trans. on Circuits and Systems for Video Technology 11(6), 688–695 (2001)
8. MPEG-7: Visual experimentation model (XM) version 10.0. ISO/IEC/ JTC1/SC29/WG11, Doc. N4062 (2001)
9. Avrithis, Y., Doulamis, A., Doulamis, N., Kollias, S.: A stochastic framework for optimal key frame extraction from mpeg video databases (1999)
10. Duda, R.O., Hart, P.E., Stork, D.G.: Pattern Classification, 2nd edn. Wiley Interscience, Chichester (2000)
11. Deerwester, S., Dumais, S., Furnas, G.W., Landauer, T.K., Harshman, R.: Indexing by latent semantic analysis. Journal of the Society for Information Science 41(6), 391–407 (1990)
12. Smeaton, A.F., Over, P., Kraaij, W.: Evaluation campaigns and trecvid. In: MIR 2006. Proceedings of the 8th ACM International Workshop on Multimedia Information Retrieval, pp. 321–330. ACM Press, New York, NY, USA (2006)

Annotation of Heterogeneous Multimedia Content Using Automatic Speech Recognition

Marijn Huijbregts, Roeland Ordelman, and Franciska de Jong

University of Twente, Dept. of Electrical Engineering,
Mathematics and Computer Science,
P.O. Box 217, 7500 AE, Enschede, The Netherlands
{m.a.h.huijbregts,ordelman,fdejong}@ewi.utwente.nl
http://hmi.ewi.utwente.nl

Abstract. This paper reports on the setup and evaluation of robust speech recognition system parts, geared towards transcript generation for heterogeneous, real-life media collections. The system is deployed for generating speech transcripts for the NIST/TRECVID-2007 test collection, part of a Dutch real-life archive of news-related genres. Performance figures for this type of content are compared to figures for broadcast news test data.

1 Introduction

The exploitation of linguistic content such as transcripts generated via automatic speech recognition (ASR) can boost the accessibility of multimedia archives enormously. This effect is of course limited to video data containing textual and/or spoken content but when available, the exploitation of linguistic content for the generation of a time-coded index can help to bridge the semantic gap between media features and search needs. This is confirmed by the results of TREC series of Workshops on Video Retrieval (TRECVID)[1]. The TRECVID test collections contain not just video, but also ASR-generated transcripts of segments containing speech. Systems that do not exploit these transcripts typically do not perform as well as the systems that do incorporate speech features in their models [13], or to video content with links to related textual documents, such as subtitles and generated transcripts.

ASR supports the conceptual querying of video content and the synchronization to any kind of textual resource that is accessible, including other full-text annotation for audiovisual material[4]. The potential of ASR-based indexing has been demonstrated most successfully in the broadcast news domain. Spoken document retrieval in the American-English broadcast news (BN) domain was even declared 'a solved problem' based on the results of the TREC Spoken Document Retrieval (SDR) track in 1999 [7]. Partly because collecting data to train recognition models for the BN domain is relatively easy, word-error-rates (WER)

---

[1] http://trecvid.nist.gov

B. Falcidieno et al. (Eds.): SAMT 2007, LNCS 4816, pp. 78–90, 2007.
© Springer-Verlag Berlin Heidelberg 2007

below 10% are no longer exceptional[8,9], and ASR transcripts for BN content approximate the quality of manual transcripts, at least for several languages.

In other domains than broadcast news and for many less favored languages, a similar recognition performance is usually harder to obtain due to (i) lack of domain-specific training data, and (ii) large variability in audio quality, speech characteristics and topics being addressed. However, as ASR performance of 50 % WER is regarded as a lower bound for successful retrieval, speech-based indexing for harder data remains feasible as long as the ASR performance is not below 50 % WER, and is actually a crucial enabling technology if no other means (metadata) are available to guide searching.

For 2007, the TRECVID organisers have decided to shift the focus from broadcast news video to video from a real-life archive of news-related genres such as news magazine, educational, and cultural programming. As in previous years, ASR transcripts of the data are provided as an optional information source for indexing. Apart from some English BN rushes (raw footage), the 2007 TRECVID collection consists of 400 hours of Dutch news magazine, science news, news reports, documentaries, educational programmes and archival video. The files were provided by the Netherlands Institute for Sound and Vision[2]. (In the remainder this collection will be referred to as *Sound and Vision* data.)

This paper reports on the setup and evaluation of the speech recognition system (further referred to as SHoUT system[3] that is deployed for generating the transcripts that via NIST will be made available to the TRECVID-2007 participants. The SHoUT system is particularly geared towards transcript generation for the kind of heterogeneous, real-life media collections exemplified by the *Sound and Vision* data that will feature in the TRECVID 2007 collection. In other words, it targets adequate retrieval performance, rather than plain robust ASR.

As can be expected for a diverse content set such as the *Sound and Vision* data, the audio and speech conditions vary enormously, ranging from read speech in a studio environment to spontaneous speech under degraded acoustic conditions. Furthermore, a large variety of topics are addresses and the material dates from a broad time period. Historical items as well as contemporary video fall within the range. (The former with poorly preserved audio; latter with varying audio characteristics, some even without 'intended' sound, just noise).

To reach recognition accuracy that is acceptable for retrieval given the difficult conditions, the different parts of our ASR system must be made as robust as possible so that it is able to cope with those problems (such as mismatches between training and testing conditions) that typically emerge when technology is transferred from the lab and applied in a real life context. This is in accordance with our overall research agenda: the development of robust ASR technology that can be ported to different topic domains with a minimum effort. Only technology that complies with this last requirement can be successfully deployed for the wide range of spoken word archives that call for annotation based on speech content such as cultural heritage data (historical archives and

---

[2] Netherlands Institute for Sound and Vision: `http://www.beeldengeluid.nl/`

[3] SHoUT is an acronym for SpeecH recognition University of Twente.

interviews), lecture collections, meeting archives (city council sessions, corporate meetings).

The remainder of this paper is organised as follows. Section 2 provides an overview of the system structure, followed by a description of the training procedure that was applied for the SHoUT-2007a version aiming at the transcription of *Sound and Vision* data (section 3). Section 4 presents and discusses performance evaluation results of this system obtained using *Sound and Vision* data and various other available test corpora.

2   System Structure

Figure 1 is a graphical representation of the system work flow that starts with speech activity detection (SAD) in order to filter out the audio that do not contain speech. The audio in the *Sound and Vision* collection contains all kinds of sounds such as music, sound effects or background noise with high volume (traffic, cheering audience, etc). As a speech decoder will always try to map a sound segment to a sequence of words, processing non-speech portions of the videos (i) would be a waste of processor time, (ii) introduces noise in the transcripts due to assigning word labels to non-speech fragments, and (iii) reduces speech recognition accuracy in general when the output of the first recognition run is used for acoustic model adaptation purposes.

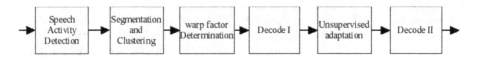

Fig. 1. Overview of the decoding system. Each step provides input for the following step.

After SAD, the speech fragments are segmented and clustered. In this step, the speech fragments are split into segments that only contain speech from one single speaker. Each segment is labelled with its corresponding speaker ID. Next, for each segment the vocal tract length (VTLN) warping factor is determined for vocal tract length normalisation during feature extraction for the first decoding step. Decoding is done using the HMM-based Viterbi decoder. In the first decoding iteration, triphone VTLN acoustic models and trigram language models are used. For each speaker, a first best hypothesis aligned on a phone basis is created for unsupervised acoustic model adaptation. For each file, the language model is mixed with a topic specific language model. The second decoding iteration uses the speaker adapted acoustic models and the topic specific language models to create the final first best hypothesis aligned on word basis. Also, for each segment, a word lattice is created.

The following sections describe each of the system parts in more detail.

2.1 Speech Activity Detection

Most SAD systems are based on a Hidden Markov Model (HMM) with a number of parallel states. Each state contains a Gaussian Mixture Model (GMM) that is trained on a class of sounds such as speech, silence or music. The classification is done by performing a Viterbi search using this HMM.

A disadvantage of this approach is that the GMMs need to be trained on data that matches the evaluation data. The performance of the system can drop significantly if this is not the case. Because of the large variety in the *Sound and Vision* collection it is difficult to determine a good set of data on which to train the silence and speech models. Also it is difficult to determine what kind of extra models are needed to filter out unknown audio fragments like music or sound effects and to find training data for those models.

The SAD system proposed by [1] at RT06s uses regions in the audio that have low or high energy levels to train new speech and silence models on the data that is being processed. The major advantage of this approach is that no earlier trained models are needed and therefore the system is robust for domain changes. We extended this approach for audio that contains fragments with high energy levels that are not speech.

Instead of using energy as an initial confidence measure, our system uses the output of our broadcast news SAD system. This system is only trained on silence and speech, but because energy is not used as a feature and most non-speech data will fit the more general silence model better than the speech model, most non-speech will be classified as silence. After the initial segmentation the data classified as silence is used to train a new silence model and a sound model. The silence model is trained on data with low energy levels and the sound model on data with high energy levels. After a number of training iterations, the speech model is also re-trained. The result is an HMM based SAD system with three models (speech, non-speech and silence) that are trained solely on the data under evaluation.

2.2 Segmentation and Clustering

Similar to the SAD system, the segmentation and clustering system uses GMMs to model different classes. This time instead of trying to distinguish speech from non-speech, each GMM represents a single unique speaker. In order to find the correct number of speakers, first the data is divided into a number of small segments and for each segment a model is trained. After this, all models are pairwise compared. If two models are very similar (if it is believed that they model the same speaker) they are merged. This procedure is repeated until no two models can be found that are believed to be trained on speech of the same speaker. In [14] this *speaker diarization* algorithm is discussed in depth.

Although this speaker diarization approach has very good clustering results, it is not very fast on long audio files. In order to be able to process the entire *Sound and Vision* collection in reasonable time, we have changed the system slightly. For the purpose it is used here (to segment the data so that we can apply VTLN and do unsupervised adaptation) a less accurate clustering will still be very helpful.

The majority of processing time is spent on pairwise comparing models. At each merging iteration, all possible model combinations need to be recomputed. The longer a file is, the more initial clusters are needed. With this, the number of model combinations that need to be considered for merging increases more than linearly. We managed to bring back the number of merge calculations by simply merging multiple models at a time. The two models A and B with the highest BIC score are merged first, followed by the two models C and D with the second highest score. If this second merge involves one of the earlier merged models, for example model C is the same model as model A, also the other combination (B,D) must have a positive BIC score. This process is repeated with a maximum of four merges at one iteration. Without performance loss (see section 4), this procedure reduces the real-time factor from 8 times to 2.7 times real-time on a 35 minute TRECVID file.

2.3   Vocal Tract Length Normalization

Variation of vocal tract length between speakers makes it harder to train robust acoustic models. In the SHoUT system, normalisation of the feature vectors is obtained by shifting the Mel-scale windows by a certain warping factor.

If the warping factor is smaller than one, the windows will be stretched and if the factor is bigger than one, the windows will be compressed. To normalize for the vocal tract length, large warping factors should be applied for people with a low voice and small warping factors for people with a high voice (note that this is typically because of how we implemented the warping factor. In the literature often big warping factors are linked to high voices instead of low voices).

In order to determine the speakers warping factors, a gaussian mixture model (referred to as VTLN-GMM) was trained with data from the Spoken Dutch Corpus (CGN) [10]. In total, speech of 200 male speakers and 200 female speakers was used. The GMM only contained four gaussians. For the 400 speakers in the training set, the warping factor is determined by calculating the feature vectors with a warping factor varying from 0.8 to 1.2 in step sizes of 0.04 and determining for each of these feature sets what the probability on the VTLN-GMM is. For each speaker the warping factor used to create the set of features with the highest probability is chosen.

After each speaker is assigned a warping factor, a new VTLN-GMM is trained using normalised feature vectors and again the warping factors are determined by looking at a range of factors and picking the one with the highest score. This procedure is repeated a number of times so that a VTL normalised speech model is created. From this point on, the warping factor for each speaker is determined by looking at a range of factors on this normalised model. The acoustic models needed for decoding are trained on features that are normalised using this method.

Figure 2 contains the warping factors of the hundred speakers of the VTLN-GMM training set split in male and female. Three training iterations were performed. At the third iteration a model with eight gaussians was trained. This model is picked to be the final VTLN-GMM.

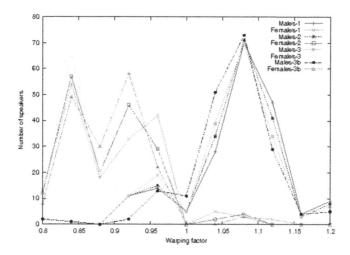

Fig. 2. The warping factor of 200 male and 200 female voices determined using the VTLN-GMM after each of its three training iterations. The female speakers are clustered at the left and the male speakers at the right.

2.4   Decoding

The decoder's Viterbi search is implemented using the token passing paradigm [6]. HMMs with three states and GMMs for its probability density functions are used to calculate acoustical likelihoods of context dependent phones. Up to 4-gram back-off language models (LMs) are used to calculate the priors.

The HMMs are organised in a single Pronunciation Prefix Tree (PPT) as described in [6,5]. This PPT contains the pronunciation of every word from the vocabulary. Each word is defined by an ID of four bytes. Identical words with different pronunciations will receive the same ID making it possible to model pronunciation variations.

Instead of copying PPTs for each possible linguistic state (the LM n-gram history), each token contains a pointer to its LM history (see figure 3). Tokens coming from leaves of the PPT are fed back into the root node of the tree after their n-gram history is updated. Only for tokens with the same LM history, token collisions will occur. This means that each HMM state of each node in the single PPT can contain a list of tokens with unique n-gram histories. These lists are sorted in descending order of the token probability scores.

The LM data are stored in up to four lookup tables. The first table contains unigram probabilities and back-off values for all words of the lexicon. The statistics for all available bigrams, trigrams and fourgrams are stored in three minimal perfect hash tables [2]. These hash tables contain exactly one occurrence in each slot of the table. This means that no extra memory is needed except for storing the hash function and for the key of each data structure, the n-gram history. This key is needed because during lookup, the hash function will map queries for non-existing n-grams to random slots. By comparing the n-gram of the query

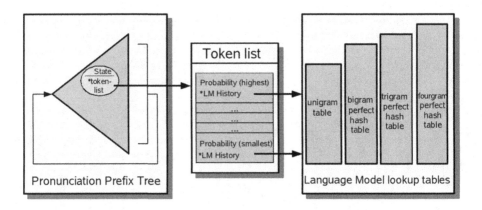

Fig. 3. The decoder uses a single Pronunciation Prefix Tree (PPT) and a 4-gram Language Model (LM). Each HMM state from the PPT can contain a sorted list of tokens. Each token from a list has a unique LM history.

to the n-gram of the found table slot, it can be determined if the search is successful. The algorithm proposed in [3,2] is used to generate the hash functions. Each n-gram table contains a backoff value so that when an n-gram probability does not exist in the hash table, the system can backoff and look up the (n-1)-gram probability of the last words of the token n-gram history.

3   System Training

3.1   SAD Model Training

The GMMs used to initialise the SAD system (see section 2.1) are trained on a small amount of Dutch broadcast news training data from the CGN corpus [10]. Three and a half hours of speech and half an hour of silence from 200 male and 200 female speakers were used. The feature vectors consist of twelve MFCC components and the zero-crossing statistic. Each GMM (silence and speech) contains 20 gaussians.

3.2   Acoustic Model Training

The acoustic models (AMs) for ASR are trained using features with twelve MFCC components. Energy is added as a thirteenth component and deltas and delta-deltas are calculated. Before calculating the MFCC features, the warping factor is determined as described in section 2.3 so that the Mel-scale windows can be shifted according to this factor.

A set of 39 triphones and one silence phone is used to model acoustics. The triphones are modelled using three HMM states. During training a decision tree is used to tie the triphone gaussians for each state. The number of triphone clusters (triphones that are mapped to the same gaussian mixture) is limited by the amount of available training data for each cluster and a fixed maximum allowed number of clusters. After clustering the triphones, the models are trained starting with one gaussian each and iteratively increasing the number of gaussians until each cluster contains 32 gaussians.

In total the acoustic models were trained on 79 hours of broadcast news, interviews and live commentaries of sport events from the CGN corpus [10]. During training 1443 triphone clusters were defined resulting in acoustic models with in total 46176 gaussians.

This data collection is the training set that is allowed to use in the primary condition of the Dutch ASR benchmark organized by the N-Best project[4].

3.3   Acoustic Model Adaptation

The clustering information obtained during speaker diarization (see section 2.2) is used to create speaker dependent acoustic models. SMAPLR adaptation ([12]) is used to adapt the means of the acoustic model gaussians. Before adapatation starts, a tree structure is created for the adaptation data. The more data is available, the deeper the tree will be. The root of the tree contains all data and the leaves only small sets of phones. The data at each branch is used to perform Maximum a Posteriori (MAP) adaptation using output of its parent nodes as prior densities for the transformation matrices. This method prevents over-adapting models with small amounts of data but it also proves to adapt models very well if more data is available. Most important, hardly any tuning is needed using this method. Only the minimum amount of data per tree node and the maximum tree depth needed to be defined. These two parameters have been tuned on broadcast news data.

3.4   Language Model Training

Broadcast news language models (LMs) were estimated from approximately 500 million words of normalised Dutch text data from various resources (See Table 1). The larger part of the text data consists of written texts (98%), of which the majority is newspaper data[5]. A small portion of the available resources, some 1 million words, consists of speech transcripts, mostly derived from the various domains of the Spoken Dutch Corpus (CGN) [10] and some collected at our institute as part of the BN-speech component of the Twente News Corpus (TwNC-speech).

Table 1. Dutch text corpora and word count based on normalised text

corpus	description	word count
Spoken Dutch Corpus	various	6.5M
Twente News Corpus (speech)	'01-'02 BN transcripts	112K
Dutch BN Autocues Corpus	'98-'05 BN teleprompter	5.7M
Twente News Corpus (written)	'99-'06 Dutch NP	513M

---

[4] http://speech.tm.tno.nl/n-best

[5] provided for research purposes by the Dutch Newspaper publisher PCM.

Table 2. OOV rates using different vocabulary selection strategies based on available training data

vocabulary	%OOV
14.6K CGN	9.04
50K general NP (99-05)	2.88
50K recent NP	2.39
51K merge recent-NP, CGN	2.33

Table 3. Language model perplexities

ID LM	PP	mixture-weigths
01 Autocues Corpus LM	389	n.a.
02 CGN comp-k (news)	797	n.a.
03 01 + TwNC-speech + 02	362	n.a.
04 Twente News Corpus (NP)	266	n.a.
05 cgn-comp-f (discussion/debates)	652	n.a.
06 mix 03-04	226	0.44 - 0.56
07 mix 03-04-05	211	0.27 - 0.55 - 0.18

For the selection of the vocabulary words we used the 14K most frequent words from the CGN corpus with a minimum word frequency of 10, a selection of recent words from 2006 newspaper (until 31/08/2006, see below) data and a selection of most frequent words from the 1999-2005 newspaper (NP) data set. The aim was to end up with a vocabulary of around 50K words that could be regarded as a more or less stable, general BN vocabulary that could later be expanded using task-specific words learned from either meta-data or via multi-pass speech recognition approaches. Table 2 shows that a selection of words from speech transcripts (CGN) augmented with recent newspaper data gave the best result with respect to OOV (2.33 %).

Trigram language models were estimated from the available data sources using modified Kneser-Ney smoothing. Perplexities of the LMs were computed using a set of manually transcribed BN shows of 7K words that date from the period after the most recent data in our text data collection. From the best performing LMs interpolated versions were created. In Table 3 perplexity results and mixture-weights of the respective models are listed. The mixture-LM with the lowest perplexity (211) was was used for decoding and contained about 8.9M 2-grams and 30M 3-grams.

3.5 Video Item Specific Language Models

For every video item in the *Sound and Vision* collection, descriptive metadata is provided containing, among others, content summaries of around 100 words. In order to minimise the OOV rate for content words in the video items, item-specific language models were created using these descriptions, a database of newspaper articles (Twente News Corpus) and an Information Retrieval (IR) system that returns ranked lists of relevant documents given a query or 'topic'.

The procedure was as follows:

1. use description of video item from metadata as a query
2. generate LM training data from top N most relevant documens given a query
3. select most frequent words
4. create topic specific language model
5. mix specific model with background language model using a fixed weight

The meta-data descriptions were normalised and stop-words were filtered out in the query processing part of the IR. For the initial experiments described in this paper, the top 50, 250 and 1000 documents from the ranked lists were taken as input for language model training. Every word from the meta-data description was added to the new vocabulary. New words from the newspaper data were only included when they exceeded a minimum frequency count (10 in the experiments reported here). Pronunciations for the new words were generated using an automatic grapheme-to-phoneme converter [11]. As no example data was available for every video item to estimate mixture weights appropriately, a fixed weight was chosen (0.8 for the NP-LM in the experiments reported here).

4   Experimental Results

4.1   Evaluation Data

For evaluation purposes we used a number of different resources. For evaluation of SAD and speaker diarization we used RT06s and RT07s conference meetings. These data do not match the training conditions as will also be the case with the *Sound and Vision* data and should provide a nice comparison. Also, from the *Sound and Vision* data, 13 fragments of 5 minutes each were randomly selected and annotated manually. To compare speech recognition results on *Sound and Vision* data with broadcast news transcription performance we selected one recent broadcast news show from the Twente News Corpus (TwNC-BN) that dates after the most recent date of the data that was used for language model training. For global comparison we also selected broadcast news data from the Spoken Dutch Corpus (CGN-BN).

4.2   Speech Activity Detection

In Table 4, the SAD error rate on RT06s and RT07s meeting data and *Sound and Vision* data are listed. On the RT06s and *Sound and Vision* data we also evaluated our BN SAD system (baseline). The results show that on both evaluation corpora we improved on the baseline BN SAD system. The SAD error on the conference meeting data is conform state-of-the-art[6].

---

[6] Because of the rules of the NIST evaluation we are prohibited from comparing our results with those of other participants. See http://www.nist.gov/speech/tests/ rt/rt2006/spring/pdfs/rt06s-SPKR-SAD-results-v5.pdf for the actual ranking.

Table 4. Evaluation results of the SAD component showing results of the BN system (baseline) and optimised results (SHoUT-2007a)

eval	baseline	SHoUT-2007a
RT06s	26.9%	4.4%
RT07s		3.3%
Sound and Vision	18.3%	10.4%

4.3 Speaker Diarization

As we did not have ground truth transcriptions on *Sound and Vision* or BN data for diarization, we could only test diarization on conference meeting data. The speaker diarization error (DER) on the conference meetings of RT07s is 11.14% on the Multiple Distance Microphone (MDM) task. The DER on the Single Distance Microphone (SDM) task, where only one single microphone may be used which is more comparible to the task in our system, is 17.28%. Both results are conform state-of-the-art[6].

4.4 Automatic Speech Recognition

Table 5 shows the evaluation results on automatic speech recognition. It contains the Word Error Rates (WER) of the systems with and without VTLN applied. For all conditions we see a stable improvement over the baseline. It can be observed that there is a substantial performance difference between the TwNC-BN set and the CGN-BN set. This may be because the TwNC-BN data is more difficult or that the audio conditions in the CGN-BN data set better match the training conditions as a large part of our acoustic training material is derived from this corpus. Note that the CGN-BN data we used for evaluation were not in our acoustic model and language model training sets. Again it shows that *Sound and Vision* data is a difficult task domain.

Table 5. Evaluation results of the ASR component on different evaluation corpora (eval) showing Word Error Rate without VTL normalization (baseline) and with normalization (SHoUT-2007a)

eval	baseline	SHoUT-2007a
TwNC-BN	32.8%	28.5%
CGN-BN	22.1%	19%
Sound and Vision	68.4%	**64.0%**

4.5 Video Item-Specific Language Models

When we compared the baseline ASR results with the ASR approach that uses video item-specific LMs we observed that for the condition that uses the 1000 highest ranked documents for estimating the item-specific LMs some video items

Table 6. Results of baseline speech recognition on *Sound and Vision* data (base) compared with speech recognition results using a video item specific language models (item-LM)

item	base	item-LM	delta
01	73.7	73.5	-0.2
02	55.9	56.3	+0.4
03	53.7	50.7	-3.0
04	74.7	73.8	-0.9
05	79.2	79.4	+0.2
06	42.6	43.2	+0.6
07	60.3	57.9	-2.4
08	62.8	64.1	+1.3
09	59.0	56.6	-2.4
10	39.1	38.7	-0.4
11	74.1	72.4	-1.7
all	61.4	60.6	-0.8

WER improved significantly (up to 3 % absolute) whereas for others the improvement was only marginal or even negative. On average WER improved with 0.8% absolute.

In Table 6 baseline ASR results are compared with the ASR approach that uses video item specific LMs. Only the results for the condition that uses the 1000 highest ranked documents for estimating the item specific LMs are shown as this condition produced the best results. The results indicate that using item specific LMs generally helps to improve ASR performance.

5   Conclusions

It is clear that ASR results on the *Sound and Vision* data leave room for improvement. When we average the results on the two types of BN data we end up with a BN-WER of 23.7% so there is a large gap between performance on BN data and performance on our target domain. The assumption is that an important part of the error is due to mismatch between training and testing conditions (speakers, recording set-ups). By implementing noise reduction techniques such as successfully applied in [1], we expect to improve system robustness on this part. On the LM level, we will fine-tune the video-specific LM algorithm and work on the lattice output of the system (rescoring with higher order n-grams).

Acknowledgements

The work reported here was partly supported by the Dutch bsik-programme MultimediaN (http://www.multimedian.nl) and the EU projects MESH (IST-FP6-027685), and MediaCampaign (IST-PF6-027413). We would like to thank all those who helped us with generating speech transcripts for the *Sound and Vision* test set.

References

1. Anguera, X., Wooters, C., Pardo, J.: Robust speaker diarization for meetings: Icsi rt06s evaluation system. In: RT 2006. LNCS, vol. 4299, Springer, Heidelberg (2007)
2. Cardenal, A., Dieguez, J., Garcia-Mateo, C.: Fast lm look-ahead for large vocabulary continuous speech recognition using perfect hashing. In: Proceedings ICASSP 2002, Orlando, USA, pp. 705–708 (2002)
3. Czech, Z.J., Havas, G., Majewski, B.S.: An optimal algorithm for generating minimal perfect hash functions. Information Processing Letters 43(5), 257–264 (1992)
4. de Jong, F.M.G., Ordelman, R.J.F., Huijbregts, M.A.H.: Automated speech and audio analysis for semantic access to multimedia. In: Avrithis, Y., Kompatsiaris, Y., Staab, S., O'Connor, N.E. (eds.) SAMT 2006. LNCS, vol. 4306, pp. 226–240. Springer, Heidelberg (2006)
5. Demuynck, K., Duchateau, J., Van Compernolle, D., Wambacq, P.: An efficient search space representation for large vocabulary continuous speech recognition. Speech Commun. 30(1), 37–53 (2000)
6. Finke, M., Fritsch, J., Koll, D., Waibel, A.: Modeling and efficient decoding of large vocabulary conversational speech. In: Proceedings Eurospeech 1999, Budapest, Hungary, pp. 467–470 (1999)
7. Garofolo, J.S., Auzanne, C.G.P., Voorhees, E.M: The TREC SDR Track: A Success Story. In: Eighth Text Retrieval Conference, Washington, pp. 107–129 (2000)
8. Gauvain, J.-L., Adda, G., Adda-Decker, M., Allauzen, A., Gendner, V., Lamel, L., Schwenk, H.: Where Are We in Transcribing French Broadcast News? In: InterSpeech, Lisbon (September 2005)
9. Nguyen, L., Abdou, S., Afify, M., Makhoul, J., Matsoukas, S., Schwartz, R., Xiang, B., Lamel, L., Gauvain, J.L., Adda, G., Schwenk, H., Lefevre, F.: The 2004 BBN/LIMSI 10xRT English Broadcast News Transcription System. In: Proc. DARPA RT 2004, Palisades NY (November 2004)
10. Oostdijk, N.: The Spoken Dutch Corpus. Overview and first evaluation. In: Gravilidou, M., Carayannis, G., Markantonatou, S., Piperidis, S., Stainhaouer, G. (eds.) Second International Conference on Language Resources and Evaluation, vol. II, pp. 887–894 (2000)
11. Ordelman, R.: Dutch Speech Recognition in Multimedia Information Retrieval. PhD thesis, University of Twente, The Netherlands (October 2003)
12. Siohan, O., Myrvol, T., Lee, C.: Structural maximum a posteriori linear regression for fast hmm adaptation. In: ISCA ITRW Automatic Speech Recognition: Challenges for the Millenium, pp. 120–127 (2000)
13. Smeaton, A.F., Over, P., Kraaij, W.: Evaluation campaigns and trecvid. In: MIR 2006. 8th ACM SIGMM International Workshop on Multimedia Information Retrieval (2006)
14. van Leeuwen, D.A., Huijbregts, M.A.H.: The ami speaker diarization system for nist rt06s meeting data. In: RT 2006. LNCS, vol. 4299, pp. 371–384. Springer, Heidelberg (2007)

A Model-Based Iterative Method for Caption Extraction in Compressed MPEG Video*

Daniel Márquez and Jesús Bescós

Grupo de Tratamiento de Imágenes, Escuela Politécnica Superior
Universidad Autónoma de Madrid, E-28049 Madrid, Spain
{daniel.marquez,j.bescos}@uam.es

Abstract. We here describe a method for caption extraction that totally works in the MPEG compressed domain. As opposed to other compressed domain methods; it does not need to refine their results in the pixel domain. It consists of two phases: first, a selection of candidate frames with captions, based on a rigorous statistical design of an AC coefficients mask; second, an extraction of caption boxes from the pre-selected set of candidate frames. Caption extraction relies on a model-based approach to obtaining the caption mask, robust enough to avoid the use of any subsequent refinement.

1 Introduction

Caption extraction in the MPEG compressed domain[1][3][4][5] generally consists of two independent phases: location of candidate frames with captions, and caption extraction from these frames. The first phase aims to avoid analysing frames without captions; it applies several global and simple frame metrics to select frame intervals that probably contain captions. The second phase, the most crucial one, first applies energy binarization to obtain a raw caption mask for each frame. Next, this mask is refined with techniques that exploit the captions persistency along time and their rectangular shape. Finally, candidate frames are refined in the pixel domain.

In section 2, we present two independent techniques to select the AC coefficients mask best suited for the detection of caption blocks. Section 3 focuses on the location of candidate frames that potentially include captions, and Section 4 details the iterative approach to caption extraction. In Section 5 we include results for both phased. Finally, Section 6 provides conclusions of our work.

2 Characterization of Caption Blocks with AC Coefficients

Captions present two main characteristics that affect the values of the AC coefficients. Firstly, they contain characters, which show step edges in some dominant directions. Secondly, they show high contrast respect to the background in order to enhance legibility. Both properties make that some AC coefficients from each caption block

---

* Work partially supported by the European Commission under its 6th Framework Programme (FP6-027685 - MESH Project) and by Spanish Institutions under projects TIN2004-07860-C02-01 and S-0505-TIC-0223.

B. Falcidieno et al. (Eds.): SAMT 2007, LNCS 4816, pp. 91–94, 2007.

systematically have higher absolute values. However, almost every researcher has handled his own subset of AC coefficients, always empirically developed [1][2][3][4][5]. In order to select a robust subset we apply two rigorous methods: one based on energy histograms and another based on Self Organization Maps. To obtain representative results we use a set of news clips extracted from the MPEG-7 Content Set (352x288 MPEG-1) and from a German news program (720x576 MPEG-2).

The energy histogram compiles the percentage of AC terms whose values are included in a data interval. As we focus on the most energetic values, we just work over the higher energy range of the histogra, (see Fig. 1). We then threshold the energy with one standard deviation of its distribution and obtain the higher energy coefficients for the reprsentative set of video clips.

A Self-Organizing Map (SOM) is a type of iterative Artificial Neural Network based on competition. The goal of using a SOM is to detect which pattern is the most common when it is trained using AC coefficient vectors extracted from caption blocks. In this sense, vectors are clustered in a two-dimensional array to provide a visual representation of the 63-dimensional input data. We then select clusters which involve a minimum 10% input vectors. Finallly, a threshold is applied to each cluster's coefficients to obtain the resulting AC coefficients mask. Fig. 2b shows a winner cluster with a SOM trained only with caption blocks.

Both methods provide the same AC coefficient mask, which emphasizes the importance of horizontal and vertical AC coefficients. This resulting AC coefficients mask is formed by $AC_{0,j}$ $j = [1, 4]$ and $AC_{i,0}$ $i = [1, 3]$.

3   Location of Candidates Frames

This phase discards large temporal frame intervals from the second phase, hence decreasing the overall computational cost. To locate them we compute energy increments between consecutive intra-coded frames, k and $k+l$:

$$\Delta E_k = E_{k+l} - E_k \, , \quad E_k = \sum_{MxN} \sum_{j=1}^{4} \left| AC_{0,j}^{k+l} \right|^2 , \qquad (1)$$

where E_K is the frame energy of an MxN blocks frame calculated just using the horizontal elements of the AC coefficients mask. Next, we apply a threshold depending on the dimensions of the frame.

4   Model-Based Iterative Method for Caption Extraction

Whenever a candidate frame is located, a binary image or caption mask should be extracted, so that captions can be analysed later by an OCR system. We propose a method that exploits a caption model based on spatial and time properties.

We observed that captions have similar constraints when considering videos with 352x288 and 720x576 frame sizes, so that we can establish a single caption-model. Considering the sum of the mean size plus a standard deviation of our data set we derived a 5x3-block minimum caption model. This establishes that every caption block should have four possible combinations of 3x2 neighbouring caption blocks (see Fig. 2).

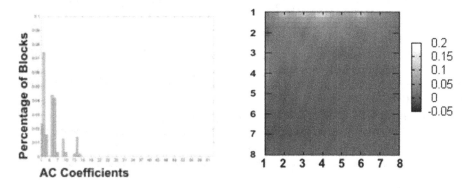

Fig. 1. Left: example of an energy histogram of the higher energy interval of a 3' 38'' clip. Right: a winner cluster unit, indicating (in light grey) the higher energy coefficients.

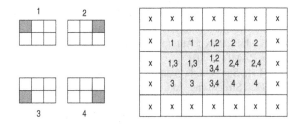

Fig. 2. Proposed caption model (right, caption blocks in grey) indicating the different 1-4 neighbour masks (left, applied over the grey block) that should fulfill each caption block.

We separately exploit the coefficients of the AC mask corresponding to horizontal and vertical frequencies, hence obtaining two different energy block images for every k frame: H_k and V_k, respectively. Firstly, we carry out an initial binarization for every H_k image with an energy threshold located one standard deviation over the mean energy. Secondly, we refine results using the vertical energies applying a threshold to every block i of the V_k image which depends on the caption model:

$$T_{k,i}^{ref} = (1+\alpha)\mu_{Vk}$$

(2)

where α is a model-adaptation coefficient and μ_{Vk} is the mean energy of V_k. Coefficient α measures how well a candidate block fulfils the caption model. Its value increases inversely to the maximum probability of the block belonging to one of the four possible neighbour combinations.

Finally, each candidate block has to remain at least one second in order to be marked definitively as a caption block.

5   Results and Conclusions

The goal of the first phase is to achieve maximum recall. We get a mean recall of 94.94 % and a precision of 61,1%. Previous works, e.g. Y. Zhang et al. [5] got a recall of 65% and a precision of 43%. D. Chen et al. [4] got a recall of 90% and a precision of 93%. They achieved a good precision; however, they miss a 10% of true candidates against our 5%.

The goal of the second phase is again to maximize recall since an OCR system fails if some caption blocks are missing. We achieve a recall of 91.8% and a precision of 64.3% with a false alarm rate of 4.6%. Previous results, e.g. S. Chun et al. [6] got a recall of 80.0% and 86.9% of precision, whereas Y. Zhang et al. [5] got a recall of 81% and a precision of 50%.

We can conclude that our system provides superior results, only working in the compressed domain and managing videos with different dimensions.

References

1. Lim, Y.K., Choi, S.H., Lee, S.W.: Text Extraction in MPEG Compressed Video for Content-based Indexing. In: Proc. ICPR (2000)
2. Crandall, D., Kasturi, R.: Robust Detection of Stylized Text-Events on Digital Video. In: Proc. 6th Int. Conf. on Document Analysis and Recognition (2001)
3. Zhong, Y., Zhang, H., Jain, A.K.: Automatic Caption Localization in Compressed Video. IEEE Transactions on PAMI (2000)
4. Chen, D.Y., Hsiao, M.H., Suh-Yin, L.: Automatic Closed Caption Detection and Filtering in MPEG Vídeos for Vídeo Structuring. Journal of Information Science and Engineering 22(5) (2006)
5. Zhang, Y., Chua, T.: Detection of Text Caption in Compressed Domain Vídeo. In: Proc. ACM Workshop on Multimedia (2000)
6. Chun, S., Kim, H., Kim, J.R., Oh, S., Sull, S.: Fast Text Caption Localization on Vídeo Using Visual Rythm. In: Proc. 5th Intl. Conf. on Recent Advances in Visual Information Systems (2002)

Automatic Recommendations for Machine-Assisted Multimedia Annotation: A Knowledge-Mining Approach

Mónica Díez and Paulo Villegas⋆

Telefónica I+D, Valladolid, Spain

Abstract. Recommender systems apply knowledge discovery techniques to help in finding associated information. In this paper, we investigate the use of association rule mining as an underlying technology for a recommender system aimed at improving the annotation process of multimedia news documents. The accuracy of these systems is very sensitive to the number of already annotated news items (the "cold-start"- problem); ontology-based semantic relations are being used to alleviate this situation.

1 Introduction

To enable advanced query, browsing and content management capabilities, multimedia systems need to be able to provide semantic metadata about the content items they include. This semantic information is hard to obtain: automatic semantic extraction systems are still error-prone, and manual annotation is very costly to produce and can also suffer from lack of consistence between annotators. A number of current research lines strive to join the outcome of automatic knowledge discovery with manual intervention of an operator, achieving thus a semi-automatic approach that intends to merge the best of both worlds.

In this paper, we follow a slightly different and complementary approach, developed within the MESH project [1], which targets multimedia news management systems. Here the mixing of automatic and manual operations is done in a strictly forward fashion, by producing automatic recommendations of possible annotation keywords. This will enable an annotator to quickly select and confirm from among the recommended concepts and keywords, thus cutting down annotation time significantly. To create the suggested keywords we use as a basis the automatically extracted metadata, available context and past history.

2 Automatic Keyword Recommender

The metadata associated to a set of structured documents, such as a collection of semantic annotations of multimedia news documents, can be successfully related

⋆ Work partially supported by the European Comission, under contract number FP6-027685 (MESH). We would also like to thank Pablo Castells and Miriam Fernández from Universidad Autónoma de Madrid for providing the customised IMDb datasets.

B. Falcidieno et al. (Eds.): SAMT 2007, LNCS 4816, pp. 95–98, 2007.
© Springer-Verlag Berlin Heidelberg 2007

to systems containing transactions, like Web usage statistics or frequent shopping list mining: where these contain pages sequentially visited or sets of products bought at the same time, our data is composed of a set of disjoint (but maybe related) concepts linked as annotations to a media content item (a news video), extracted automatically via multimodal analysis of the media.

All these datasets can be viewed as transactions. Thus, our framework can be a good candidate to experiment with some of the technologies that have been developed within association mining for rule extraction.

Association mining is one of the important sub-fields in data mining, where rules that imply certain association relationships among a set of items in a transaction database are discovered. The efforts of most researchers focus on discovering rules in the form of implications between itemsets, which are subsets of items that have adequate supports (i.e. occur frequently enough).

The motivation for using association mining within the context of keyword recommendations is twofold. First, rules extracted from association mining can improve the automatic annotation process of news documents (through better understanding of the semantic concepts inferred from previous multimedia analysis); second, the estimation of further concepts can be considered by reusing annotation history, via the design of adaptive recommender systems.

A first prototype of an AKR (Automatic Keyword Recommender) component was thus developed and integrated in our news media framework, meant to be used for the management of multilingual selections of collections of news video items, related audio broadcasts and text excerpts.

Initial annotations are automatically produced from the content using statistical pattern recognition approaches and multimodal analysis based on modeled news ontologies. The keyword recommendation approach can then complement that automatic analysis by providing additional suggestions (which probably could never be produced by media analysis), to be validated by a human annotator through an specialised interface.

This first version was based on some of the principles from the Apriori algorithm [2], which will find, given an initial set of transactions, association rules that will predict the occurrence of an item based on the occurrences of other items in the transaction.

The pseudocode of this implementation is shown below in Table 1, where A_k denote the annotation used for filtering, R_k denote the k-recommendations returned by the algorithm, C_k denote the set of candidate k-itemsets, and F_k represents the set of frequent k-itemsets.

3   Performance Analysis: The Cold-Start Problem

To test and analyze the performance of the implemented keyword recommender we need a huge real-life transactional dataset made up of reasonable sets of keywords. Our own multimedia news framework does not contain yet a big enough set of annotated news, so we complemented the tests with a dataset from the

Table 1. Keyword recommendation generation of the KwdRecom algorithm

```
1: k = 1
2: F_k = { Find A_k| support count ≥ minsup threshold }
3: do
4:    k = k + 1
5:    C_k = {Generate candidate itemsets for (F_{k-1}) }
6:    for each transaction t (news document)
7:    C_t = {Identify all candidates / C_k ∈ t }
8:    for each candidate itemset c in C_t
9:      support(c) ++
10:   end for
11:   end for
12:   F_k = {Extract C_t| support count ≥ minsup threshold }
13:   R_k = {Collection of k − itemsets| annotation A_pattern takes place}
14:   R_k = {Eliminate from R_k those annotations that are already present}
15:while F_k is not empty
16:Return { U {R_k}}
```

Internet Movie Database (IMDb) [3], composed of a large list of movies with their associated set of explanatory keywords.

Experimental results carried out on this first version of the recommender showed that sometimes few or no recommendations are produced, because the correlation between the concepts involved is low. This problem decreases the overall efficiency and, much more important, the user confidence on the recommender. Although no recommender system can work without some initial information, the final efficiency is defined by the capacity to manage with a smaller number of examples.

This situation, commonly faced by recommender systems, is known as the cold-start problem [4], and appears whenever recommendations are required for new items for whom little or no information has yet been acquired. Our system cannot discover co-occurrence because there is not enough previously annotated semantic concepts upon which to base any correlations.

This fact has triggered our work to define a new and enhanced AKR, which can use other available information in addition to pure statistical co-occurrences. Two types of such additional information are currently being tried:

- One approach has been to look at the hierarchical structure of the annotated media, trying to find relationships between annotations across the hierarchy.
- The other is to look at the semantic structure of the annotations themselves, taking advantage of the fact that they belong to a pre-defined ontology. This is commented in the next section.

4   Ontologies: A Good Combination

Ontologies are defined by establishing relationships between linguistic terms that specify domain concepts at different abstraction levels, being structured as a set of entities, attributes, relationships and axioms [5]. Thus, an ontology can provide a variety of concepts and its relationships from the working domain or scenario, defined from the news document context or the past history.

Taking into account the recommendations produced by the frequency itemset mining approach described in the first section, but trying to overcome its start-up problems, we are currently developing a mixed system that can take advantage of the knowledge defined in an ontology of the semantic concepts used in the annotation process. We are aiming both at lessening the cold-start problem and at an improved guarantee of the quality of the recommendation.

In this approach, an *is-a* hierarchy of relationships driven by the ontology entities is held, so that e.g. higher level entities could be used to infer broader concept interests, increasing the possibilities of finding a recommendation. Other types of relations as given by the ontology structure are also being investigated.

The confidence of the final recommendation is formulated through a correlation between the co-occurrence frequency of the concepts obtained as recommendations from the association mining algorithm, and the interest of those concepts based on the context defined for the news document being processed.

When a specific concept gets affected by the cold-start problem and no recommendations can be retrieved, some super-class entities or concepts derived from ontology-based relationships are used as input in the recommendation algorithm, broadening the activity field of the search. An exponentially decreasing fractional interest is used for those related classes and concepts.

5   Conclusions

In this paper, an algorithm for automatic suggestion of keywords is described, developed within an application for annotating multimedia news documents. Empirical evaluation of a first approach based on association-mining techniques showed that performance decreased in those cases in which little knowledge was available due to scarcity of previous annotations. In order to improve performance, ontologies are being used to make up for the lack of initial information.

References

1. MESH: Multimedia semantic syndication for enhanced news services,
   http://www.mesh-ip.eu
2. Agrawal, R., Srikant, R.: Fast algorithms for mining association rules in large databases. In: Bocca, J.B., Jarke, M., Zaniolo, C. (eds.) VLDB. Proceedings of the 20th International Conference on Very Large Data Bases, Santiago, Chile, pp. 487–499 (September 1994)
3. IMdb: The internet movie database. http://www.imdb.com
4. Maltz, D., Ehrlich, K.: Pointing the way: active collaborative filtering. In: CHI 1995. Proceedings of the SIGCHI conference on Human factors in computing systems, New York, NY, USA, pp. 202–209 (1995)
5. Guarino, N., Poli, R.: Formal ontology in conceptual analysis and knowledge representation. International Journal of Human and Computer Studies 43(5/6) (1995)

Video Summarisation for Surveillance and News Domain*

Uros Damnjanovic, Tomas Piatrik, Divna Djordjevic, and Ebroul Izquierdo

Department of Electronic Engineering, Queen Mary, University of London,
London E1 4NS, U.K.
{uros.damnjanovic, tomas.piatrik, divna.djordjevic,
ebroul.izquierdo}@elec.qmul.ac.uk

Abstract. Video summarization approaches have various fields of application, specifically related to organizing, browsing and accessing large video databases. In this paper we propose and evaluate two novel approaches for video summarization, one based on spectral methods and the other on ant-tree clustering. The overall summary creation process is broke down in two steps: detection of similar scenes and extraction of the most representative ones. While clustering approaches are used for scene segmentation, the post-processing logic merges video scenes into a subset of user relevant scenes. In the case of the spectral approach, representative scenes are extracted following the logic that important parts of the video are related with high motion activity of segments within scenes. In the alternative approach we estimate a subset of relevant video scene using ant-tree optimization approaches and in a supervised scenario certain scenes of no interest to the user are recognized and excluded from the summary. An experimental evaluation validating the feasibility and the robustness of these approaches is presented.

Keywords: Spectral clustering, ant-tree clustering, scene detection, video sumarization.

1 Introduction

With new video material being created every day, the users need to spend more and more time viewing the content even though they might not be interested into all its aspects. Therefore efficient browsing and access to relevant video items is a crucial application in modern multimedia systems. Not only that new material is created everyday, but also new applications that utilize the creation and distribution of multimedia content and its metadata are emerging. For example WWW applications make the process of video publishing and sharing available to everyone. Video on demand, news broadcasting and syndication, personal video archiving, surveliance and security camera tracking are just some of the examples of exiting applications which are working with huge video databases. Not only time requirement issues but also storage issues have to be taken into account when working with videos. Storing whole video

---

* The research leading to this document has received funding from the European Community's Sixth Framework Programme. However, it reflects only the author's views, and the European Community is not liable for any use that may be made of the information contained therein.

B. Falcidieno et al. (Eds.): SAMT 2007, LNCS 4816, pp. 99–112, 2007.

when just a small part is used waists a lot of disk space resulting in decrease of application efficiency. Conventional multimedia systems are still not capable to overcome above mentioned constraints. For example, most existing video players are offering only fast forward and backward option, leaving the user without the possibility to efficiently search the video. Contrary, modern multimedia applications need to be efficient, providing the user only with pieces of content that are of interest. However the definition of what is interesting may vary from user to user, or from application to application making the problem more difficult. But the essence of the approach is to present to the user as much interesting information as possible in the smallest possible time interval. In this aspect, video summaries can be seen as short but highly informative representation of the content, presented to the user who wants to have fast overview of the available information.

Looking at the video from bottom to top, essential part of a video are frames, which are further organized into shots [1-2]. The next level in the video structure is a scene, defined as a group of shots with similar semantic content. Shots can still be seen as low level of video organization, while scenes represent more semantic level of video organization. After the scenes are found one need to choose which part of the scene actually contains the information important to the user and present it. This can be done either as key frame presentation, or as a video skim.

Research efforts put in the problem of video summarization and scene detection resulted in large amount of available literature on this topic. Most of the approaches focuses on either detecting the scenes within the video or on defining some importance measure which is the used for selecting parts of the video which will be presented to the user. Main problem in both approaches is still to connect the low level representation of the video with high level semantic representation. For example problem of scene detection should sometimes group together shots that are visually toatally different. Using only low level representation of video this is very hard task to do. Also, automatic detection of informative parts of a video can be very hard task having in mind subjective nature of importance definition. In [3] authors presented the tool that utilizes MPEG-7 visual descriptors and generates a video index for summary creation. The resulting index generates a preview of the movie and allows non-linear access to the content. This approach is based on hierarchical clustering and merging of shot segments that have similar features and neighboring them in the time domain. In [4] Rasheed and Shah construct a shot similarity graph, and use graph partitioning normalized cut for clustering shots into scenes. In a similar approach [5] authors proposed a novel way to assess clustering quality using "eigengap" measure. Video summarization based on the optimization of viewing time, frame skipping and bit rate constraint is described in [6]. For a given temporal rate constraint the optimal video summary problem is defined as finding a predefined number of frames that minimize the temporal distortion. Otsuka et. al in [7] presented their video browsing system that uses audio to detect sport highlights by identifying segments with mixture of the commentators excite speech and cheering. Video motion analysis can be used for creating video summaries as in [8]. In this approach Wang et al. showed that by analyzing global/camera motion and object motion is possible to extract useful information about the video structure. Motion vectors are also used in [9] for demonstrating semantic inference by using the MPEG-7 low-level motion activity features. In [10] motion based method for video summarization and scene detection is achieved by analyzing the temporal variations of some coefficients of the 2D affine model.

In this paper we present two frameworks for efficient video summarization. One based on spectral clustering methods for temporal segmentation of the video followed by representative key frame extraction based on the motion analysis of detected scenes. In this approach we assume that scenes that are rich in motion are more interesting to the user that static ones. It is a logical choice for news and surveillance video domain, having in mind that usually events and highlights are related to some dynamic process within videos. For example, in news video domain, showing the actual event is much more interesting to the user than presenting anchor person speaking about the event. Motion analysis of scenes results in key-frames that are rich in the local motion opposed to the global motion. Finally, we propose the approach for analyzing the structure or resulting clusters, in order to define importance measure for every part of the scene based on the statistical analysis of eigenvectors used in clustering process. We show that is possible to have information about the motion activity in the scene, without extracting motion vectors from MPEG stream, and to use this information to create video skims that will be presented to the user. In the alternative approach, ant-tree clustering with automatic partitioning of video scenes is used for video summarization. It is combined with semi-supervised approach for recognizing non-relevant scenes (e.g. anchor person for the news domain and static indoor or outdoor scenes for the surveillance video). The remaining clusters and their relevance in regards to the timeline is used for obtaining video summaries. We compare the obtained video skims with the manually generated ground truth in both domains.

Structure of the paper is as follows. In section 2 the spectral clustering approach for building video summaries is presented. This also incorporates the eingen vector analyses and definition of importance measure for every part of the scene. Section 3 presents the ant-tree approach for building video summaries together with scene recognition. In section 4 the experimental results are depicted by comparing the generated summaries with ground truth data. Section 5 presents some conclusion.

2 Spectral Clustering for Video Summarization

We present a video summarization approach based on spectral normalized cut algorithm. Since spectral clustering methods have proved to be efficient in detecting block structure of a similarity matrix for some dataset, in this section we show how this block structure can be formed over a set of video frames, and then used to detect scenes in a video. As mentioned scenes are neighboring video segments with similar visual content. The task of detecting scenes can be seen as one of clustering frames based on their visual properties and temporal position. Once the scenes are detected one needs to present each scene to the user. The goal is to find a small set of frames that provide the core information about the whole scene. Each frame is then ranked based on the motion activity, and finally video summary is created. The ranking is based on the fact that the structure of the similarity matrix of eigenvectors corresponds to the structure of the changes in the video.

2.1 Spectral Clustering

Spectral clustering has its origin in a spectral graph partitioning,. a popular algorithm in high performance computing. Now days, spectral clustering has many applications in machine learning, exploratory data analysis, computer vision and speech processing.

The success of such algorithms depends heavily on the choice of the metric, but this choice is generally not treated as part of the learning problem. Spectral algorithms use information contained in the eigenvectors of data affinity matrix to detect structures.

Given a set of data points, the pair wise similarity matrix is defined as matrix W with elements w_{ij} representing a measure of similarity between points i and j. Spectral clustering techniques make use of the spectrum of the data similarity matrix to cluster the points. The matrix mechanics is closely related to the more general singular value decomposition. Square matrices have an eigenvalue/eigenvector equation with solutions that are the eigenvectors x_λ and the associated eigenvalues λ. The main idea behind all spectral clustering algorithms is similar. Initially a pair-wise similarity has to be defined and a similarity matrix built. Eigen decomposition of the similarity matrix results in eigenvalues and eigenvectors which are finally used to cluster the dataset.

The set of data points are denoted by V, with cardinality $|V| = N$. For each pair of points $i, j \in V$, a similarity $w_{ij} = w_{ji} \geq 0$ can be viewed as weight of the undirected edge of a graph G over V. The matrix $W = \lfloor w_{ij} \rfloor$ plays the role of *"real valued"* similarity matrix for G. Let d_i represent a degree of node I:

$$d_i = \sum_{j \in I} w_{ij} \tag{1}$$

And let the volume of a set $A \subset V$ be defined as:

$$VolA = \sum_{i \in A} d_i \tag{2}$$

Let D be a $N \times N$ diagonal matrix with values d_i, $i \in [1, N]$ on its diagonal. Then Laplacian matrix of the graph G is defined as:

$$L = D - W \tag{3}$$

The set of edges between two disjoint sets $A, B \subseteq V$ is called *edge cut* or in short *cut* between A and B:

$$cut(A, B) = \sum_{i \in A, j \in B} w_{i,j} \tag{4}$$

Once the specific eigenvalues and its corresponding eigenvectors are found, membership of each point from the dataset is determined by investigating specific entries of eigenvectors. Every entry of eigenvector corresponds to exactly one point from the dataset. By comparing the entry to some value that is chosen to be the splitting point, membership of a point to one or the other partition is determined. A set of clusters $C = \{C_1, C_2, ... C_k\}$ defines partitioning of V into nonempty mutually disjoint subsets. In the graph theoretical paradigm a clustering represents a *multiway cut* in the graph G. After creation of a specific graph structure, the objective function can be easily defined. Usually this is some measure that describes the relations between two or more separate clusters. The clustering problem can be seen as a problem of finding partitions of a dataset in a way that similarities inside clusters are large, and similarities between different clusters are small. Most of the algorithms can be thought of as consisting of three stages:

- **Pre-processing:** This is a form of normalization of the similarity matrix W in order to avoid ill conditioned adjacency matrix.
- **Spectral Mapping:** Top k eigenvectors of the pre-processed similarity matrix are computed. Each data point is represented by the value of a specific component in the aforementioned eigenvectors.
- **Post-processing/Grouping:** A grouping algorithm clusters the data based on the respective values of eigenvectors.

Normalized Cuts algorithm, *Ncut*, was first introduces by Shi and Malik in [11] as a heuristic algorithm aiming to minimize the *Normalized Cut* criterion. Originally this approach was created to solve the problem of perceptual grouping in the image data. The normalized cut between two sets $A, B \subseteq I$ is defined as:

$$NCut(A, B) = Cut(A, B)\left(\frac{1}{VolA} + \frac{1}{VolB}\right) \tag{5}$$

The algorithm consists of minimizing (5) by solving the generalized eigensystem:

$$L x = \lambda D x \tag{6}$$

Ncut algorithm focuses on the second smallest eigenvalue of (6) and its corresponding eigenvector, λ^L and x^L respectively. In [12] it is shown that when there is a partitioning A, B of V such that

$$x^L = \begin{cases} \alpha, i \in A \\ \beta, i \in B \end{cases} \tag{7}$$

then A, B is the optimal cut and the value of the cut itself is λ^L. This result represents the basis of spectral segmentation by normalized cuts. After the dataset is divided in two groups, the binary clustering algorithm can be run recursively.

In [12] it is shown that multway cut approach gives better results then recursively applying binary clustering, until k clusters are reached. In a mulitwaycut approach instead of using only one eigenvector of (6), top k eigenvectors are used, where k is predefined number of clusters. Let matrix X be $N \times k$ matrix, created by stacking the top k eigenvectors in columns. Each row of X corresponds to a point in the dataset and is represented in a k-dimensional Euclidian space. Finally, k clusters are obtained by applying the K-means algorithm over the rows of X.

2.2 Video Summarization with Spectral Clustering

The first step in our algorithm is to create the similarity matrix. Instead of analyzing the video on a key frame level, we are using a predefined ratio of frames in order to stimulate the similarity matrix block structure. For surveillance videos similarity between all frames is high resulting in the similarity matrix which has almost all entries close to 1. Any change that happens in the video will result in a local change of the block structure of the similarity matrix.

In most cases events occurring in the surveillance video are short making the extraction of important events from the whole set of frames a difficult task. The emphasis in the summarization of surveillance videos is to detect small and short changes in the video, and to present them to the user based on importance levels. On the other side news videos are formed of clusters that are significantly different. The main task in the news summarization is to properly cluster scenes, and then to analyze clusters in order to find most informative representatives.

The similarity matrix W is made of pair wise similarities $w_{i,j}$ between two frames i and j defined as:

$$w_{i,j} = \exp\left(-\frac{d(i,j)^2}{2\sigma^2}\right)$$

(8)

Where $d(i,j)$ is a distance is over the set of low- level feature vectors and σ is scaling parameter. Biger the σ is, higher the sensitivity of the similarity measure. After creating the similarity matrix and solving the generalized eigensystem from (6), the next step is to determine the number of clusters in the video. This number is then passed to a classical k-means algorithm. Automatic determination of the number of clusters is not a trivial task. In [13] is shown that every similarity matrix have a set of appropriate number of clusters depending on the choice of the parameter σ used in equation (8). For automatic detection of number of clusters for fixed σ we use the results of matrix perturbation theory [14]. The matrix perturbation theory states that the number of clusters in a dataset is highly dependent on the stability of eigenvalues/eigenvectors determined by the eigengap δ:

$$\delta_i = |\lambda_i - \lambda_{i+1}|$$

(9)

With λ_i and λ_{i+1}, being two consecutive eigenvalues of (6). The number of clusters k is then found by searching for the maximal eigengap over a set of eigenvalues:

$$k = \left\{ i \Big| \delta_i = \max_{j \in [1,N]} (\delta_j) \right\}$$

(10)

After the right number of cluster is determined, k-means algorithm is initialized with k equidistant rows of the matrix X (defined in section 2.1). Results of the k-means algorithm are clusters that give importance information for various applications Fig. 1. clusters which came from the k-means are used in the decision process, which will actually perform the summarization of the video, Fig. 2. Decision process analyzes the structure of the clusters and decides how to present them to the user. Let $\tau_{i,j}$ be the length of the j-th continuous sequence of cluster i as in Fig. 2, and let $\sum_{j \in [1, n_i]} \tau_{i,j} = T_i$ be the total length of the cluster i, where n_i is the total number of disconnected continuous segments of the cluster i. T_v is the length of the whole video, T_s is the length of the resulting summary and $\delta_{m,n}^{(p)}$ is the Euclidean distances between centroids of m-th and n-th cluster in p-dimensional feature space.

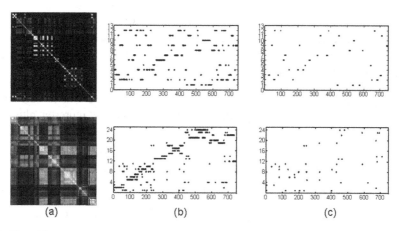

Fig. 1. The video summarization process, the top row represents news domain example, the bottom row surveillance video example. a) Similarity matrix. (b) Result of the k-means clustering. Cluster indicators are plotted over the time axis. Every point on the horizontal axis corresponds to the fixed time interval. (c) Final summary structure. For the news videos, long continuous segments are used to build the summaries. In the surveillance domain, long segments are removed and only short sequences are used in building the summary.

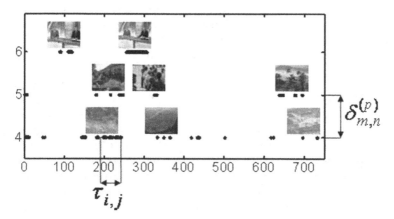

Fig. 2. Plot of cluster indicators over the time axis for news videos. Long clusters are used to give short segments that will be used for building the summary. Parameters $\tau_{i,j}$ and $\delta_{m,n}^{(p)}$ are used in the analysis of clusters. $\tau_{i,j}$ is the length if the continuous segment, while $\delta_{m,n}^{(p)}$ is the distance between centroids of two clusters in p-dimensional feature space.

The cluster analysis for the surveillance videos search for clusters with short continuous segments using the following rule:

if $\tau_{i,j} < T_{tr}$ the sequence i of the cluster j contains important information.

Or in the case of the news domain the rule is as follows:

if $\tau_{i,j} > T_{tr}$ the sequence i of the cluster j contains important information.

Where T_{tr} is experimentally determined threshold. Motivation for the decision rules is that in the surveillance videos long scenes correspond to static scenes where nothing important is actually happening in the video. While for the news videos long scenes usually contain information that is important to the user.

After important clusters are extracted, next step is to decide how to present them to the user. The main task of the video summarization is to save time for browsing the video. Time saving is achieved by choosing short segments that represent clusters. Let T_{si} be the duration of each cluster i representative segment. With representative segment we mean part of the each cluster that will be used to build the summary. Duration T_{si} in the surveillance video is set to a fixed value, while in the case of news video it is proportional to the length of the cluster. To calculate the duration T_{si} for each cluster, distances between cluster centroids are found. If $\delta_{m,n}^{(p)} < \delta_{tr}$, then clusters m and n will contribute together to the overall duration of the summary T_s with: $T_{smn} = T_{sm} = T_{sn}$. Now, T_{si} is calculated using the following equation:

$$T_{si} = T_s \cdot \frac{T_i}{T_v} \tag{11}$$

The last step in our algorithm is to find representative segments of the specified duration for each cluster. We propose simple method for motion activity estimation that is based on the statistical theory of spectral clustering. The authors in [15] showed a connection between spectral clustering and Markov random walks where pair wise similarities are seen as flows in Markov random walks within a probabilistic framework for properties of eigenvectors and eigenvalues. In the case of the ideal dataset with high similarities within the same clusters and with sum of inter-cluster similarities close to zero, resulting eigenvectors will be piecewise constant. In other words, eigenvector values on the same side of a cut will have almost same values. In our approach we studied properties of datasets which resulted in non-constant eigenvectors. By applying statistical analysis of eigenvector entries, it is possible to get the insight of the motion activity level. We use statistical analysis of eigenvectors on scene level to extract representatives for each interesting segment. The algorithm searches for the segments with largest values of accumulated absolute change between consecutive eigenvector entries. Areas with high change in the eigenvector entries correspond to the regions with high motion activity levels. If t_r is the start time of a representative scene and $\Delta E_i = |E_{i+1} - E_i|$ is the absolute difference between consecutive entries of eigenvectors, then t_r can be found by using the following formula:

$$t_r = t_i \left| \sum_{i \in (i, i+T_{si})} \Delta E_i = \max_{i \in [1,N]} \sum_{j \in (j, j+T_{si})} \Delta E_j \right. \tag{12}$$

Segments extracted using equations (12) are finally used to build video summaries, which are then presented to the user.

3 Ant-Tree Clustering for Video Summarization

Many researchers in computer science have been inspired by real ants [16] and have defined artificial ants paradigms for dealing with optimization or machine learning problems [18-20]. Previous models involve the ability of ants to sort objects [20-21] or to build a colonial odor [22] and mechanical structures by a self-assembling behavior [23]. In this section, we present a video summarization approach based on ant-tree clustering algorithm. We model the ability of ants to build live structures with their bodies [24] in order to discover, in a distributed and unsupervised way, a tree-structured organization and summarization of the video data set.

3.1 Ant-Tree Clustering

The ant-tree clustering method was inspired by self-assembling behavior of African ants and their ability to build chains by their bodies in order to link leaves together [23]. The main principles of the algorithm are depicted in Fig. 3: each ant represents a node of the tree to be assembled, i.e. a data to be clustered. On the basis of a root (initial support node) on which the tree will be built, ants will gradually fix themselves until all ants are attached to the structure.

The movement and fixing of ants in a position depends on the similarity value of the data and on the local neighborhood of the moving ants. Similarity measure between two data points is denote by $sim(i, j)$ which gives, for a couple of data points $(x_i, x_j), i, j \in [1, N]$, a value in the range [0, 1] where N is the number of data points. Thus, for each ant a_i the following concepts are defined:

- the outgoing link of a_i is the link that a_i can maintain toward another ant;
- the incoming links of a_i are the links that the other ants maintain towards a_i, these bonds can be the legs of the ant;
- the data x_i is resented by a_i;
- There are two thresholds one for higher- similarity $T_{sim}^{H}(a_i)$ and one for lower-similarity (dissimilarity) $T_{sim}^{L}(a_i)$, which are locally updated by a_i.

At each step, an ant a_i will connect itself or move according to the similarity with its neighbors. While there is still a moving ant a_i, we simulate its action according to the movement. The first ant is directly connected to the support. Next, for each ant a_i, the two following cases need to be considered.

The first case is when ant a_i is placed on the support; if a_i is similar enough to a^+ so that $sim(a_i, a^+) > T_{sim}^{H}(a_i)$, where a^+ is the ant which is the most similar to a_i among the ants already connected to the support; then a_i is moved towards a^+ in order for both ants to be clustered in the same subtree, i.e. the same cluster.

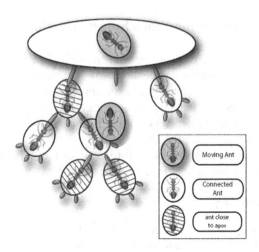

Fig. 3. Building of a tree with artificial ants

Otherwise if a_i is dissimilar enough to a^+ that is $sim(a_i, a^+) < T_{sim}^L(a_i)$, then it is connected to the support and it becomes a representative ant of a new subtree. This means that the new sub-tree is created, ant its ants will be as dissimilar as possible to representative ants in other sub-trees connected to the support. Finally, if a_i is not simi-lar or dissimilar enough to a^+, the thresholds is updated in order to allow a^+ to be more tolerant and to increase its probability of being connected the next time it will be considered (in the meantime, other ants will have changed the tree). The updating pro-cedure of the proposed threshold for improving summarization is present in section 3.2.

The second case is when a_i is on top of another ant denoted by a_{pos}; if there is a free incoming link for a_{pos} and if a_i is similar enough to a_{pos}, that is if the follow-ing holds true $sim(a_i, a^+) > T_{sim}^H(a_i)$ and at the same time dissimilar enough to ants connected to a_{pos}, then a_i is connected to a_{pos}. In this case, a_i represents the root of a new sub-tree or sub-cluster below a_{pos}. Its dissimilarity with other ants directly connected to a_{pos} is such that sub-clusters of a_{pos} will be well "separated" from each other (while being similar to a_{pos}).

Otherwise, a_i is randomly moved toward a neighbor node of a_{pos} and its thresh-olds are updated in the same way as described in the previous case. So, a_i will move around within the graph, in order to find the optimal location. The algorithm ends when all ants are connected.

3.2 Video Summarization by Ants

To obtain a video summary, an ant tree is built so that nodes correspond to video frames and the edges are what needs to be discovered in order to summarize the

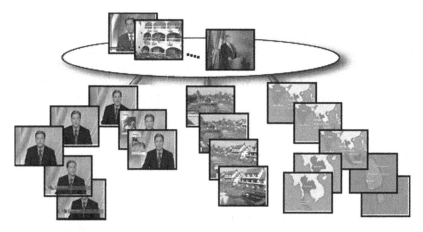

Fig. 4. Tree structuring of video frames by ant-tree clustering

content. One should notice that this tree will not be strictly equivalent to a dendogram as used in standard hierarchical clustering techniques. Here each node corresponds to one data point, while in the case of dendograms data points correspond to leaves.

A fundamental step in our video summarization approach is to create the similarity matrix and organize video frames into the tree-structure using ant-tree clustering method as depicted in Fig. 4. Results of the ant-tree algorithm are clusters which are used in the decision process for creating video summaries.

In order to increase quality of clustering and for automatic detection of a number of clusters, the following definition of threshold updating is proposed:

$$T_{sim}^{H}(a_i) = T_{sim}^{H}(a_i) \cdot \frac{\alpha - 1}{\alpha}, \quad T_{sim}^{L}(a_i) = T_{sim}^{L}(a_i) \cdot \frac{\beta}{\gamma} \tag{13}$$

$$\beta_{(t)} = \frac{\max \delta_{m,n}^{(p)} - \min \delta_{m,n}^{(p)}}{\max \delta_{m,n}^{(p)}}, \ t \in [0, R] \tag{14}$$

where α represents sensitivity, $\delta_{m,n}^{(p)}$ is distance between support frames of two clusters m, n in p-dimensional feature space, R is iteration number of the clustering approach and γ is used for normalization. The higher α is, the more sensitive the clustering algorithm to changes among frames. The main idea of our approach is to iterate the clustering process until optimal number of clusters is found. The parameter $\beta_{(t)}$ is initialized for first run of ant-tree algorithm and a new $\beta_{(t+1)}$ is obtained after all frames are connected. The optimal partitioning is achieved when $\beta_{opt} = |\beta_{(t)} - \beta_{(t-1)}| \leq \beta_{tr}$. Finding the β_{opt} is also important in the stage of calculating the duration T_{si} (defined in 2. 2) of each cluster that will be used to build the summary. If some clusters are too similar to each other ($\delta_{m,n}^{(p)}$ is very small), then frames will be put together and will contribute to the overall duration of summary as one scene. Duration of each representative segment of video is calculated using equation (11). In order to present most important information of video content to the user in short time, the following procedure is applied. We defined certain scenes of no interest to the user according to the type of video. In the news domain it is the scene an "anchor person" and in the surveillance domain statical "indoor" and "outdoor" scene. After all frames are clustered, and the above mentioned scenes are recognized we excluded clusters contained these scenes from the summary using constraints.

4   Experimental Evaluation

For experimental evaluation in the news domain we used two 20 minutes long news videos, and in the surveillance domain two 20 minute surveillance videos (one depicting an inside room with people entering and leaving and the other depicting an outdoor parking). Low-level features are computed on the frame level using one frame per second sample ratio. We used MPEG-7 color layout features for representing each selected frame.

Spectral parameter σ from (8) is experimentally chosen in such a way that the number of clusters corresponds to the specific application. When choosing the σ values, we create the similarity matrix and examine the resulting eigenvalues. Specifically we examine the number of clusters in created similarity matrix. For example setting σ high in the news domain will result in few clusters, while in the surveillance domain it result in only one cluster. Different values of σ come from different structure of the specific domains. In news video domain, high level of visual changes of frames is significantly bigger then in the surveillance domain. In the surveillance doman, small visuall changes lead to small changes of the simialrtiy measure that are hard to detect. For the news domain σ is set to 20, while for the surveillance domain σ is set to5. Modification of values for parameter σ influence changes in sensitivity of the similarity measure and the ability to detect small changes in the video. The spectral clustering approach initializes k-means clustering and the final clustering is obtained after 10 to 15 iterations.

In the case of ant-tree based summarization after empirical evaluations $\beta_{(0)} = 0.1$ and parameter α from (13) has different values in respect of the type of changes present in the particular domain. For news videos we set $\alpha = 100$ since the events are depicted with dynamic scenes with large variations in content. For the surveillance videos the sensitivity is defined with higher granularity (one person or a car moving) and therefore we set $\alpha = 1000$.

The spectral method gives better performances in case of the surveillance domain (87% of events are detected) since it can detect regions with high motion activity, while the ant-tree clustering approach detects 74 % of events. In the news domain where anchor person precedes most of the stories (events) the scene recognition properties of the ant-tree approach lead to higher performances 86.2% of correctly detected event comparing to 82% for the spectral approach. In this case most of the relevant scenes have high-motion activity; hence this feature is not discriminative enough of its own.

Table 1. Results of the video summarisation

Method	Video type	Number of detected events	Number of missed events	Total length of videos	Total length of summaries
Spectral	News	24	5	40 min	3 min
	Surveillance	27	4	40 min	2 min
Ant-tree	News	25	4	40 min	3 min
	Surveillance	23	8	40 min	1,5 min

Overall the experimental results show that our proposed summarization approaches abridged the original video with a compression factor up to 16:1 while capturing most of user defined relevant scenes.

5 Conclusion

We have shown that spectral and ant-tree clustering approaches can efficiently be used for summarizing videos. In the first case the main reason is that high similarity between consecutive frames in video sequence strengthens the block structure of the similarity matrix in spectral clustering. Furthermore, without the need for extra processing time the structure of the video can be efficiently mapped to eigenvectors and eigenvalues, and used for discovering important segments of the video in both domains. In the latter approach for video summarization, ant-tree clustering, the emphasis is on scene recognition. The approach generates a set of representative shots and extracts the tree structure of a video sequence. In case of surveillance video, where scenes are very similar to each other, ant-tree based summarization shows lack of consideration of motion activity within scenes. Our future work will encompass the advantages of approaches, joining together motion activity and scene recognition properties underlying the individual approaches.

References

1. Calic, J., Izquierdo, E.: Towards real time shot detection in the MPEG compressed domain. In: Proceedings of the Workshop on Image Analysis for Multimedia Interactive Services (2001)
2. Yeo, L.B., Liu, B.: Rapid scene analysis on compressed video. IEEE Transactions on Circuits & Systems for Video Technology 5, 533–544 (1995)
3. Lee, J., Lee, G.G., Kim, W.Y.: Automatic video summarizing tool using MPEG-7 descriptors for personal video recorder. IEEE Transaction on Consumer Electronics 49, 742–749 (2003)
4. Rasheed, Z., Shan, M.: Detection and Representation of scenes in videos. IEEE Transactions on Multimedia 7, 1097–1105 (2005)
5. Odobez, J., Gatica-Perez, D., Guillemot, M.: Video shot clustering using spectral methods. In: 3rd Workshop on Content Based Multimedia Indexing (CBMI) (2003)

6. Li, Z., Schuster, G.M., Katsaggelos, A.K.: MINMAX optimal video summarization. IEEE Transactions on Circuits and Systems for Video Technology 15, 1245–1256 (2005)
7. Osuka, I., Radharkishnan, R., Siracusa, M., Divakaran, A., Mishima, H.: An enhanced video summarization system using audio features for personal video recorder. IEEE Transactions on Consumer Electronics 52, 168–172 (2006)
8. Wang, Y., Zhang, T., Tretter, D.: Real time motion analysis towards semantic understanding of video content. In: Conference on Visual Communications and Image Processing (2005)
9. Peker, K.A., Alatan, A.A., Akansu, A.N.: Low level motion activity features for semantic characterization of video. IEEE International Conference on Multimedia and Expo 2, 801–804 (2000)
10. Peyrard, N., Bouthemy, P.: Motion-based selection of relevant video segments for video summarization. Multimedia Tools and Applications, pp. 259-276 (2005)
11. Shi, J., Malik, J.: Normalized cuts and image segmentation. IEEE Transactions on PAMI 22, 888–905 (2000)
12. Alpert, C., Khang, A., Yao, S.: Spectral partitioning: The more eigenvectors, the better. Discrete Applied Mathematics (1999)
13. Manjunath, B.S., Salembier, P., Sikora, T.: Introduction to MPEG-7. John Willey & Sons, New York, NY (2002)
14. Zheng, X., Lin, X.: Automatic determination of intrinsic cluster number family in spectral clustering using random walk on graph. In: ICIP 2004. International Conference on Image Processing, vol. 5, pp. 3471–3474 (2004)
15. Meila, M., Shi, J.: A random walks view of spectral segmentation. AI and Statistic (2001)
16. Holldobler, B., Wilson, E.O.: The Ants. Springer, Heidelberg (1990)
17. Dorigo, M., Di Caro, G.: Ant Algorithms for Discrete Optimization. Technical Report, pp. 98–10 (1999)
18. Dorigo, M., Stutzle, T.: Ant Colony Optimization. MIT Press, Cambridge (2004)
19. Bonabeau, E., Dorigo, M., Theraulaz, G.: Swarm Intelligence: From Natural to Artifical Systems. Oxford University Press, Oxford (1999)
20. Lumer, E., Faieta, B.: Diversity and Adaptation in Populations of Clustering Ants. In: 3tll Conference on simulation and adaptive behavior-: from animals to animats, pp. 501–508 (1994)
21. Kuntz, P., Snyers, D., Layzell, P.: A stochastic heuristic for visualizing graph clusters in a hi-dimensional space prior to partitioning. Journal of Heuristic 5 (1999)
22. Labroche, N., Monmarche, N., Venturini, G.: A new clustering algorithm based on the chemical recognition system of ants. In: Proceedings of the 15th European Conference on Artifical Inteligence (2002)
23. Azzag, N., Monmarch, H., Slimane, M., Venturini, G., Guinot, C.: Antree: a new model for clustering with artificial ants. In: IEEE Congress on Evolutionary Computation, pp. 8–12 (2003)
24. Lioni, A., Sauwens, C., Theraulaz, G., Deneubourg, J.L.: The dynamics of chain formation in Oecophylia longinoda. Journal of Insect Behavior 14, 679–696 (2001)

Thesaurus-Based Ontology on Image Analysis

Sara Colantonio[1], Igor Gurevich[2], Massimo Martinelli[1],
Ovidio Salvetti[1], and Yulia Trusova[2]

[1] Institute of Information Science and Technologies, Italian National Research Council,
via G. Moruzzi 1, 56124 Pisa, Italy
{Sara.Colantonio, Ovidio.Salvetti,
Massimo.Martinelli}@isti.cnr.it
[2] Dorodnicyn Computing Center, Russian Academy of Sciences,
Vavilov str. 40, 119333 Moscow, Russian Federation
{igourevi,ytrusova}@ccas.ru

Abstract. The paper is devoted to the development of an ontology of the
domain "Image processing, analysis, recognition, and understanding" based on
the existing image analysis thesaurus. Such an ontology could be used to
support a wide range of tasks, including automated image analysis, algorithmic
knowledge reuse, intelligent information retrieval, etc. Main steps and first
results of the ontology development process are described.

Keywords: image analysis, thesauri, ontologies.

1 Introduction

Since the beginning of the 1990s, research on ontology is becoming increasingly
widespread in the computer science community. Ontologies are used to support a
great variety of tasks in many research areas, such as knowledge management and
organization, natural language processing, databases and knowledge bases, digital
libraries, geographic information systems, information retrieval and so on.

The work presented in this paper is devoted to the development of an ontology of
the domain "*I*mage processing, *A*nalysis, recognition and understanding"(shortened in
IA) based on an existing image analysis thesaurus [1]. The main goals of the ontology
are to structure domain knowledge and to unify a terminology of the domain. In
addition to providing a common vocabulary and shared understanding of the given
domain, the ontology can be used to support a range of tasks, including (a) ontology-
driven image analysis; (b) algorithmic knowledge reuse; (c) intelligent information
retrieval. Some examples of image processing ontologies have been reported in the
literature (e.g., [2, 3]), though many of them were developed for solving a specific
task and only include image processing concepts needed for a certain problem.

The main contribution of our work is the development of an ontology which will
cover all important aspects of IA (main concepts, their properties and relations). As a
basis for building the ontology we used the Image Analysis Thesaurus (IAT) [1]
which is being developed at the Scientific Council "Cybernetics" of the Russian
Academy of Sciences (RAS) and detailed later at the Dorodnicyn Computing Center

B. Falcidieno et al. (Eds.): SAMT 2007, LNCS 4816, pp. 113–116, 2007.

of the RAS. IAT provides a rich terminology of IA domain, which makes it a good starting point for the ontology of the given subject domain. The paper presents preliminary steps of reengineering the existing IAT into an ontology aiming at making it more useful and widely used.

2   Image Analysis Thesaurus

2.1   General Overview

IAT is intended for systematization of ill-structured and changing terminology in IA domain. An existing version of IAT contains about 2000 terms. This collection of terms continues to grow, since the thesaurus is regularly updated to reflect changes in the given domain. The concepts are represented by descriptors, which are distributed among five functional categories: *objects* (image types, image elements), *tasks*, *methods*, *instruments* (operators), and *attributes* (of objects, tasks, methods, and instruments). The *system* of semantic categories of descriptors, linked together by the hierarchical, conceptual and pragmatic relations, simulates the image analysis domain and defines the conceptual model of the knowledge base in this domain.

IAT was developed in accordance with international standards on thesauri construction [4]. The thesaurus was implemented in Visual Fox Pro 7.0 (VFP7.0). Detailed description of the IAT program module is presented in [1].

2.2   Semantic Problems

Thesauri are powerful tools for information indexing, classification, and retrieval, but the standard thesaurus relationships – hierarchical, associative and equivalence relations - are not differentiated enough to support more complicated tasks, for example, intellectual decision making and ill-structured problem solving.

With respect to intelligent processing needs it is evident that a more sophisticated knowledge organization system is required. It is necessary to refine standard thesaurus relationships into more specific kinds of relations such as "*hasSubtask*", "*hasProperty*", "*methodFor*", etc. The relationships in an ontology are more differentiated and explicitly named (see Table 1). The specification of rules and constraints are presented, so that they reflect the context of the domain for which the knowledge has to be represented [4].

Table 1. The IAT relationships compared with relationships of future ontology

Image Analysis Thesaurus	Image Analysis Ontology
edge	Edge
Narrower term: edge pixel	*<hasComponent>* edge pixel
Narrower term: roof edge	*<includesSpecific>* roof edge
median filtering	median filtering
Broader term: non-linear smoothing method	*<hasProperty>* non-linearity
Related term: image smoothing	*<methodFor>* image smoothing
Related term: image smoothing method	*<performedByInstrument>* median filter
Related term: median filter	*<methodFor>* noise reduction
Related term: noise reduction	*<hasAttribute>* window size
Related term: window size	

3 Reengineering the IAT into an Image Analysis Ontology

An ontology may take a variety of forms but it necessarily includes a vocabulary of terms and some specification of their meaning. IAT contains the most important terms and their definitions in the IA domain. It helps to avoid building the ontology of the given domain starting from scratch. By reengineering the thesaurus into the ontology we mean the process of transforming the information contained in the IAT in order to encode it in more widely used format and refine it as needed.

Converting IAT into an ontology is a multi-step procedure. In [4] the following main steps of the reengineering process were identified: (1) definition of an ontology structure; (2) extraction of all information from the thesaurus; (3) manually editing and refinement of the ontology. First results obtained on each stage are described below.

3.1 Defining an Ontology Structure

Automated problem solving is an important task in the domain of processing and understanding the information represented in the form of an image. Image processing, and, in most cases, image analysis and recognition can be reduced to applying the system of transformations on an image. Usually, several processing steps have to be combined to obtain the solution. At each of these stages a problem of selecting the most relevant method and its best parameters appears. Syntactic combining is not sufficient for the efficient solution of image analysis and understanding problems; the possibility of semantic combining is necessary.

The Image Analysis Ontology (IAO) is aimed at describing the knowledge about available image processing techniques for intelligent image analysis. IAO will contain concepts and relationships that are required to describe various methods, tasks that can be solved by these methods as well as different image types, image elements and their characteristics. The following 5 top-level classes of concepts were identified: "Tasks", "Methods", "Instruments", "Objects" and "Attributes". The relationships among these classes are as follows (see Fig. 1): a task can be solved by alternative methods; a method is performed by some instruments; each instrument is characterized by a set of attributes and objects describing input and output data.

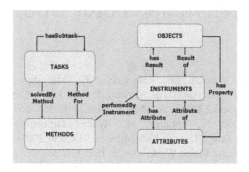

Fig. 1. Examples of relations between classes

3.2 Extraction of All Information from Existing Thesaurus

The information contained in IAT was converted into OWL (Web Ontology Language), a language that can be used to: (i) formalize a domain by defining classes and properties of those classes; (ii) define individuals and assert properties about them, and (iii) reason about these classes and individuals to the degree permitted by the formal semantics of the OWL language.

Since the manual conversion of IAT into OWL is a routine and very time-consuming, we wrote a semi-automatic customized transformation procedure able to fill better the gap between the syntactic and the semantic formats. We created some *virtual* tables (*views*) to collect and relate the information contained in the thesaurus and then we exported an XML document obtained grouping such views. Afterwards, we imported the XML document into an XML database, specifically the eXist XML native database. Finally, a sequence of three queries, written with the XQuery language was executed. All terms of the IAT were converted to classes (concepts). The hierarchical IAT relationships were used to form the class' hierarchy, which constitutes the basic taxonomy of the ontology.

3.3 Editing and Refining the Ontology

The result of the previous step was edited by using the Protégé editor in order to check the ontology consistence. Currently, the ontology is under refinement for further improvements, e.g. introducing new relations and new properties.

4 Conclusions

Future work will focus on the refinement of the ontology structure. One of the important tasks will be to define a full set of relations between concepts needed for problem solving as well as different properties of classes. This way, our ontological approach will have the advantages to contain more precise and deeper semantics, and assure reusability and share-ability.

Acknowledgments. The work was partly supported by the Russian Foundation for Basic Research (pr. nrs. 06-01-81009, 06-07-89203, 07-07-13545), by the Program of the Presidium of the RAS "Fundamental Problems of Computer Science and Information Technologies" (nr. 2.14), by MUSCLE NoE – FP6-507752 and by the CNR-RAS bilateral project.

References

1. Beloozerov, V.N., Gurevich, I.B., Gurevich, N.G., Murashov, D.M., Trusova, Y.O.: Thesaurus for Image Analysis: Basic Version. Pattern Recognition and Image Analysis: Advances in Mathematical Theory and Applications 13(4), 556–569 (2003)
2. Hudelot, C., Maillot, N., Thonnat, M.: Symbol Grounding for Semantic Image Interpretation: From Image Data to Semantics. In: ICCVW 2005. 10th IEEE International Conference on Computer Vision Workshops, p. 1875 (2005)
3. Asirelli, P., Little, S., Martinelli, M., Salvetti, O.: MultiMedia metadata management: a proposal for an infrastructure. In: Proc. of SWAP 2006, Pisa, CEUR (December 18-20, 2006)
4. Soergel, D., Lauser, B., Liang, A., Fisseha, F., Keizer, J., Katz, S.: Reengineering Thesauri for New Applications: the AGROVOC Example. Journal of Digital Information 4(4) (2004)

A Wavelet-Based Algorithm for Multimodal Medical Image Fusion

Bruno Alfano[1], Mario Ciampi[2], and Giuseppe De Pietro[2]

[1] IBB-CNR, Via S. Pansini, 5 - 80131 Naples, Italy
{bruno.alfano@ibb.cnr.it}
[2] ICAR-CNR, Via P. Castellino, 111 - 80131 Naples, Italy
{mario.ciampi,giuseppe.depietro}@na.icar.cnr.it

Abstract. Medical images coming from different sources can often provide different information. So, combining two or more co-registered multimodal medical images into a single image (image fusion) is an important support to the medical diagnosis. Most of the used image fusion techniques are based on the Multiresolution Analysis (MRA), which is able to decompose an image into several components at different scales. This paper presents a novel Wavelet-based method to fuse medical images according to the MRA approach, that aims to put the right "semantic" content in the fused image by applying two different quality indexes: variance and modulus maxima. Experimental tests show very encouraging results in terms of both quantitative and qualitative evaluations.

1 Introduction

Extraordinary evolution of image acquisition technology enables physicians to deal with several kinds of images for the medical diagnosis. As matter of fact, the diversification of the typology of sensors for acquiring medical images (such as CT, PET and MRI scanners) has granted the possibility of consulting simultaneously a range of multimodal images of the same body part of a patient. Such images provide different and often complementary contents, e.g. CT images supply anatomical information, (for example if a tumour is penetrated in bone structures), whereas PET images deliver functional information (such as determining the real extension of a tumour).

Therefore, only one type of image is often not sufficient to make a physician able to individuate possible pathologies, but he needs to consult simultaneously different images. For this reason, many several algorithms have been studied to the aim of fusing multimodal medical images.

Image fusion is the process of extracting significant information from multiple images and synthesizing them in the same image. In literature, it is well established that the Multiresolution Analysis (MRA) [1] is the approach that best suits image fusion, for its ability to decompose an image in more images at different resolutions in order to analyze them separately. Wavelet decomposition is the method that best fits the MRA approach regarding images, because it is able to separate an image in low- and high-frequency contents [2].

B. Falcidieno et al. (Eds.): SAMT 2007, LNCS 4816, pp. 117–120, 2007.

This paper presents a novel Wavelet-based algorithm to blend two co-registered multimodal medical images trying to extract salient information from them.

2   Proposed Approach

The complete fusion scheme of the algorithm proposed is shown in Fig. 1.

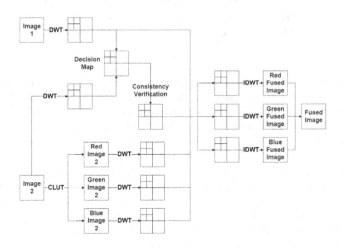

Fig. 1. Fusion scheme proposed

The algorithm consists in a window-based technique that combines low-frequency components (approximations) and high-frequency components (details) of images respectively using the variance maximum and the modulus maximum as quality indexes, which consider in the first case the change of the contrast of pixels and in the second case the edges as significant information.

To date, many other Wavelet-based fusion techniques [3,4,5] have been proposed. The solution presented in this paper differs from those ones for two basic reasons: i) it is a completely selective-based approach (for instance, approximation coefficients are not combined by averaging them), and ii) it colors one of the images before the fusion process start and it faces the problem of combining the RGB channels separately in order to enable a physician to distinguish in the fused image if a pixel is sourced from the first image or the second one.

The whole process can be organized in five steps that are described afterwards in details. It is assumed that both the sources are grey-scale images.

1. Image 2 is colored by means of a Colour Look-up Table (CLUT), a technique that transforms a grey-scale pixel into a RGB pixel by consulting a known table. Then, by fixing a level of decomposition, 2-D Discrete Wavelet Transforms (DWTs) decompose Image 1 and Image 2 with its RGB channels. Thus, five matrixes of coefficients are created.

2. Variances of coefficients contained into sliding windows that scroll decompositions of Image 1 and Image 2 are computed and compared. Named as
S the source image, as W the size of the window, as K the decomposition
level and as a the matrix of approximation coefficients, the variance can be
defined as follows:

$$var_S^{(K)}(W) = \frac{1}{|W|} \sum_{(i,j) \in W} \left(a_S^{(K)}(i,j) - \frac{1}{|W|} \sum_{(i,j) \in W} a_S^{(K)}(i,j) \right)^2. \quad (1)$$

Approximation coefficients of the fused image are all the central pixels of
the windows with the maximum value of the variance. Specifically, a binary
image (named decision map), in which values 0 and 1 indicate if the final
coefficient will be extracted from the first or the second image, is created.
3. With regards to detail coefficients, sums of moduli of coefficients contained
into a sliding window of size W are computed as following:

$$mod_S^{(K)}(W) = \sum_{(i,j) \in W} |d^{(K)}(i,j)|, \quad (2)$$

where d is the matrix of detail coefficients. The decision map is completed
by considering all the central pixels of the windows with modulus maximum.
4. A consistency verification of the decision map is performed, that is a sliding
window of size W scrolls it in order to avoid the presence of noise, characterized by the choice of extracting a pixel from one image, while the most
of the neighbour pixels (that is, all the pixels inside the window except the
central pixel) come from the other one. Only in this case, the algorithm flips
the value of such a pixel in the decision map.
5. The RGB channels of the decomposed fused image are figured extracting
coefficients from the Wavelet decompositions of Image 1 and of the RGB
channels of Image 2 according to the verified decision map. Then, 2-D Inverse
DWTs are applied to resulting RGB channels and finally the obtained images
are composed into one in order to produce the final RGB fused image.

3 Experimental Results

In this section the proposed approach is tested with another technique that
averages approximation coefficients. We have evaluated results by using two
different quality indexes: Mutual Information [6], often used to assess also the
registration of medical images, and the one proposed by Wang and Bovik [7].

We have implemented these algorithms in C++ language and exploited two
open source libraries: ITK [8], for processing medical images compliant to the DI-
COM [9] standard, and VTK [10], for visualizing and rendering images. Datasets
considered for the test are two series of multimodal images of 10 slices (CTs
and PETs). The resolution of CTs is of 512x512 pixels, whereas for PETs is of
128x128 pixels. For this reason, a resizing of PET images has to be preliminarily

	Mutual Information		Wang and Bovik	
Slice	Mean	Variance	Mean	Variance
1	0.383356	0.395590	0.383717	0.403864
2	0.385227	0.401047	0.384714	0.406014
3	0.392685	0.413141	0.388363	0.407864
4	0.391598	0.413616	0.387095	0.406453
5	0.392148	0.412260	0.387202	0.403259
6	0.394681	0.414405	0.389981	0.404458
7	0.399621	0.415722	0.387454	0.400452
8	0.385982	0.405891	0.384140	0.396303
9	0.402599	0.419474	0.389815	0.401747
10	0.400288	0.418346	0.382649	0.396750

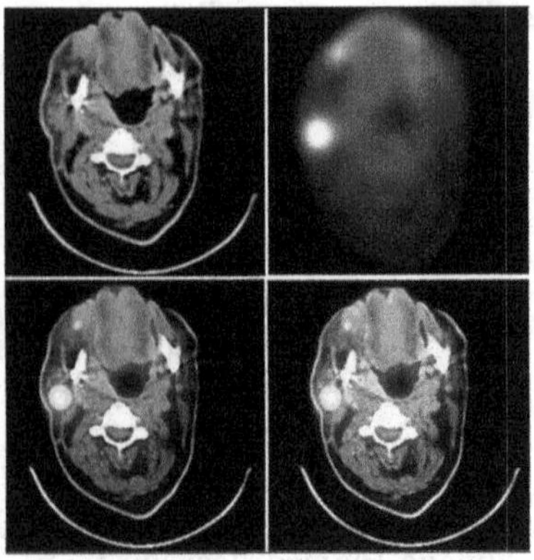

Fig. 2. Numerical results and clockwise a CT, a PET, fusions with variance and mean

performed. Images are decomposed by Daubechies Wavelets level 20 with two levels of decomposition and the size of the sliding window chosen is 5x5.

Figure 2 reports both quantitative and qualitative evaluations. As shown, both Mutual Information and Wang and Bovik's quality index provide encouraging results as well as the resulting image computed with the proposed approach is more accurate than the one figured with the classical mean-based criterium and both characteristics of source images are distinguishable.

References

1. Piella, G.: A General Framework for Multiresolution Image Fusion: From Pixels to Regions. Information Fusion 4, 259–280 (2003)
2. Chipman, L.J., Orr, T.M., Lewis, L.N.: Wavelets and Image Fusion. IEEE Transactions on Image Processing 3, 248–251 (1995)
3. Berbar, M.A., Gaber, S.F., Ismail, N.A.: Image Fusion Using Multi-Decomposition Levels of Discrete Wavelet Transform. Visual Information Engineering, 294–297 (2003)
4. Pajares, G., de la Cruz, J.M.: A Wavelet-based Image Fusion Tutorial. PR(37) 9, 1855–1872 (2004)
5. Qu, G., Zhang, D., Yan, P.: Medical Image Fusion by Wavelet Transform Modulus Maxima. Optics Express 9(4), 184 (2001)
6. Pluim, J.P.W., Maintz, J.B.A., Viergever, M.A.: Mutual Information Based Registration of Medical Images: A Survey. IEEE Transactions on Medical Imaging 22(8), 986–1004 (2003)
7. Wang, Z., Bovik, A.C.: A Universal Image Quality Index. IEEE Signal Processing Letters (2002)
8. http://www.itk.org/
9. http://medical.nema.org/
10. http://public.kitware.com/VTK/

Context-Sensitive Pan-Sharpening of Multispectral Images

Bruno Aiazzi[1], Luciano Alparone[2], Stefano Baronti[1], Andrea Garzelli[3],
Franco Lotti[1], Filippo Nencini[3], and Massimo Selva[1]

[1] Institute of Applied Physics "Nello Carrara", IFAC-CNR, 10 via Madonna del
Piano, 50039 Sesto F.no (FI), Italy
{b.aiazzi,s.baronti,f.lotti,m.selva}@ifac.cnr.it
[2] Department of Electronics and Telecommunications, University of Florence,
3 via S. Marta, 50139, Florence, Italy
alparone@lci.det.unifi.it
[3] Department of Information Engineering, University of Siena, 56 via Roma, 53100,
Siena, Italy
{garzelli,nencini}@dii.unisi.it

Abstract. Multiresolution analysis (MRA) and component substitution (CS) are the two basic frameworks to which image fusion algorithms can be reported when merging multi-spectral (MS) and panchromatic (Pan) images (Pan-sharpening). State-of-the-art algorithms add spatial details derived from the Pan image to the MS bands according to an injection model. The capability of the model to describe the relationship between the MS and Pan images is crucial for the quality of fusion results. Although context adaptive (CA) injection models have been proposed in the framework of MRA, their adoption in CS schemes has been scarcely investigated so far. In this work a CA injection model already tested for MRA algorithms is evaluated also for CS schemes. Qualitative and quantitative results reported for IKONOS high spatial resolution data show that CA injection models are more efficient than global ones.

1   Introduction

Image fusion has become important in the remote sensing community because of the existing physical constraints between spatial and spectral resolution. This makes desirable the spatial enhancement of poor-resolution multispectral (MS) data or the spectral enhancement of high ground resolution data having poor spectral resolution (as a limit case, a single broadband panchromatic image).

A typology of simple and fast well established algorithms for data fusion is denoted as component substitution (CS). This family of methods is based on a transformation of all the original bands in a new vectorial space. Then, one of the transformed components (usually the smooth intensity **I**) is *substituted* by the panchromatic image, **Pan**, before the inverse transformation is applied. This procedure is equivalent to *inject*, i.e. add, the difference between the sharp panchromatic and the smooth intensity images into the resampled MS bands [1].

B. Falcidieno et al. (Eds.): SAMT 2007, LNCS 4816, pp. 121–125, 2007.

A widespread CS technique is the Gram-Schmidt (GS) spectral sharpening. This method has been invented by Laben and Brover in 1998 and patented by Eastman Kodak [2]. GS is presently adopted in the ENVI-IDL tool for image processing. Following GS basic concepts, a modified GS algorithm called Gram-Schmidt Adaptive (GSA) has been obtained, in which the smooth intensity $\mathbf{I}$ is computed as a weighted average of the MS bands.

A second family of data fusion methods is based on multiresolution analysis (MRA). MRA algorithms decompose each image into a series of bandpass channels in the spatial frequency domain. The techniques based on MRA substantially perform an injection, with suitable weights, of $zero - mean$ high-frequency spatial components taken from Pan image into resampled versions of the MS bands. These methodologies have been considered to overcome the problem of spectral distortion that can be usually noticed in CS fusion results. In particular, MRA methods based on generalized Laplacian pyramids (GLP) have the property that they can take care of the magnitude of the Fourier transform (MTF) of the optical point spread function of the imaging system. Actually, if the highpass filter used to extract the high spatial frequency components from the Pan image is taken such as to approximate the complement of the MTF of the MS band to be enhanced, then the high frequency components can be restored. A further goal of an advanced fusion method is to increase spectral information by unmixing the coarse MS pixels through the sharp Pan image. This task requires a suitable model establishing how the highpass information coming from the Pan have to be inserted into the MS bands. In this perspective, remarkable improvements have been obtained for MRA techniques by considering local context adaptive (CA) models instead of simple global models.

The aim of this work is to present the key points of image fusion based on CS and MRA and to evaluate if CA injection models suitable for MRA methods can be useful also for CS techniques. This is done by defining a general scheme that is capable of modeling both CS and MRA fusion methods. In particular, this scheme is applied for improving the performances of the GSA method when a set of local factors weighting the high frequency details is considered. The results obtained are compared with a state-of-art MRA technique [3], i.e. the GLP-CA-MTF algorithm [4]. This paper reports in sect. 2 the proposed general scheme. Quality issues are considered in sect. 3 with a qualitative and quantitative evaluation on an IKONOS data set.

2   General Image Fusion Scheme

Let us consider the general fusion scheme reported in Fig. 1. We examine the usual case in which a multispectral sensor collects information into four frequency bands: the blue band $\mathbf{B}$, the green band $\mathbf{G}$, the red band $\mathbf{R}$ and the near infrared band $\mathbf{NIR}$. Moreover, we denote as $\mathbf{MS}_i$, $i = 1, \ldots, 4$, the generic multispectral band taken from the considered set. The symbol $\sim$ means that the generic band is expanded to the same spatial scale of the full resolution Pan image $\mathbf{Pan}$, while the symbol $\wedge$ indicates the generic band at the end of the fusion process. Thus,

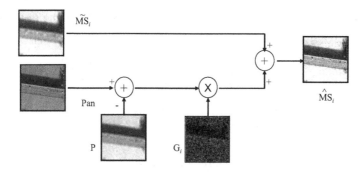

Fig. 1. General scheme representing both MRA and CS family fusion methods

the block diagram of Fig. 1 can be synthesized by the following formula, where $\mathbf{G}_i$ is a weight matrix which modulates the high spatial frequencies to be injected into each band according to the considered model (local or global) and $\mathbf{P}$ is a low spatial frequency image differently computed changing the family of fusion methods:

$$\hat{\mathbf{MS}}_i = (\mathbf{Pan} - \mathbf{P}) \cdot \mathbf{G}_i + \tilde{\mathbf{MS}}_i, \quad i = 1, \ldots, 4 \tag{1}$$

Concerning the GSA technique, it is defined by (2):

$$\mathbf{P} = \mathbf{I} = \hat{w}_1 \cdot \tilde{\mathbf{B}} + \hat{w}_2 \cdot \tilde{\mathbf{G}} + \hat{w}_3 \cdot \tilde{\mathbf{R}} + \hat{w}_4 \cdot \tilde{\mathbf{NIR}} + b \tag{2}$$

where the weights $\{\hat{w}_i\}$ are obtained by means of a linear regression algorithm [5] in order to minimize the MSE between a linear combination of the multispectral bands and the full resolution Pan image reduced to the spatial scale of the original MS bands. The constant parameter b is an offset image introduced in the model to consider an additive component that may arise from the acquisition process because of the sensor spectral responses.

Regarding the MRA family, $\mathbf{P}$ is a low-pass spatial filtered version of the **Pan**. Consequently, it can be denoted as $\mathbf{P} = \mathbf{Pan}_F$.

The injection model represented by the weight matrix $\mathbf{G}_i$ is performed by calculating suitable statistics on the entire image in the global case or on a 11×11 sliding window of the current pixel in the local case. The adopted statistics are based on the covariance factors between the MS bands and the Pan image, and are reported in Table 1 for the considered algorithms.

3   Results and Comparisons

Quantitative and qualitative results have been assessed on a data set of four MS bands and a single Pan acquired by high resolution IKONOS satellite. Quantitative assessments are obtained by degrading the spatial resolution of all original images by a factor of 4 and performing fusion on the reduced images [6]. The comparison of fused with original images allows global synthetic scores as Q4 [7],

Table 1. Paradigm of the fusion methods represented in Fig. 1

Method	w_i	Filtered Pan	g_i	Injection Model
GSA	$\hat{w}_i$	no	$\frac{cov(\mathbf{Pan}_F,\tilde{\mathbf{MS}}_i)}{var(\mathbf{P}_F)}$	**global**
GSA-CA	$\hat{w}_i$	no	$\frac{cov(\mathbf{Pan}_F,\tilde{\mathbf{MS}}_i)}{var(\mathbf{P}_F)}$	**local**
MRA	nd	yes	$\frac{cov(\mathbf{I},\tilde{\mathbf{MS}}_i)}{var(\mathbf{I})}$	**global**
MRA-CA	nd	yes	$\frac{cov(\mathbf{I},\tilde{\mathbf{MS}}_i)}{var(\mathbf{I})}$	**local**

SAM and ERGAS [8] to be computed. The score values are reported in Table 2 and show that the CA strategy improves both GSA and MRA. MRA-CA results the best algorithm, closely followed by GSA-CA.

Fig. 2 shows a 256 × 256 image of the area of Toulouse, France. The full-resolution Pan image is reported in Fig. 2(a) while the original true-color MS image, expanded to the scale of Pan, is reported in Fig. 2(b). Fusion results are reported in figures 2(c-f), also in true color composition.

Qualitative evaluation is more difficult. All the methods are *good* and their relative variations can be fully appreciated only when flickering couples of images on the screen. GSA and GSA-CA images are the sharpest ones, while MRA and

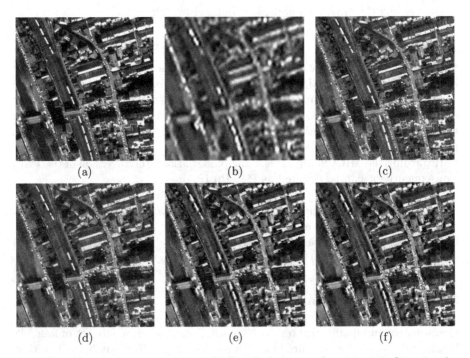

(a)   (b)   (c)

(d)   (e)   (f)

Fig. 2. 256 × 256 color compositions of full-scale fusion results at 1 m pixel spacing for the IKONOS image. (a): original Pan image; (b): original MS bands (4 m) re-sampled to Pan image scale(1 m); (c): GSA; (d): GSA-CA; (e): MRA; (f): MRA-CA.

Table 2. Average cumulative quality scores between original 4 m MS and fused images obtained from 16 m MS and 4 m Pan. EXP denotes 16 m MS resampled to a 4m size.

	EXP	GSA	GSA-CA	MRA	MRA-CA
Q4	.630	.921	.926	.922	.928
SAM (*degrees*)	4.85	2.96	2.87	2.94	2.85
ERGAS	5.94	2.76	2.69	2.73	2.66

MRA-CA are more spectrally faithful and this explains the results of Table 2. The introduction of the CA strategy has the effect of modulating the contrast in the fused images depending on the CA function response.

4   Concluding Remarks

The general scheme proposed for CS and MRA based fusion methods allow an easy comparison between different fusion strategies. An adaptive injection strategy of the spatial details extracted from the Pan image has been evaluated for two efficient fusion algorithms based on CS and MRA analysis, respectively. Context adaptive injection is capable of improving the quality of images fused with both CS and MRA methods. Future work will explore different context adaptive models capable of improving the performances of image fusion algorithms.

References

1. Tu, T.M., Su, S.C., Shyu, H.C., Huang, P.S.: A new look at IHS-like image fusion methods. Information Fusion 2(3), 177–186 (2001)
2. Laben, C.A., Brower, B.V.: Process for enhancing the spatial resolution of multispectral imagery using pan-sharpening. Technical Report US Patent # 6,011,875, Eastman Kodak Company (2000)
3. Alparone, L., Wald, L., Chanussot, J., Thomas, C., Gamba, P., Bruce, L.M.: Comparison of pansharpening algorithms: outcome of the 2006 GRS-S data-fusion contest. IEEE Trans Geosci. Remote Sensing 45(10) (October 2007)
4. Aiazzi, B., Alparone, L., Baronti, S., Garzelli, A., Selva, M.: MTF-tailored multiscale fusion of high-resolution MS and Pan imagery. Photogramm. Eng. Remote Sens. 72(5), 591–596 (2006)
5. Ross, S.M.: Introduction to Probability and Statistics for Engineers and Scientists. Elsevier Academic Press, Burlington, MA (2004)
6. Wald, L., Ranchin, T., Mangolini, M.: Fusion of satellite images of different spatial resolutions: assessing the quality of resulting images. Photogramm. Eng. Remote Sens. 63(6), 691–699 (1997)
7. Alparone, L., Baronti, S., Garzelli, A., Nencini, F.: A global quality measurement of Pan-sharpened multispectral imagery. IEEE Geosci. Remote Sensing Lett. 1(4), 313–317 (2004)
8. Ranchin, T., Wald, L.: Fusion of high spatial and spectral resolution images: the ARSIS concept and its implementation. Photogramm. Eng. Remote Sens. 66(1), 49–61 (2000)

Semantic Annotation of 3D Surface Meshes Based on Feature Characterization

Marco Attene, Francesco Robbiano, Michela Spagnuolo, and Bianca Falcidieno

Istituto di Matematica Applicata e Tecnologie Informatiche - Genova, CNR, Italy

Abstract. In this paper we describe the main aspects of a system to perform non-trivial segmentations of 3D surface meshes and to annotate the detected parts through concepts expressed by an ontology. Each part is connected to an instance in a knowledge base to ease the retrieval process in a semantics-based context. Through an intuitive interface, users create such instances by simply selecting proper classes in the ontology; attributes and relations with other instances can be computed automatically based on a customizable analysis of the underlying topology and geometry of the parts.

1 Introduction

Digital 3D shapes have a fundamental role in important and diverse areas such as product modeling, medicine, virtual reality and simulation, and their impact on forthcoming multimedia-enabled systems is foreseen to grow significantly. In the latest years we have assisted to an impressive growth of online repositories of 3D shapes [3,1,2,4] which reveals the importance of making these resources more accessible and easy to share and retrieve.

The Stanford repository is one of the earliest and widely used [4] and it maintains simple records of 3D shape models: *core* data – the geometry – plus a brief textual description. More recently, the increase of the number of shapes called for more intelligent searching methods; among them we can cite the first significant results proposed by the Princeton Shape Benchmark [2], which supports searching by geometric similarity starting from queries defined by sketching or by example. A different perspective is adopted by the AIM@SHAPE Shape Repository [3], which is based on a formal organization of 3D models enriched with metadata that make it possible to search for content also in terms of knowledge *about* the represented shapes. Most repositories attempt to ease the retrieval process by associating each shape to a coarse category (e.g. vehicles, aircraft, humans). Note that, since such association is typically a manual operation, a finer subdivision would be a too demanding and subjective task. Nonetheless, the ever-growing size of 3D repositories is making the retrieval hard even in a single bounded category. Thus, characterizing shapes of a given domain is becoming more and more important, and specific annotation approaches that require minimal human intervention must be devised.

B. Falcidieno et al. (Eds.): SAMT 2007, LNCS 4816, pp. 126–139, 2007.

To achieve this goal, the structural subdivision of an object into subparts, or *segments*, has proven to be a key issue. At a cognitive level, in fact, the comprehension of an object is often achieved by understanding its subparts [5,6]. For instance, if we consider an object which has a *human face* it is likely that the object is a *human being* (the presence of a subpart influences the interpretation of the whole object), and if we want to retrieve a human which is *strong* we can search for humans whose arms'*volume* is big (here the quantitative characterization of a part influences the qualitative interpretation of the whole). Furthermore, a proper subdivision of the shape would allow users to access directly subparts: users will be able not only to search, but also to retrieve *legs*, *noses* and *heads* even if the repository was originally intended for whole human body models. Such a possibility is extremely important in modern design applications: creating original shapes from scratch, in fact, is a time-consuming operation and it requires specific expertise, hence re-using parts of 3D shapes is recognized to be a critical issue.

Clearly, the retrieval of 3D objects within a repository can be significantly improved by annotating each shape not only as a whole, but also in terms of its meaningful subparts, their attributes and their mutual relations. The possibility to semantically annotate shape parts may have a relevant impact in several domains. An application that we consider particularly significant, for example, is the creation of *avatars* in emerging *MMORPGs* (Massive Multiplayer Online Role-Playing Games) and in online virtual worlds such as Second Life [7]. Currently the avatar design is done from scratch or through the personalization of a predefined amount of parameters (e.g. shape of the body, skin, hair, eyes, clothes). Producing an original avatar through this approach, however, is becoming more and more difficult due to the exponential *demographic* growth in the aforementioned virtual worlds (currently, in Second Life there are more than 7 million residents, each one with his/her avatar). In our view, the possibility to browse and search huge repositories of virtual characters and their parts would end up in a potentially infinite number of avatars, obtained by combination and further personalization of the different parts. Annotated parts might also be exploited by proper interpretation rules to allow their retrieval based on high-level characterizations (e.g. retrieve "heads of Caucasian adult male" or "heads of children", or "heads of black women").

In general, both the extraction and the annotation of the subparts are characterized by an inherent context dependance: the kind of geometric analysis used to detect the segments, as well as the *interpretation* of the segments, can significantly vary in different contexts. A nearly-cylindrical object can be annotated as a *finger* in the domain of *human bodies*, as a *piston* in the domain of *car engines*, and may be not detected at all in another domain in which this kind of features is not interesting. Thus, it is important to devise annotation approaches which are both general (i.e. they must not depend on any particular context) and flexible (i.e. they must easily adapt to any particular context).

1.1   Overview and Contributions

In this paper we tackle the aforementioned part-based annotation problem by presenting a flexible and modular system called the *ShapeAnnotator*. We use surface meshes to represent 3D shapes, and ontologies to describe the annotation domains. By exploiting results from Computer Graphics, Vision and Knowledge Technologies, the ShapeAnnotator makes it possible to load a 3D surface mesh and a domain ontology, to define the meaningful shape parts, to annotate them properly and to save the result in a knowledge base. The ShapeAnnotator makes use of the following solutions:

- A *multi-segmentation* framework to specify complex and heterogeneous surface segments (the *features*);
- A set of functionalities called *segmentmeters* to calculate geometric and topological characterizations of the segments;
- An ontology module which allows the user to browse the domain ontology and to create instances *describing* the features;
- A mechanism to *teach* the system how instance properties can be computed automatically based on segmentmeters.

2   Related Work

To our knowledge, existing literature does not deal with any framework to support feature-based annotation processes for 3D shapes. The two main tasks addressed in our work are the *segmentation* and the *annotation* of 3D shapes. A lot of research that deals with the integration of these two aspects is related to traditional visual media, images and videos in particular. A variety of techniques has been developed in image processing for content analysis and segmentation, and the available annotation techniques are mainly based on the computation of low-level features (e.g. texture, color, edges) that are used to detect the so-called regions-of-interest [8,9,10]. For video sequences, keyframe detection and temporal segmentation are widely used [11].

Although the general framework of content analysis and annotation developed for images or videos may be adapted to 3D shapes, the two domains are substantially different. For images and videos, indeed, the objective is to identify relevant objects in a scene, with the inherent complexity derived by occlusions and intersections that may occur. In the case of 3D, information about the object and its features is complete, and the level of annotation can be therefore more accurate. A peculiarity of 3D shapes is that geometric measures of the single segments, such as bounding box length or best-fitting sphere radius, are not biased by perspective, occlusions or flattening effects. Therefore these measures can be directly mapped to cognitive properties (roundness, compactness, ...).

2.1   Keyword-Based and Ontology-Based Annotation

Generally speaking, the purpose of annotation is to create correspondences between objects, or segments, and conceptual tags. Once an object and/or its parts are annotated, they can easily match textual searches. Stated differently, advanced and semantics-based annotation mechanisms support content-based retrieval within the framework of standard textual search engines.

The two main types of textual annotation are *keyword*-driven and *ontology*-driven. In the first case users are free to tag the considered resources with any keyword they can think of, while in the second case they are tied to a precise conceptualisation. The trade-off is between flexibility and meaningfulness. In fact, in the case of free keyword annotation users are not forced to follow any formalized scheme, but the provided tags have a meaning just for themselves: since no shared conceptualisation is taken into account, the association of the tag to a precise semantic interpretation can be only accidentally achieved. Well-known examples of this kind of annotation for 2D images are *FLICKR* [12] and *RIYA* [13].

In the *ontology*-driven annotation, the tags are defined by an ontology. An ontology is a formal, explicit specification of a shared conceptualization of a domain of knowledge, and expresses the structuring and modeling of that particular domain [14,15]. Since the conceptualisation is shared, there is no freedom in the selection of tag names, but this is rewarded with a common understanding of the given tags eligible for selection. Moreover, the shared conceptualisation can also be processed by computer applications, opening up challenging opportunities for the further enrichment of search results with inference mechanisms [16]. In M-*OntoMat-Annotizer* [17] the user is allowed to highlight segments (i.e. regions) of an image and to browse specific domain ontologies in order to annotate parts of the former with specific instances of the latter. Similarly, *Photostuff* [18] provides users the ability to annotate regions of images with respect to an ontology and publish the automatically generated metadata to the Web.

The work presented in this paper falls in the class of the ontology-driven annotation approaches. Specifically, we tackle the problem of annotating shapes belonging to a specific category which is described by an ontology (i.e. human bodies, cars, pieces of furniture, ...). Each of these ontologies should conceptualize shape features characterizing the category, their attributes and their relations. In a *human body*, for example, *head*, *arms* and *legs* are relevant concepts, relations such as *arm* is_a *limb* hold, and attributes such as the *size* of the *head* are helpful to infer higher-level semantics (e.g. ethnic group, gender, age-range).

2.2   Segmentation of Polygonal Meshes

Given an ontology that describes a given class of shapes, the optimal solution for annotating 3D models would be to use a shape segmentation algorithm able to automatically detect all the features conceptualized by the ontology. This approach is far from being feasible, as existing segmentation algorithms hardly target semantic features and usually follow a pure geometric approach. Recent surveys of these methods can be found [19] and [20].

Much of the works tackling the segmentation problem with the objective of *understanding* a shape [21] are inspired by studies on human perception. For example there are theories that received a large consensus [22,5] and that indicate how shapes are recognized and mentally coded in terms of relevant parts and their spatial configuration, or structure.

In another large class of methods, the focus is mainly on the detection of geo-metrically well-defined features. These segmentations do not produce natural features but patches useful for tasks that require different and possibly non-intuitive schemes (e.g., approximation, remeshing, parameterization) [23,24,25]. Generally speaking, this approach is feasible when the features have some formal structure that can be associated to a mathematical formulation. In natural domains, for example human body models, there is no clue on how to define relevant features, and only few methods in the literature tackled a semantics-oriented segmentation in these kind of domains [26].

3   Multi-segmentation and Part-Based Annotation

In the case of 3D shapes, the identification of relevant features is substantially different from the corresponding 2D case. For 2D images, segmentation algo-rithms are not always considered critical to define features for annotation; on a flat image, in fact, useful features may be even sketched by hand [17]. In contrast, a 3D shape may be very complex and drawing the boundary of a feature might become a rather time-consuming task, involving not only the drawing stage, but also rotating the scene, translating and zooming in and out to show the portions of the surface to draw on. Moreover, while on a 2D image a closed curve de-fines an inner area, in 3D this is not always true. Hence, for the 3D case, using segmentation algorithms to support feature detection is considered a mandatory step.

Nevertheless, the huge amount of different and specialized works on mesh segmentation indicates that satisfactory results are missing. The majority of the methods used in Computer Graphics are not devised for detecting specific features within a specific context, as for example is the case of form-feature recognition in product modeling and manufacturing. The shape classes handled in the generic segmentation contexts are broadly varying: from virtual humans to scanned artefacts, from highly complex free-form shapes to very smooth and feature-less objects. Moreover, it is not easy to formally define the meaningful features of complex shapes in a non-engineering context and therefore the capa-bility of segmentation methods to detect those features can only be assessed in a qualitative manner [20].

Hence, our proposition is that, due to intrinsic limitations, no single algorithm can be used to provide rich segmentations, even within a single domain. This motivates the introduction of a theoretical framework for working with multi-segmentations, that allow for a much more flexible support for semantic seg-mentation. The intuition behind multi-segmentation is that a meaningful shape segmentation is obtained by using in parallel a set of segmentation algorithms and by selecting and refining the detected segments.

Most segmentation algorithms proposed in the literature [20] strive to subdi-vide the surface into non-overlapping patches forming an exhaustive partitioning of the whole model (Figure 1(b), (c) and (d)). Our proposition is that even this assumption is too restrictive: following the claim that the segmentation has to reflect the cognitive attitude of the user, the detected parts do not necessarily have to constitute a partition of the model, as some segments may overlap, and some surface parts may not belong to any significant segment at all. Therefore, it is often possible to design a proper technique for the identification of a particular class of features [27,23] and, if there is the need to identify features of different classes, it is possible to use different segmentation algorithms and take the features from all of their results. In some cases, moreover, there is an intrinsic *fuzziness* in the definition of the boundaries of a feature (i.e., in a human body model the neck may be considered part of both the head and the torso). This is another reason to avoid the restriction of using a sharp partitioning of the whole to identify all the relevant segments.

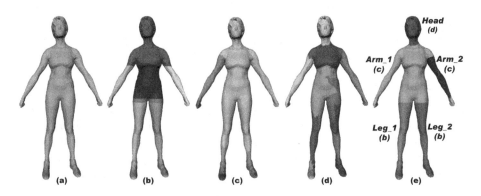

Fig. 1. An original mesh (a) has been partitioned using different segmentation algorithms: [28] in (b), [27] in (c) and [23] in (d). Only the most relevant features taken from (b), (c) and (d) have been selected and annotated in (e).

Due to these observations, we introduce the concept of *multi-segmentation* of a 3D surface represented by a triangle mesh, and say that in a multi-segmented mesh, the results of several segmentation approaches may overlap (e.g. {(b),(c),(d)} in Figure 1). When a multi-segmented mesh is interpreted within a specific context, some of the segments can be considered particularly *meaningful*. Such meaningful segments (i.e. the features) can be annotated by specific conceptual tags describing their meaning within the context. In this paper, we refer to an *annotated mesh* as to a multi-segmented mesh in which some of the segments have been annotated (e.g. Figure 1 (e)).

4   The Shape Annotator

Having established *what* is an annotated mesh, it remains to explain *how* to produce it out of an existing triangle mesh. In principle, an expert in a particular domain should be able to identify significant features and to assign them a specific meaning. As an example, an engineer should be able to look at a surface mesh representing an engine and identify which parts have a specific mechanical functionality. Unfortunately, to the best of our knowledge, today there is no practical way to transform such expertise into usable content to be coupled with the plain geometric information.

To bridge this gap, we defined an annotation pipeline and developed a prototype graphical tool called the *ShapeAnnotator*. This tool has been specifically designed to assist an expert user in the task of annotating a surface mesh with semantics belonging to a domain of expertise.

After loading a model and a domain ontology, the first step of the annotation pipeline is the feature indentification, i.e. the execution of segmentation algorithms to build the multi-segmented mesh. Once done, from the resulting multi-segmented mesh interesting features can be interactively selected. Each interesting feature can then be annotated by creating an instance of a concept described in the ontology. Optionally, the system may be also programmed to automatically compute attributes and relations among the instances to significantly enrich the resulting knowledge base.

4.1   Feature Identification

In order to identify surface features, the ShapeAnnotator provides a set of segmentation algorithms. Our prototype has a plugin-based architecture so that it is possible to import proper segmentation algorithms according to the requirements of the specific class of features. In the current implementation, we have chosen a number of algorithms that cover quite a wide range of feature types. In particular, it is possible to capture:

- *Planar features* through a clustering based on a variational shape approximation via best-fitting planes [24];
- *Generalized tubular features* with arbitrary section computed by the *Plumber* algorithm introduced in [27];
- *Primitive shapes* such as planes, spheres and cylinders through a hierarchical segmentation based on fitting primitives [23];
- *Protrusions* extracted through shape decomposition based on Morse analysis using the height function, the distance from the barycenter and the integral geodesics [28].

Using these tools, it is possible to roughly capture features and also to refine them through morphological operators. Up to now the following operators are available:

- Growing a segment by adding a strip of triangles to its boundary;
- Shrinking a segment by removing a strip of triangles from its boundary;
- Merging two segments.

It is also possible to remove a segment or to add a new segment from scratch (i.e., a single triangle), and edit it through the above operators.

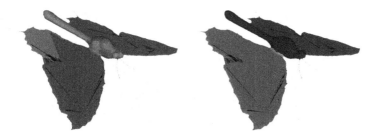

Fig. 2. Definition of non-trivial features starting from a raw segmentation. On the left, the HFP algorithm [23] could not capture the features due to degenerate mesh elements. On the right the unprecise features computed have been edited to obtain a better segmentation.

These functionalities make it possible to refine raw segmentations and properly define useful non-trivial features within few mouse clicks, as shown in Figure 2. Further examples are shown in Figure 6.

4.2 Manual Annotation

To annotate the features, the user may select proper conceptual tags within a domain of expertise formalized as an OWL [29] ontology. This choice offers a number of advantages, including the non-negligible fact that OWL is supported by popular ontology editors [30]. Strictly speaking, for the current functionalities of the ShapeAnnotator, a simpler language could be sufficient, as long as the user is prompted with the chance of selecting among relevant concepts; the choice of OWL, however, has been driven by the potential evolution of the ShapeAnnotator, which is foreseen to become more *intelligent* in the sense of providing inference functionalities (see Section 5).

Non trivial ontologies may be huge [31], and effective browsing facilities are fundamental to reduce the time spent to seek the proper concept to instantiate. In our approach, the ontology is depicted as a graph in which nodes are classes and arcs are relations between them (see Figure 3, left).

Browsing the ontology consists of moving along paths in the graph, which means jumping from a concept to another across relations. The navigation may be customized by the user and, while the simplest way of browsing is across relations of type `subClassOf` or `superClassOf`, it is possible to select any combination of properties that will be shown by the browser (see Figure 3, middle). Once a proper concept has been identified, the ShapeAnnotator provides the possibility to create an instance, which means providing a URI (i.e., a unique name) and setting the value of the properties (attributes and relations) defined in the ontology for the class being instantiated (see Figure 3, right).

Each instance may be modified in a second step in order to make it possible to define non-trivial relations between instances of the knowledge base (i.e., `myHead isAdjacentTo myNeck`).

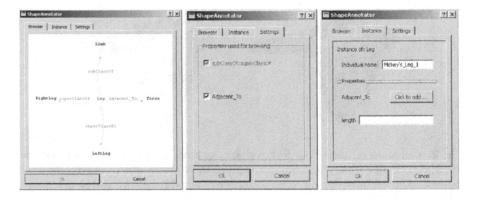

Fig. 3. The ontology browser, the selection of navigation settings and the creation of an instance.

4.3   Automatic Annotation

Currently, our system requires the user to manually select the concepts to instantiate; for attributes and relations between instances, however, there is the possibility to *tell* the ShapeAnnotator how these properties can be calculated without the user intervention. The ShapeAnnotator, in fact, comes with a set of functionalities to measure geometric aspects of shape parts (i.e. bounding box length, radius of best-fitting cylinder, ...) and to establish topological relations among the parts (i.e. adjacency, containment, overlap, ...). We call these functionalities *segmentmeters*.

Though segmentmeters work independently of any ontology, the user may define their *interpretation* within each specific domain of annotation. Namely, the user may establish a set of *connections* between topological relations and conceptual relations (e.g. "segment adjacency" ↔ is_connected_to) and between calculated values and class attributes (e.g. "radius of best-fitting cylinder" ↔ through_hole :: radius).

After having established such connections, the instance properties are transparently computed by the system. For example, when annotating a reverse engineered mechanical model, a part may be manually annotated as a Through_hole, while its parameter radius is automatically computed by the ShapeAnnotator as the radius of the cylinder that best fits the geometry of the part; if two adjacent segments are manually annotated as instances of the class stiffener, the relation is_adjacent_to is automatically set to conceptually link the two instances. An example of connection is shown if Figure 4.

Furthermore, there is the possibility to combine some segmentmeters within formulae to be connected to specific attributes. When annotating a human body model, for example, the user may want to tell the ShapeAnnotator something like "for each instance of the class head set the attribute size with the length of the great circle of the segment's best-fitting sphere". This will be accomplished

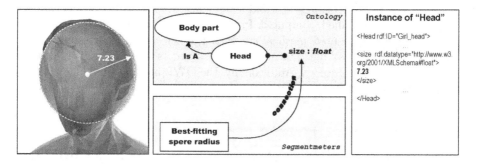

Fig. 4. The attribute *size* of the instance *Girl_head* is automatically set to the value 7.23 because this is the value computed by the connected segmentmeter

by connecting the formula $2 * \pi * best\_fitting\_sphere\_radius$ to the attribute head :: size.

Since we believe that modularity is crucial to provide a flexible annotation framework, we made our system able to load additional segmentmeters implemented externally as plug-ins, just as we have done for the segmentation algorithms.

In our prototype the connections can be established through a proper dialog in which all the segmentmeters are shown as buttons. Currently, they belong to the following two groups:

– **Topological relations between segments** consisting of *adjacency, overlap, disjointness* and *containment*;
– **Geometric aspects of a segment** consisting of *oriented bounding box length, width* and *height, best-fitting sphere radius, best-fitting cylinder radius.*

By clicking on a segmentmeter button, a list of properties defined in the domain ontology is shown and the user may select some of them to establish connections. The list of properties shown is filtered so that only admissible connections can be selected; this avoids, for example, the connection of a property with more than one segmentmeter, or between non-compatible segmentmeters and ontology properties (e.g. "segment adjacency" ↔ through_hole :: radius).

The connections can be established either before the creation of instances or afterwards. In the former case, for each newly created instance the properties are computed on the fly based on the existing connections; in the latter case, the values of the properties of the existing instances are (re)computed according to the newly established connections.

To allow their easy reuse when annotating several models in the same domain, the connections can also be saved as an XML-based file and loaded later on.

4.4 Resulting Knowledge Base

The result of the annotation process is a set of instances that, together with the domain ontology, form a knowledge base. Each instance is defined by its URI, its type (i.e., the class it belongs to) and some attribute values and relations that might have been specified/computed. In its current version, the ShapeAnnotator saves the multi-segmented mesh along with the selected, and possibly edited features as a single PLY file. The instances are saved as a separate OWL file that imports the domain ontology. Additionally, the OWL file contains the definition of two extra properties:

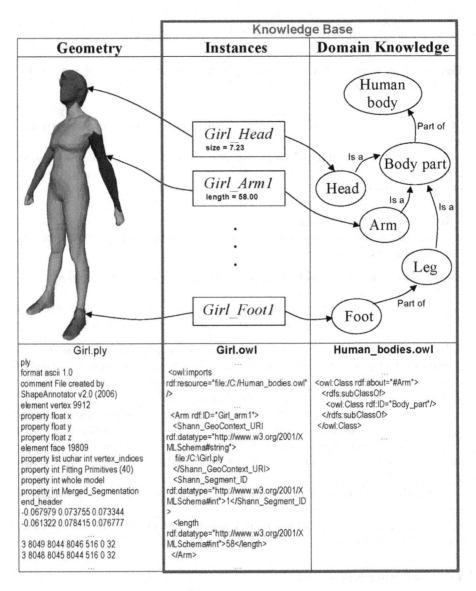

Fig. 5. The instances along with their specific relations represent a formal bridge between geometry and semantics

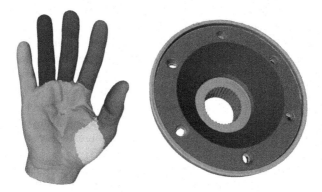

Fig. 6. Examples of segmentation of a natural and an artificial shape. The bright color on the hand surface indicates that the corresponding part is an overlap of segments (in this case the thumb and the palm).

- `ShannGeoContextURI`, whose value is the URI of the multi-segmented mesh (typically the path to the PLY file saved by the ShapeAnnotator);
- `ShannSegmentID`, whose value is an index that specifies a segment in the multi-segmented mesh.

All the instances produced during the annotation pipeline are automatically assigned values for the above two properties, so that the link between semantics and geometry is maintained within the resulting knowledge base (see Figure 5).

Note that the OWL files produced by the annotation of several models in the same domain can constitute a unified knowledge base; this would contain all the instances describing the models in terms of their meaningful parts, allowing unprecedented levels of granularity for query formulation. Once an instance has been located, for example, it is possible to retrieve the geometry of the corresponding part and, possibly, to extract it without the need to download and process the whole model.

5   Conclusions and Future Research

This paper tackles the problem of providing useful semantic annotations to 3D shapes. We have discussed the key aspects of the subject, and shown that simple keywords attached to a whole shape do not provide enough information to answer complex queries. Thus, we have illustrated how to decompose the shape into interesting features within the multi-segmentation framework, and introduced the annotation pipeline to attach a formal semantics to the features and the whole shape. We have pointed out that the process is unfeasible using only state-of-the-art approaches. Conversely, we have described our novel ShapeAnnotator tool that makes it possible to annotate 3D shapes through few mouse clicks using the pipeline proposed. The introduction of segmentmeters along with their context-based interpretation represents a first step towards automatic annotation methods for the 3D domain.

In future developments, we plan to treat also lower-dimensional features (i.e. curves and points) with a twofold benefit: they will be eligible for annotation, just as the other segments, and they will be useful to edit other segments (e.g. a curve may be used to split a segment into two subsegments).

In its current version, the ShapeAnnotator has minimal inference capabilities which have been implemented just to provide a flexible browsing of the ontology. This means that input ontologies are assumed to be mostly asserted; if not, the user can use an offline reasoner to produce the inferred parts. Future developments are targeted to this aspect, and internal inference capabilities are foreseen. Besides simple deductions on the input ontology, inference will also be used to (partially) automate the whole annotation pipeline. Although the process can be completely automated in rather few domains, in many others the user might be required to contribute only to disambiguate few situations.

Acknowledgements

This work is supported by the AIM@SHAPE Network of Excellence (EC FP6 IST Contract N. 506766). Thanks are due to all the members of the Shape Modeling Group of the CNR-IMATI-GE, and to the people that, through discussions and suggestions, made it possible to develop the Shape Annotator. *This work was conducted using the Protégé resource, which is supported by grant LM007885 from the United States National Library of Medicine.*

References

1. [online]: AIM@SHAPE shape repository (2004), http://shapes.aimatshape.net/
2. [online]: 3D CAD browser (2001), http://www.3dcadbrowser.com/
3. Shilane, P., Min, P., Kazhdan, M., Funkhouser, T.: The princeton shape benchmark. In: SMI 2004. Proceedings of the Shape Modeling International 2004, pp. 167–178. IEEE Computer Society, Washington, DC, USA (2004)
4. [online]: The Stanford 3D Scanning Repository.
   http://graphics.stanford.edu/data/3dscanrep
5. Marr, D.: Vision - A computational investigation into the human representation and processing of visual information. W. H. Freeman, San Francisco (1982)
6. [online]: Network of excellence AIM@SHAPE, EU FP6 IST NoE 506766 (2004), http://www.aimatshape.net
7. [online]: Secondlife. http://www.secondlife.com
8. Hermes, T., Miene, A., Kreyenhop, P.: On textures: A sketch of a texture-based image segmentation approach. In: Decker, R., Gaul, W. (eds.) Classification and Information Processing at the Turn of the Millenium. Proc. 23rd Annual Conference Gesellschaft für Klassifikation e.V. 1999, March 10-12, 2000, pp. 210–218 (2000)
9. Xiaohan, Y., Ylä-Jääski, J., Baozong, Y.: A new algorithm for texture segmentation based on edge detection. Pattern Recognition 24(11), 1105–1112 (1991)

10. Pavlidis, T., Liow, Y.T.: Integrating region growing and edge detection. IEEE Trans. Pattern Anal. Mach. Intell. 12(3), 225–233 (1990)
11. Browne, P., Smeaton, A.F.: Video retrieval using dialogue, keyframe similarity and video objects. In: ICIP 2005. International Conference on Image Processing (2005)
12. [online]: Flickr. http://www.flickr.com/
13. [online]: Riya. http://www.riya.com/
14. Gruber, T.R.: Towards principles for the design of ontologies used for knowledge sharing. In: Guarino, N., Poli, R. (eds.) Formal Ontology in Conceptual Analysis and Knowledge Representation, Deventer, The Netherlands, Kluwer Academic Publishers, Dordrecht (1993)
15. Guarino, N.: Formal ontology and information systems. In: Guarino, N. (ed.) Formal Ontology in Information Systems, pp. 3–18. IOS Press, Amsterdam (1998)
16. Saathoff, C.: Constraint reasoning for region-based image labelling. In: VIE 2006. 3rd IEE International Conference of Visual Information Engineering. Special Session on Semantic Multimedia (2006)
17. Petridis, K., Anastasopoulos, D., Saathoff, C., Timmermann, N., Kompatsiaris, I., Staab, S.: M-OntoMat-Annotizer: Image annotation. linking ontologies and multimedia low-level features. In: Gabrys, B., Howlett, R.J., Jain, L.C. (eds.) KES 2006. LNCS (LNAI), vol. 4251, pp. 2006–2010. Springer, Heidelberg (2006)
18. [online]: Photostuff. http://www.mindswap.org/2003/photostuff/
19. Shamir, A.: Segmentation and shape extraction of 3D boundary meshes. In: Eurographics 2006 - State of the Art Reports, pp. 137–149 (2006)
20. Attene, M., Katz, S., Mortara, M., Patanè, G., Spagnuolo, M., Tal, A.: Mesh segmentation - a comparative study. In: SMI 2006. Proceedings of the IEEE International Conference on Shape Modeling and Applications 2006, p. 7. IEEE Computer Society, Washington, DC, USA (2006)
21. Attene, M., Biasotti, S., Mortara, M., Patane, G., Falcidieno, B.: Computational methods for understanding 3D shapes. Computers&Graphics 30(3), 323–333 (2006)
22. Biederman, I.: Recognition-by-Components: A theory of human image understanding. Phicological Review 94, 115–147 (1987)
23. Attene, M., Falcidieno, B., Spagnuolo, M.: Hierarchical mesh segmentation based on fitting primitives. The Visual Computer 22(3), 181–193 (2006)
24. Cohen-Steiner, D., Alliez, P., Desbrun, M.: Variational shape approximation. In: SIGGRAPH 2004. ACM SIGGRAPH 2004 Papers, pp. 905–914. ACM Press, New York, NY, USA (2004)
25. Zhang, E., Mischaikow, K., Turk, G.: Feature-based surface parameterization and texture mapping. ACM Transactions on Graphics 24(1), 1–27 (2005)
26. Mortara, M., Patané, G., Spagnuolo, M.: From geometric to semantic human body models. Computers & Graphics 30(2), 185–196 (2006)
27. Mortara, M., Patanè, G., Spagnuolo, M., Falcidieno, B., Rossignac, J.: Plumber: A method for a multi-scale decomposition of 3D shapes into tubular primitives and bodies. In: SM 2004. Proceedings of the ACM Symposium on Solid Modeling, pp. 339–344. ACM Press, New York, NY, USA (2004)
28. Biasotti, S.: Computational Topology methods for Shape Modelling Applications. PhD thesis, University of Genoa, Italy (2004)
29. [online]: OWL web ontology language guide (February 2004), W3C Recommendation http://www.w3.org/tr/2004/rec-owl-guide-20040210/
30. [online]: The Protégé ontology editor (2006), http://protege.stanford.edu
31. [online]: The NCI cancer ontology (2003),
    http://www.mindswap.org/2003/cancerontology

3D Classification Via Structural Prototypes

Silvia Biasotti, Daniela Giorgi, Simone Marini, Michela Spagnuolo,
and Bianca Falcidieno

CNR-IMATI, Genova, Italy
{silvia,daniela,simone,michi,bianca}@ge.imati.cnr.it

Abstract. We describe a 3D shape classification framework, and discuss the performance of selective and creative prototypes extracted from structural descriptors.

1 Introduction

Shape classification methods establish the membership of an unknown query shape in one of a set of classes, thus inferring semantic information about the query model. In this paper, we discuss the role of 3D prototypes for shape classification. Prototypes are embedded in a general, dissimilarity-based classification framework. The flow is illustrated in Fig. 1. For each class, a small set of *prototypes* is defined (a); a query is classified at run-time by matching its descriptor vs. the subset of prototypes (b), thus reducing the search space. Prototypes can be defined in either a *selective* or a *creative* manner. In the first case, one or a few class members are chosen to represent the whole class. In the second case, new descriptors are generated. We focus on testing and comparing selective and creative prototypes; in particular, we discuss the creative prototypes in [1].

2 Dissimilarity Based 3D Shape Classification

Given a database D with n models classified in disjoint classes, we compute a descriptor S_i for each model and consider a dissimilarity measure d between

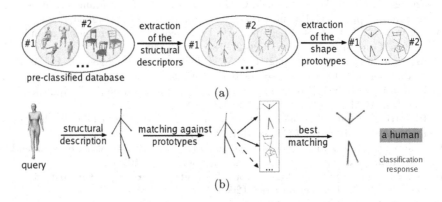

Fig. 1. Classification using prototypes extracted from structural descriptors

B. Falcidieno et al. (Eds.): SAMT 2007, LNCS 4816, pp. 140–143, 2007.

descriptors [2]. d is used to derive a query-to-class *membership measure* $\tilde{d}$, so that a query is classified by selecting the class which minimizes $\tilde{d}$.

We analyze two classification schemes. The first is based on the Nearest Neighbor (NN) rule. Let $N = \{1, \ldots, n\}$, and let $N_k \subset N$ be the set of indices corresponding to the models in a class C_k. The distance $\tilde{d}(Q, C_k)$ is then: $\tilde{d}(Q, C_k) = \min_{i \in N_k} d(S_Q, S_i)$. The second rule uses a set R of t *shape prototypes* $\{P_i\}$, $i = 1, \ldots, t$, with $t \ll n$. When considering a single prototype P_k for each class C_k, the membership measure $\tilde{d}$ is defined as the distance between the query descriptor and the class prototype: $\tilde{d}(Q, C_k) = d(S_Q, P_k)$. A similar rule is applied when $t > 1$ prototypes are considered: $\tilde{d}(Q, C_k) = \min_{i \in T_k} d(S_Q, P_i)$, with T_k the set of indices of the t prototypes of the class C_k.

3   Selective and Creative Prototypes

We choose *selective prototypes* according to their degree of similarity to other models, called *eccentricity*. For each descriptor $S \in C_k$, the eccentricity is: $avg_{C_k}(S) = \frac{\sum_{R \in C_k} d(S, S_R)}{|N_k|}$, with $|N_k|$ the number of objects in C_k. We can choose as selective prototypes either those minimizing this value (*inner* models), or those maximizing it (*boundary* models), or *average* models, see Sec. 4.

Creative prototypes summarize, in a new descriptor, the relevant shape features of a class. The creative prototypes used in this paper (cfr. [1] for details) are extracted from structural descriptors coded as attributed graphs, in our case Extended Reeb Graphs (ERGs) with the geodesic function [3]. For each class, we select a seed model and match it against the remaining class members. We store the resulting editing operations, and apply a subset of them to the seed, that is transformed in a new descriptor (the creative prototype), that is still an attributed graph. In Fig. 2(a) we show some chair models, their ERG descriptors, and the corresponding prototype (right). Fig. 2(b) shows the attributes, i.e. shape parts, associated with the nodes of the structural prototype. The latter is a composition of characteristic class features; notice the rear part and the legs.

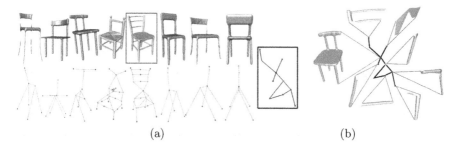

(a)   (b)

Fig. 2. (a) The prototype of a set of chairs and (b) the attributes of its nodes

4   Evaluation

We use the 400 models of the SHREC'07 (http://aimatshape.net/event/SHREC) Watertight Track, plus 80 external queries. We compute the classification rate, i.e. the percentage of queries correctly classified. The values are 85%, 61% and 66% for NN, selective and creative prototypes, respectively. The prototypes-based schemes show lower performances than NN, but their added value is the reduction of comparisons. Fig. 3 compares creative, selective, and random proto-types. The abscissa is the ratio between the number of queries and the size of the training set. The three plots refer to 1 prototype (most eccentrical one), 2 proto-types (2 most eccentrical) or 3 prototypes (most, lowest and average eccentrical) per class. Creative prototypes perform better than selective ones: they better exploit the information in a class, being built by considering all class members.

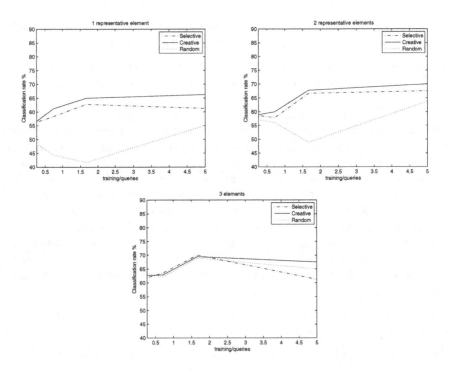

Fig. 3. Classification rates using 1, 2 or 3 creative, selective, and ramdom prototypes

Using classification as a preliminary step in a retrieval pipeline can mini-mize the number of comparisons and false positives. In the example of Fig. 4, retrieving in the classical way yields the results in Fig. 4(top). If the query is beforehand classified, the search can be restricted to the top-ranked classes, thus considering a small percentage of data and also discarding some false positives (Fig. 4(bottom)). Fig. 5 plots the percentage of false positives within the first

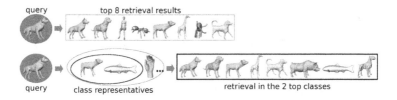

Fig. 4. Improvement of the retrieval process when pre-classification is involved

20 items ranked vs. the number of classes considered. Without pre-classification, the percentage is 58%; performing classification first, it reduces to 15%, 38.75% and 33.75%, resp. using NN, a selective and a creative prototype. Creative prototypes perform better than selective ones. A few classes are enough to obtain good retrieval results, but precision improves only if top-ranked classes are considered.

Finally, we plan to tune the number of prototypes on the class complexity.

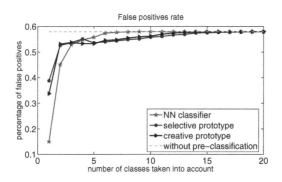

Fig. 5. False positive percentage, averaged over 80 queries against 400 models

Acknowledgements

Work developed in the CNR activity (ICT-P03), the EU NoE "AIM@SHAPE" (http://www.aimatshape.net) and the MIUR-FIRB project "SHALOM".

References

1. Marini, S., Spagnuolo, M., Falcidieno, B.: Structural shape prototypes for the automatic classification of 3D objects. Computer Graphics and Applications 27 (2007)
2. Pekalska, E., Duin, R.P.W., Paclik, P.: Prototype selection for dissimilarity-based classifiers. Pattern Recognition 39, 189–208 (2006)
3. Biasotti, S., Marini, S., Spagnuolo, M., Falcidieno, B.: Sub-part correspondence by structural descriptors of 3D shapes. Computer-Aided Design 38, 1002–1019 (2006)

Post-processing Techniques for On-Line Adaptive Video Summarization Based on Relevance Curves

Víctor Valdés and José M. Martínez

Grupo de Tratamiento de Imágenes
Escuela Politécnica Superior, Universidad Autónoma de Madrid, Spain
{Victor.Valdes,JoseM.Martinez}@uam.es

Abstract. This paper presents a group of post-processing techniques aimed to on-line adaptive video summary generation based on video analysis curves, named relevance curves obtained by different approaches (e.g., extraction of visual features, semantic features, rate-distortion curves). The developed techniques can be applied to improve the quality of the generated video summaries, control the summaries length and constitute a way to generate summaries with independence of the approach taken to generate the relevance curves that are used as basis for summary generation.

1 Introduction

Video summarization is a key enabling technology for different applications ranging from video browsing (via keyframes, storyboards or video skims) to content adaptation via transmoding (e.g., video to slide show for a limited performance terminal or network).

Video summarization has been studied during the last years by many researchers and different taxonomies for video summarization have been proposed from different points of view:

- o from the point of view of analysis techniques can be grouped in off-line, that is, optimization techniques that are applied to the whole available video sequence [1] or on-line, applying the analysis to the video frame-by-frame or GoP-by-GoP-[2].
- o from the point of view of analysis outcome techniques can provide unconnected keyframes [3] or video skims from the original video[4].
- o from the point of view of presentation [5] (e.g. a set of keyframes can be presented as a story-boards or as a videoposter).

In our work we define a video summary taxonomy based on the analysis outcome: keyframes (the summary is created by selecting a predefined number of isolated keyframes), video skims (the summary is created dropping frames without the requirement of maintaining the original timeline) and video segments (the summary is composed by complete fragments –segments- of the original video without dropping any frame in the subsegment).

Most of the summarization techniques are based on some activity [6][7] or global optimization [4] curve that outputs video summaries (video skims in our taxonomy).

B. Falcidieno et al. (Eds.): SAMT 2007, LNCS 4816, pp. 144–157, 2007.

This may yield to additional problems if on-line (real-time) working mode is required: need of recoding the selected frames (versus cutting of available -closed- GoPs), variations in the presentation speed with respect to the recording one, ... Therefore, for the applications where video segments are required (either for allowing real-time summarization or not changing the presentation speed –e.g. creation of trailers-), video segments (in our taxonomy) are the best output type.

The on-line summarization approach provides adaptation mechanisms not only for improving the efficiency of the summarization task but to enable concepts such as the generation of video summaries of broadcasted content as it is being received or, in the future, to allow watching video summaries from stored video content as they are being generated.

In this paper we propose a set of post-processing techniques that allow the on-line summarization into video segments from relevance curves. In order not to loss any information, an optional technique is incorporated providing some fast-forward viewing of the unselected video segments incorporating them as video skims.

The paper is structured as follows: in section 2 a state of the art about video summarization techniques is presented; in section 3 an overview of the developed summarization module is shown; Section 4 depicts the proposed post-processing techniques for video summarization while section 5 presents some results of those techniques for an specific example; Finally section 6 presents the conclusions and future work.

2   State of the Art

In the last years there have been a lot of developments covering different approaches to video summarization paving the ground to applications such as fast content browsing, transmission and retrieval of audiovisual content. Most part of the existing techniques are aimed to work off-line, without the requirements imposed by the real-time operation mode which are mainly the lack of information about the incoming video (including length and or other extractable characteristic such as visual information, shot length or any other semantic information), the inability to eliminate from the produced video abstract any frame or video segment which has been already outputted and, of course, the limited processing time. Some of the further explained techniques can not completely or partially applied to real time summarization approaches.

According to [8] existing work in video summaries generation can be categorized in three classes:

- o   Sampling based: In this case, the keyframe extraction is performed by sampling the original audiovisual content uniformly or randomly from the original video [9][10]. This approach could be easily applied in video skims generation by outputting shots of video from the original content but it is the most simple method and has an important disadvantage in the arbitrariness of the obtained summarized or skimmed video.
- o   Shot based: In this kind of summary generation each shot represents a continuous period and some keyframes are extracted from each shot. The extraction of this keyframes can be based in several features such as motion or colour [3], colour clustering or global motion[11][12]. This kind of techniques could be easily extrapolated to video skim generation by taking a

piece of video per each shot in the original sequence but the application to real-time processing is more complicated as the length and characteristics of each shot can not be known a priori.

o   Segment based: This is a more sophisticated approach in which the key-frames selection is performed over higher-level units known as segments. These segments can be, for example, different scenes in a movie each of them composed by several shots. In video summarization targeting to key-frame extraction, the sets of frames are clustered and then, from each segment, representative frames are extracted depending on different criterions [13][14]. The clustering approach provides a way to avoid redundancy in video summaries by grouping similar shots or frames. Once more this approach can not be applied to real-time summarization because clustering of frames can not be performed without the whole audiovisual content available.

Regarding video-skimming specifically oriented approaches, two main categories can be differentiated:

o   Highlight oriented: In this kind of video skimming the video skim is composed by a set of specific relevant parts of the original video for example movie trailers or sport highlights [15]. This kind of video skimming could be performed in real-time if the characteristics of the highlights can be clearly defined and are fixed. If the selection of this highlight must be performed by selecting between several parts of a video this technique can not be applied in real-time.

o   Summary oriented: This kind of video skimming is oriented to give a summarized version of the original video by selecting a set of fragments of the original video along its entire length providing an overview of the whole video [16].

In opposition to the keyframe extraction techniques, video skimming involves a higher level of complexity due to the different ways in which it is possible to deal with the non relevant considered segments of video. The direct way to deal with that is just dropping the parts of the video which are not considered as important [17] but a more complex possibility is to fast-playback those less relevant video segments in opposition with the more relevant parts played at normal speed [7]. The advantage of this approach is that there is no risk of completely losing an important piece of a video (in the worst case it is played quickly) but it adds complexity to the summary generation process in which the speed and transition of fast-forward shots of video must be considered.

Regarding the features in which the video summarization is based there are lots of possibilities. The shot change detection is a common starting point in the most part of video summarization techniques as it is commonly used to segment the videos in shots before performing further analysis. Other typical features are motion activity [17], textual information or audio [16]. Other specific domain approaches make use of several specialized features (i.e. person detection or gesture analysis[18]) but can not be considered as common techniques.

3   Overview of the On-Line Summarization Module

On-line summarization allows the generation of partial or complete summaries of video sequences while they are being captured (e.g., for surveillance applications), contributed (e.g., for showing previews from a large number of reporters' contributions) or broadcasted (e.g., for creating hot news summaries for a premium service). The generation of on-line summaries requires visual content analysis is also performed on-line: both shot boundary detection [19] as well as visual features extraction. The proposed architecture, that takes also into account user preferences information, is presented in Fig. 1.

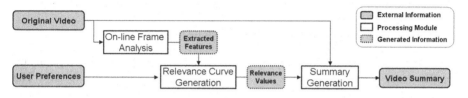

Fig. 1. Proposed on-line video summarization system

The summary generation process starts with an on-line frame analysis which outputs a set of extracted features which, together with the user preferences are applied in the Relevance Curve Generation module to create the Relevance Curve that will drive the selection of frames. This frame selection process is performed in the Summary Generation module and depends on the length of the desired summaries and the values in the Relevance Curve to control the adaptive thresholds used to drop or select frames. The way in which the thresholds are set to control the output rate and the different post-processing techniques aimed to generate visually pleasant summaries will be described in section 4.

4   Post-processing Techniques for Segments Generation

4.1   Basic Frame Selection Process

The Relevance Curve synthesizes all the analysis or optimization algorithms applied to the video in order to indicate the (semantic) features to maintain in the summary and is the main measurement applied in current summarization decisions. Fig. 2 shows a simplistic approach example for frame selection in which the frames with a Relevance Curve value higher than a user-defined threshold are included in the output summary and those with lower Relevance Curve value are not included.

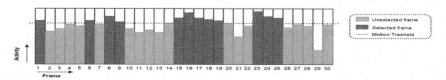

Fig. 2. Threshold based frame selection

The threshold value must be selected depending on the studied content, mean Relevance Curves values for different video segment types (e.g., sports, interviews, live coverage) and what kind of video segments it is desired to keep in the video summary (e.g., anchorman segments, high motion activity segments). Section 5 shows a case of threshold value selection to preserve the 'footage report' parts of news bulletin videos using a Relevance Curve based on motion activity.

This kind of approach can produce undesired results such as non-smooth output videos. In the example shown in Fig. 2 it can be noted that for a period of 30 frames, about one second of video, some isolated frames and two groups of 3 and 5 frames would be added to the output summary. If this happens in several parts of a video this effect could produce abrupt and unpleasant summaries (we should remember that we are targeting the creation of video segments summaries, not keyframe summaries).

4.2 Eliminating Isolated Frames and Very Short Segments

The first measure applied to avoid this kind of undesired effect is not using directly the Relevance Curve value but a Score measure calculated within the OLSM. The Score values will depend on the Relevance Curve value for each frame, Score value of the previous frame and the shot boundaries in the video sequence. Taking into account the Score of previous frame will smooth the original Relevance Curve shown in Fig. 2. The *Score* measure for frame i will be calculated as:

$$Score(i) = Score(i-1) \cdot SF + RelevanceCurve(i) \cdot (1-SF) \tag{1}$$

where *SF -ScoreFade-* is between 0 and 1 which specifies the weight of the previous frame Score in the calculation of the Score value for current frame.

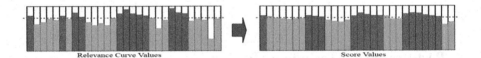

Fig. 3. Score value calculation

Fig.3 shows how the calculation of the *Score* value can smooth the relevance curve. If this *Score* curve is taken into account to decide which frames to select or drop instead of taking the Relevance Curve directly the undesired effects in the output can be, in part, avoided but not completely eliminated.

Another unpleasant effect is caused by shot boundaries, as if they are not taking into account, it is possible that the segment contents a reduced number of an ending or starting shot, but creates a subjective unpleasant effect.

4.3 Eliminating Short Segments Including a Shot Boundary

To take shot boundaries into account in order to decide which frames are going to be included and which not, a shot change related parameter should be included in the Score measure. The Score calculation (1) is reformulated as follows:

$$Score(i)=Score(i-1) \cdot SF + (RelevanceCurve(i) + SCB) \cdot (1 - SF) \qquad (2)$$

where *SCB –Shot Change Bonus-* is a configurable parameter which is added to the *Score* value when a shot change is detected, and zero if not. Note that this value can be both positive or negative to increase or decrease the possibilities of including in the output summaries frames close to shot changes but it is recommended to choose a value with the same order of magnitude as *RelevanceCurve* values in order to keep both *SCB* and *RelevanceCurve* values significative in the *Score* calculation. As the *Score* value is calculated taking into account previous *Score* values a very high *SCB* values can affect long sequences of frames. Fig. 4 shows the effects of high positive *SCB* value (a) which will increase the chances of the frames following a shot change to be selected for the output summary, or high negative *SCB* value (b) which decreases the chances of the shot change following frame of being selected.

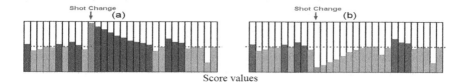

Fig. 4. *SCB* effects on *Score* value

Even with application of the explained measures, it would be still quite frequent to find sets of frames with *Score* value over the threshold interleaved with groups of frames with lower *Score* value. This will cause that there will be still shorts segments that will produce unpleasant effects in a video segments summary.

4.4   Fixing a Minimum Size of Video Segments

In order to avoid small video segments, Blocks of Frames (BoF) have been considered. Each frame *Score* value is stored in a buffer and, when this buffer is full, the mean *Score* value for all the frames in the buffer is calculated and this BoF is then dropped or added to the final result depending on the relevance threshold which, in this case, should be set taking into account previous knowledge about the kind of content and the *Score* values the frames can reach (section 5 shows an example for news bulletins).. The size of the BoF should be selected taking into account the time needed by the user to perceive the shown content. Fig. 5 shows an example of frame block division, mean *Score* value calculation, and selection of BoF with mean *Score* value over a threshold (note that in a real example the number of frames per block should be much higher in order to get smooth results).

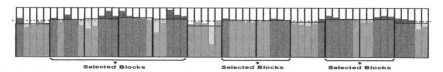

Fig. 5. Frame block selection

By the application of this BoFs selection mechanism, the obtained output provides a much pleasant experience to the final user. The effects of the *SCB* value can be still considered and the selection of BoF following a shot change can still be penalized or an incentive can be given to them depending on the *SCB* value.

4.5 Optimizing BoF Positioning

As a disadvantage of the block selection method, it is possible to find examples such as the one shown in Fig. 6 (a) in which, independently of the selection or dropping of the last considered BoF, the next block is situated with a displacement equal to the size of each block resulting in a positioning in the following frame to the end of the previous block. This can produce that, as shown in Fig. 6 (a), fragments of the original video with a high *Score* value are not selected due to the fixed positioning of each block without being centered in the high *Score* group of frames. In order to avoid this problem the number of dropped frames each time a block mean *Score* is under the relevance threshold is defined as a configurable parameter. Fig. 6 (b) shows the effect of this block dropping policy. In this example each time the mean *Score* of a block of frames is under the *Score* threshold, 3 frames are dropped and the block is positioned three frames ahead. When the block mean *Score* is higher than the *Score* threshold the whole block of frames is dumped to the output result and the next block is considered to start at the end of the just dumped block.

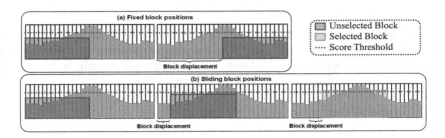

Fig. 6. Block positioning policies

There also exists another undesired effect produced in the frame block dumping. Fig. 7 shows an example in which blocks of 3 seconds have been selected to produce the output. The objective is to give the user time enough to perceive the content of a summary shot, but the example shows the effect of finding a shot change inside the frame block. In this case, as the shot is located 1 second from the beginning of the block, if this is dumped the result will be composed by a shot of 1 second and another one of 2 seconds. In this case the shot length could be too short and the user would receive a confusing piece of summary.

To avoid this problem it is possible to force the dumping of the frame buffer just when it reaches a shot change. In the Fig. 7 example the frame buffer will be filled until it reaches the shot change (1 second of content) and then decide if this content must be written in the output summary. This decision will depend on the mean relevance value for this frame period and minimum number of frames to write in the output which is configurable and is used to avoid writing of too short shots in the output summary.

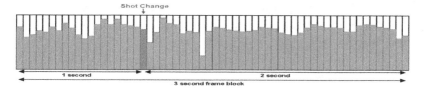

Fig. 7. Frame block including shot change

4.6 Optional Inclusion of Dropped Frames

In the previous subsections, the process for frame selection to compose the video segments summary has been explained. In this section the (optional) process to deal with the unselected frames is depicted, allowing the insertion of video skims of the unselected video segments into the summary.

The first, and most typical, approach is to drop the non selected frames. In this case there is a risk of losing shots of the original content which could be relevant for the user. In any kind of video summary generation part of the original content information is going to be lost. The decision about what information is going to be kept and which is going to be lost depends of the computer's decision algorithms criteria which can be different to the user criteria (although guided by user preferences). To minimize the possibility of losing relevant information a fast-forward mechanism can be used in the summary generation. This mechanism is applied to the to-be-dropped frames: instead of eliminating each frame in the discarded frame block the playing speed of the discarded fragments of the original video is increased by dropping only part of its frames. To avoid abrupt changes in the playing speed of the video, the dropping rate is increased uniformly until it reaches a constant maximum dropping [7]. This mechanism allows maintaining at least a part of all the visual information contained in the original video. Fig. 8 shows an example of increasing the dropping rate showing in green the frames added to the output summary and, in red, the discarded frames. If the next frame block is discarded the dropping rate continues in the maximum value and if the next block is selected to be included in the video summary the dropping rate is set to 0. The speed at which the dropping rate increases and the maximum dropping rate are parameters the user can select if this functionality is activated.

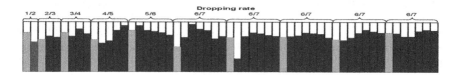

Fig. 8. Increasing dropping rate

4.7 Control of Summarization Length

Working in an on-line mode makes impossible to target to a summarization fixed length, as we don't know what will come next. Nevertheless, it is possible to target to a summarization length defined here as Time Ratios. We will consider the *resulting-TimeRatio* as the relation between the length of the output summary and the length of the original video.

$$resultingTimeRatio = \frac{length(summary)}{length(original)} \tag{3}$$

This ratio can be considered as well as the relation between the number of written frames in the summary and the original number of frames. As described in section 4.1 the frame selection process depends on several factors such as the definition of a *Score* threshold to select or delete frames, the size of the frame blocks or the different choices for shot change bonus value, etc. We will define the *objectiveTimeRatio* as the desired value we would like to obtain for the *resultingTimeRatio*. It is an objective value which can not be always reached. A value for the *Score* threshold should be chosen considering the *objectiveTimeRatio* and taking into account what kind of content it is going to be preserved in the output summary. Fig. 9 summarizes the *resultingTimeRatio* by changing the *Score* threshold and applying the summary generation algorithm to 10000 frames of two news bulletin videos studied in section 5 in which the relevance curve is calculated considering the motion activity on each shot of the video (the rest of the parameters such as frame block size, *SCB*, etc. is fixed in this experiment). It can be seen that the *resultingTimeRatio* decreases as the relevance threshold increases but the exact relation between these two parameters depends on the specific content of each video.

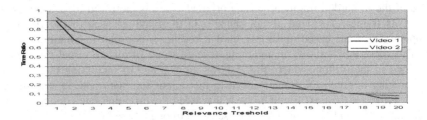

Fig. 9. ResultingTimeRatio reduction

A video with higher relevance curve will result on a higher *resultingTimeRatio* summary in opposition with a lower relevance curve video from which a very compact summary would be obtained. To avoid the differences obtained in the *resulting-TimeRatio* due to each video specific particularities a *resultingTimeRatio* dynamic control has been designed which will allow to target specific values for the output video.

To gain control over the *resultingTimeRatio* the score threshold applied to select or drop blocks of frames will not be fixed. It will be dynamically recalculated while the summarized video is being generated. For this purpose three values must be defined:

o   *BaseThreshold*: This value is the base value for the score threshold. It is the starting value from which to calculate the corrections. If this value is decreased more frames will be included in the output summary and vice versa.

- o *MaxThreshold:* The maximum value that the score threshold can reach after corrections.
- o *MinThreshold*: The minimum value that the score threshold can reach after corrections.

The *correctedTreshold* is calculated each time that frames are dropped or added to the resulting summary. For that calculation it is necessary first to calculate the *CurrentTimeRatio* as the ratio obtained until current moment:

$$CurrentTimeRatio = \frac{writtenFrames}{readenFrames} \qquad (4)$$

where *readenFrames* is the total of frames read from the original content until the moment and *writtenFrames* the number of frames from the total read that have been included in the resulting summary.
The *CF –correctingFactor-* is calculated as follows:

$$CF = [abs(t \arg etTimeRatio - CurrentTimeRatio)]^{reactionFactor} \qquad (5)$$

where *targetTimeRatio* is the desired *resultingTimeRatio* and *reactionFactor* is a value in the interval (0 1] which affects to the intensity of the threshold correction (higher correction when lower *reactionFactor* values).
The final *resultTreshold* value is calculated as follows when the *CurrentTimeRatio* is higher than the objective *TimeRatio:*

$$resultThreshold = BaseThreshold + (maxThreshold - BaseThreshold) \cdot CF \qquad (6)$$

When the *CurrentTimeRatio* is lower than the target *TimeRatio* the frame selection threshold must be decreased and it is calculated as follows:

$$resultThreshold = BaseThreshold - (BaseThreshold - minThreshold) \cdot CF \qquad (7)$$

Fig. 10 summarizes the *resultingTimeRatio* values obtained processing different number of frames from the beginning of the same video and with different *reactionFactor* values. Note that during the 100 first frames no correction is performed in the score threshold to provide some 'warp up' time before beginning to calculate *CurrentTimeRatio* statistics. The big increases and falls in the graphs are due to local particularities of the test video. It can be noted as well that with higher number of frames the *resultingTimeRatio* is closer to the target and is less sensible to high relevance curve periods on the original video.

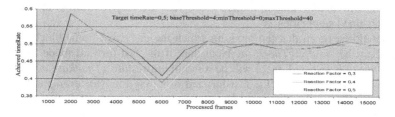

Fig. 10. Resultingtimerate for different number of processed frames

The differences in the obtained *resultingTimeRatio* depending on the *reactionFactor* are clearly visible in Fig. 11, in which several target rates are chosen for different *reactionFactor* values, and where it is shown that it is possible to obtain a resultingTimeRatio closer to the targetTimeRatio with lower reactionFactor values.

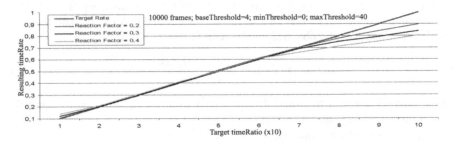

Fig. 11. ReactionFactor effect on resultingTimeRatio

It could seem that the selection of a high reactionFactor value is better because the resultingTimeRatio values are closer to the desired ones. Nevertheless Fig. 12 shows how the resultThreshold values changes with the different reactionFactor values. It can be seen that for low *reactionFactor* values the *resultThreshold* values have bigger and more abrupt changes.

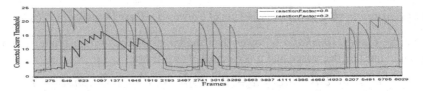

Fig. 12. Corrected *Score* threshold with different reactionFactor values

This will produce better rate results but the quality of the obtained video summary will be lower because the sudden changes in the frame selection threshold will produce periods of time in which every frame will be selected for the summary and periods in which every frame will be dropped with independence of their associated motion activity values.

The chosen *reactionFactor* must be selected carefully to avoid arbitrary summary generation but maintaining acceptable control over the *resultingTimeRatio* value. The mean, maximum and minimum threshold values are very relevant in the final quality of the obtained summaries.

It should be noted that the mode in which the discarded frame blocks are treated will affect the *resultingTimeRate*. Note that, if the fast-forward mode for discarded frame blocks is selected, there will be a minimum *resultingTimeRate* which could be achieved corresponding to the fast-forward dropping rate (for example if the maximum dropping rate is 9 of every 10 frames the minimum possible obtainable *resultingTimeRate* will be 0'1).

5   Results for News Bulletins

The proposed techniques have been tested for news bulletins videos within the MESH project[20]. The objective was to show more frames when the news footage is presented (usually more motion activity –e.g. shots changes, camera motion, scene activity, closed captions insertions, …-) than when the anchorman is present (less activity).

Of course, for real multimodality summaries it is required to take into account the audio, in order not to lose all (or the most important) parts of the speech. Nevertheless, for browsing purposes it is enough to have the visual impression that will raise the attention to a more detailed inspection of the news bulleting that raised our interest by the "visual" clue.

Obtained results are good regarding the objective of isolating parts of the news bulletins in which the presenter is speaking or an interview is being performed from those parts in which news reports are shown on screen. The motion activity can be used to differentiate between the fragments of video in which the presenter is speaking on screen (the ones with the lowest mean motion activity) and the fragments of video corresponding to reports (higher motion activity values) so this is the measure which has been applied to calculate the relevance curve for each frame in the video. Table 1 shows the results of the summarization process with different *targetTimeRates* using a frame block size of 120 frames, *BaseThreshold*=4, *MinThreshold*=0, *MaxThreshold*=40 and *reactionFactor* = 0.4. The selection of these parameters aims to preserve the report periods of time in the news bulletins as it is considered that these parts provide more visual information to generate the summary than other kind of content. Table 1 shows that the *resultingTimeRate* has a precision of about 85% and 97% for *targetTimeRate* values of 0,2 and 0,4 respectively.

Table 1. Summarization Results

	Frames	*TimeRate*	%Report	%Others
Original news bulletin	20000	1	39,3	60,7
Summary: target *TimeRate=0,2*	4634	0,23	86,4	13,6
Summary: target *TimeRate=0,4*	8309	0,41	75,3	24,7

The objective of preserving the report periods of the news bulletin is fulfilled: starting from a 39,3% of report time in the original 20000 frames the summaries achieve 86,4% for a *targetTimeRate* of 0,2 and 75,3% of report time for the 0,4 *targetTimeRate*. The maximum percentage of report time in summaries with increasing *resultingTimeRates* will decrease as it tends to the original video content percentages. These results prove the feasibility of the motion activity information to differentiate between presenter, interviews and other parts of the news bulletins and generate representative summaries from a visual point of view. The addition of additional visual features, together with audio, textual information and classification information will provide useful information to generate better quality video summaries or semantics-driven video summaries. Sample video summaries are available at:
http://www-gti.ii.uam.es/publications/PostProcessingTechniquesForOnLineAdaptive VideoSummarizationBasedOnRelevanceCurves/

6 Conclusions

Current results demonstrate that of the proposed post-processing techniques increase subjective quality of the generated on-line video summaries. The main innovation so far are the techniques implemented in order to avoid too short or abrupt sequences of frames whilst maintaining most of the relevant information and, at the same time, providing control over the length of the obtained video summaries (considering the handicap of on-line operating mode). Initial experiments have been successfully carried out with news bulletins videos generating video summaries based on the motion-activity of the videos but the proposed techniques are equally applicable to any kind of relevance curve with independence of the way it is calculated (motion activity, people presence, rate-distortion curves...) allowing the generation of different kind of videos summaries just varying the curve to deal with by the isolation of the frame/shot score calculation process and the summary generation itself.

Current work within the OLSM is mainly focused on two aspects: the relevance curves generation and the video summaries edition.

Work on relevance curves generation is mainly related with the evaluation and integration of more visual descriptors (e.g., camera motion, face detection, number of objects, closed caption presence) in the relevance curve calculation process and the study of different methods to calculate this relevance curve fusing the different individual curves, considering also multimodal features (e.g., speech detection, silence detection, speaker recognition, text detection) and their fusion.

Work on video summaries edition focus on creating different editions for the same summary content taking into account different presentation styles (e.g., insertion of maps and closed caption as thumbnails during the summarization of the footage video content, anchorman in a small video window overlayed over the footage coverage summary).

Future work includes the design of subjective evaluation[5] of the resulting summaries both in terms of understanding of the main information within the original video and the aesthetic (editorial style) of the summary.

Acknowledgements

Work partially supported by the European Commission under the 6th Framework Program (FP6-027685 - Mesh). This work is also partially supported by the Spanish Government under project TIN2004-07860-C02 (MEDUSA) and by the Comunidad de Madrid under project S-0505/ TIC-0223 (ProMultiDis-CM).

References

1. Rao, Y., Mundur, P., Yesha, Y.: Automatic video summarization for wireless and mobile environments. In: IEEE International Conference on Communications 2004, vol. 3, pp. 1532–1526 (June 2004)
2. Bescos, J., Martinez, J.M., Herranz, L., Tiburzi, F.: Content-driven adaptation of online video. In: Proc. CBMI 2007 (in press)

3. Zhang, H.J., Wu, J., Zhong, D., Smoliar, S.W.: An integrated system for content-based video retrieval and browsing. Pattern Recognition 30(4), 643–658 (1997)
4. Li, Z., Schuster, G.M., Katsaggelos, A.K., Gandhi, B.: Rate-Distorsion Optimal Video Summary Generation. IEEE Transactions on Image Processing 14(10), 1550–1560 (2004)
5. Christel, M.G.: Evaluation and User Studies with Respect to Video Summarization and Browsing. In: Proc. of SPIE, vol. 6073, pp. 196–210 (2006)
6. Lagendijk, R.L., Hanjalic, A., Ceccarelli, M., Soletic, M., Persoon, E.: Visual search in a? SMASH system. In: Proc. ICIP 2006, pp. 671–674 (2006)
7. Peker, K.A., Divakaran, A.: Adaptive fast playback-based video skimming using a compressed-domain visual complexity measure. In: Proc. ICME 2004, vol. 3, pp. 2055–2058 (2004)
8. Li, Y., Lee, S.-H., Yeh, C.-H., Jay Cuo, C.-C.: Techniques for Movie Content Analysis and Skimming. IEEE Signal Processing Magazine 23(2), 79–89 (2006)
9. Mills, M.: A magnifier tool for video data. In: Proc. ACM HCI 1992, pp. 93–98 (1992)
10. Taniguchi, Y.: An intuitive and efficient access interface to real-time incoming video based on automatic indexing. In: Proc. ACM Multimedia 1995, pp. 25–33 (1995)
11. Zhuang, Y.T., Rui, Y., Huang, T.S., Mehrotra, S.: Adaptive key frame extraction using unsupervised clustering. In: Proc. ICIP 1998, pp. 866–870 (1998)
12. Toklu, C., Liou, S.P.: Automatic keyframe selection for content-based video indexing and access. In: Proc. SPIE, vol. 3972, pp. 554–563 (2000)
13. Girgenohn, A., Boreczky, J.: Time-constrained keyframe selection technique. In: Proc. ICMCS 1999, pp. 756–761 (1999)
14. Yeung, M.M., Yeo, B.L.: Video visualization for compact presentation and fast browsing of pictorial content. IEEE Transactions on Circuits, Systems and Video Technology 7(5), 771–785 (1997)
15. Xiong, Z., Radhakrishnan, R., Divakaran, A.: Generation of sports highlights using motion activity in combination with a common audio feature extraction framework. In: Proc. ICIP 2003, vol. 1, pp. I-5–I-8(2003)
16. Huanz, Q., Lou, Z., Rosenberg, A., Gibbon, D., Shahraray, B.: Automated generation of news content hierarchy by integrating audio, video, and text information. In: Proc. ICASSP 1999, vol. 6, pp. 3025–3028 (1999)
17. Li, B., Sezan, I.: Event detection and summarization in American football broadcast video. In: Proc. SPIE, vol. 4676, pp. 202–213 (2002)
18. Ju, S.X., Black, M.J., Minneman, S., Kimber, D.: Summarization of video-taped presentations: Atomatic analysis of motion and gestures. IEEE Transactions on Circuits Systems and Video Technology 8(5), 686–696 (1998)
19. Bescos, J.: Real-time shot change detection over online MPEG-2 Video. IEEE transactions on Circuits and Systems for Video Technology 14(4), 475–484 (2004)
20. IST MESH project, http://www.mesh-ip.eu/

A Constraint-Based Graph Visualisation Architecture for Mobile Semantic Web Interfaces

Daniel Sonntag and Philipp Heim

German Research Center for Artificial Intelligence
66123 Saarbrücken, Germany
`sonntag@dfki.de, heim@interactivesystems.info`

Abstract. Multimodal and dialogue-based mobile interfaces to the Semantic Web offer access to complex knowledge and information structures. We explore more fine-grained co-ordination of multimodal presentations in mobile environments by graph visualisations and navigation in ontological RDF result structures and multimedia archives. Semantic Navigation employs integrated ontology structures and leverages graphical user interface activity for dialogical interaction on mobile devices. Hence information visualisation benefits from the Semantic Web. Constraint-based programming helps to find optimised multimedia graph visualisations. We report on the constraint-formulisation process to optimise the visualisation of semantic-based information on small devices and its integration in a distributed dialogue system.

1 Introduction

For every specific type of information there are certain categories of visual representations that are more suitable than others. The use of a graph for the visualisation of information has the advantage that it can capture a detailed knowledge structure. Therefore graphs are suitable for conveying semantic relations between individual information items and for providing an understanding of the overall information structure. We aim to display information that stem from semantic RDF[1] structures and explore if the implicit graph structure in the RDF data can be used for the knowledge visualisation process, especially on mobile devices. By additional graph presentations of answers in a linguistic question answering scenario, the user would become more engaged in the dialogue, navigate through the incrementally presented result space, and would be encouraged to pose follow-up questions in natural language. The challenge we address is the intuitive navigation in a semantically organised information space on small interaction devices. Using RDF structures for graph representations can improve the users' understanding of certain information pieces and the relations between these pieces. The second aim is to produce evidence for this hypothesis by implementing and evaluating mobile Semantic Web interfaces and applying direct structure mapping from RDF graphs toward their multimedia visualisations.

---

[1] Resource Description Framework, *http://www.w3.org/RDF/*

B. Falcidieno et al. (Eds.): SAMT 2007, LNCS 4816, pp. 158–171, 2007.

Fig. 1. SMARTWEB's main graphical user interface (left) and semantic navigation interactions (centre and right). By clicking on a certain vertex of the graph, the user can change the *focus point* for the fisheye distortion. With a click on the red arrow, the user can change an *active instance*. Clicking again on an active instance is an interaction form to ask for additional detailed multimodal information. When the structure of the dynamic graph changes, a new optimal layout is computed server-side. A further click on the *Ergebnis* (result) node results in displaying the information: **5:3** n. E., 1:1 n. V. (**1:1, 0:0**), *Ereignis* (incidence) reveals red card for player *Cufre*, for example.

In our most recent dialogue system project SMARTWEB [1], we try to provide intuitive multimodal access to a rich selection of Web-based information services; especially the handheld scenario is tailored toward multimodal interaction with ontological knowledge bases and Semantic Web services [2]. Since the application domain we have in mind is football, the knowledge base covers facts about football events, players, matches, etc., the user can ask for. The main scenario[2] we modelled is that a football fan is in Berlin to visit the 2006 FIFA World Cup. Holding the personal digital assistant (PDA) in one hand, she could, for example, ask questions like *How many goals has Michael Ballack scored this year?* or *How did Germany play against Argentina in the FIFA worldcup?* The summarised answer to the last question, SMARTWEB provides and synthesises, is *5 Spiele* (5 matches). This is presented along with textual material *Argentinien-Deutschland, (1:3) 8.6.1958 Gruppenspiel (group match), (0:0)*

---

[2] A scenario presentation provided by Deutsche Telekom can be downloaded at *http://smartweb.dfki.de/SmartWeb_FlashDemo_eng_v09.exe*

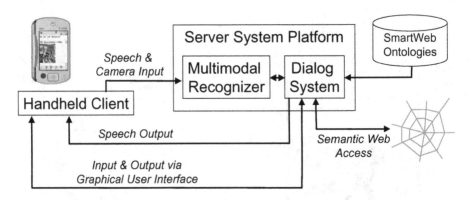

Fig. 2. The distributed SMARTWEB dialogue system architecture with handheld client

16.7.1966 Gruppenspiel, (3:2) 29.6.1986 Finale (1:0) 8.7.1990 Finale, (5:3) 30.6.2006 Viertelfinale (quarter final), and multimedia material such as images and videos, as can be seen in figure 1. Figure 1 shows the PDA interaction device, the graphical user interface [3], and the semantic navigation possibilities for manipulative interfaces with overview, zoom, filter, and details on demand functionality. To offer an overview and details on demand on small screen size at the same time, we use a fisheye distortion view in combination with automatic graph node placement.

The partners of the SMARTWEB project share experience from earlier dialogue system projects [4,5,6,7]. We followed guidelines for multimodal interaction, as explained in [8] for example, in the development process of our first demonstrator system [9] which contains the following assets: *multimodality*, more modalities allow for more natural communication, *encapsulation*, we encapsulate the multimodal dialogue interface proper from the application, *standards*, adopting to standards opens the door to scalability, since we can re-use ours as well as other's resources, and *representation*. A shared representation and a common ontological knowledge base eases the data flow among components and avoids costly transformation processes [10], which applies to the visualisation process, too. The general SMARTWEB handheld architecture to meet the requirements can be seen in figure 2, the dialogue system as well as the multimodal speech and camera recogniser (for face orientation) are explained in further depth in [2].

In this contribution we report on the use of RDF metadata to arrange information pieces in automatically layout graphs with respect to their semantic relations extracted from RDF results obtained from our knowledge servers. Humans themselves may encode information based upon its meaning [11], at least users feel familiar with this way of information arrangement. The text is structured as follows: chapter 2 and 3 describe the interaction and navigation possibilities, in chapter 4 we report on the integration process, chapter 5 presents related work, and in chapter 6 we conclude by further motivating the use of ontologies and Semantic Web data structures [12] for multimodal interaction design and implementation, and in particular, for visualising graph-like information spaces on mobile PDA devices. Our graph visualisation should provide an answer to the question how conceptual data models facilitate the generation of semantic navigation structures on mobile devices.

2   Interaction Possibilities

Basically, we want to allow the user to send requests to various information services that are linked by a Semantic Web framework. The user should be able to pose questions in natural language to get multimodal answers to be presented. Subsequently, semantic navigation (section 3) allows for iterative information retrieval by browsing and navigating through self-organising and highly interactive graph structures—as multimodal dialogue system functionality. We group the interaction modalities into three major classes:

- auditory: speech input and output
- graphical: all modalities that serve as input on the screen: touch or stylus input, the keyboard, and for output the graphical display itself
- haptic: device buttons and the cursor joystick

We focus on the touch screen stylus input and the graphical display output. The active graph node, *1. GER-ARG* in figure 1(centre), is called the *focus point*. All direct node interaction possibilities, such as changing an active ontology instance (e.g. a football game instance), can be done on the focus point. The focus point covers multiple similar ontology instances (e.g. of the same type), according to the calculated best mapping from the result structure toward the visual graph structure. Every ontology instance has information slots (figure 1(right)) and relations to other ontology instances which are represented by relation nodes. We calculate the best initial set of active relations. In addition, the user should be able to control which node relations are active and which are not. Starting from this initial setting, the user can change the focus point by clicking other instance or relation nodes.

3   Semantic Navigation

In [13] the use of multiple and distinct ontologies to support modelling, integration, and visualisation of personalised knowledge is discussed. *Semantic Navigation* is thereby defined as a way to build up and navigate views according to the logical organisation given by topic ontologies. This definition accentuates the need for content visualisation and navigation in heterogeneous ontologies, and for authoring or extraction needs. In the context of mobile interfaces and browsing in ontological answer structures, we focus on semantic navigation that helps to (1) access semantic information quickly, (2) allow for intuitive interaction, (3) allow the user to build an own cognitive map due to dynamic exploration of an unknown information space through graph interactions on mobile device

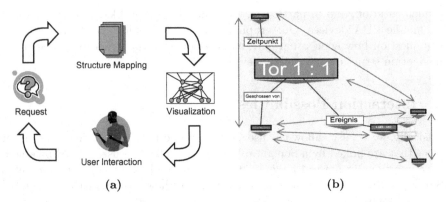

Fig. 3. (a) Interaction and visualisation loop. (b) Uniform vertices distribution.

displays. To reach this goal the RDF data must somehow be filtered, simplified, and summarised to be appropriate for the display on small devices and navigation. In addition, each graph to be displayed and its layout must be calculated according to the representation and presentation constraints, named *structure mapping* and *visualisation*, respectively. Figure 3(a) shows the interaction and visualisation loop. For example, examining the centre display of figure 1, every exchange of the active instance as user interaction initiates a new interaction and visualisation loop to calculate a new graph and its layout. In our framework we rely on an integrated ontology model where the visual graph can be extracted from result RDF structures in a straightforward way. We use all domain instances according to the ontology assertions, whereby structure mapping reduces the number of nodes needed for representing the RDF data to better fit on the small screen. The instances are grouped according to their incoming and outgoing slots (contextual structure), and every group is represented by one graph node. By sharing the same contextual structure, groups of instances provide the capabilities to be represented by one node in the graph structure. [3]

3.1   Automatic Graph Structures and Graph Layout

Central idea of this approach is to address automated layout by using constraint processing techniques [14] to represent and process causal design principles and perceptual aesthetic criteria about human visual abilities for structuring and organising information. The three central constraints for our automatic graphs are:

1. All vertices must be inside a fixed space on the handheld.
2. Vertices must not overlap.
3. Related vertices must be placed next to each other.

---

[3] In the integrated ontology framework this means to specify PDA slot preferences for e.g. match instances: (i) *referee*, (ii) *weather*, (iii) *spectators*, and so on (figure 1(right)). The constraint-based visualisation step optimises the layout in such a way that five ontological soccer match instances are grouped into one graph node (figure 1(middle)).

In addition to these constraints for the automatic graph layout, a number of aesthetic criteria should be considered as far as possible [15]. For display on a small mobile device, we take four aesthetic criteria into consideration:

1. Avoid edge crossings.
2. Keep edge lengths uniform.
3. Distribute vertices uniformly.
4. Keep vertices conform with user expectations.[4]

The design of an aesthetically pleasing layout is characterised as a combination of a general search problem in a finite discrete search space and a constraint satisfaction optimisation problem (CSP) [16]. The task is to find positions for all instance vertices satisfying the three constraints, as well as the four aesthetic criteria which should be considered as far as possible. The computation of the solution has to be fast to guarantee real-time interactivity.[5]

A CSP is defined by a set of variables, $X_1, X_2, ..., X_n$, and a set of constraints, $C_1, C_2, ..., C_m$, whereby each variable X_i has a nonempty domain D_i of possible values. Each constraint C_i involves some subset of the variables and specifies the allowable combinations of values for that subset. A CSP state is defined by an assignment of values to some or all of the variables, $\{X_i = v_i, X_j = v_j, ...\}$. In a *complete assignment* every variable is mentioned, and a CSP *solution* is a complete assignment that satisfies all the constraints. Fortunately, the problem to satisfy the constraints mentioned above can be modelled by a simple kind of CSPs; it involves only discrete variables that have finite domains for vertex positions. A position consists of discrete x- and y- coordinate numbers. The domains for these coordinates are restricted by the width and height of the handheld display. If the maximum domain size of any variable in a CSP is d, then the number of possible complete assignments is $O(d^n)$, where n is the number of variables. In the worst case, therefore, we cannot expect to solve finite-domain CSPs in less than exponential time that is, exponential in the number of variables, but fortunately the number of variables (nodes) is rather low in our domain. We chose a *refinement model* [6] over an *perturbation model* [7], because the latter corresponds

---

[4] To provide consistent visual encodings and smooth transitions between consecutive displays.

[5] A time limit of three seconds on producing a satisfying layout solution is set. If the constraints are not satisfied within that limit, we assume the constraints to be inconsistent. A way to handle inconsistent constraints is described in section 3.2.

[6] Variables are initially unconstrained; constraints are added as the computation unfolds, progressively refining the permissible values of the variables. Solving CSPs is based on removing inconsistent values from variables' domains until the solution is found (by *forward checking*, to look at each unassigned variable Y that is connected to X by a constraint and deleting from Y's domain any value that is inconsistent with the value chosen for X. If a partial solution violates any of the constraints, backtracking is performed.

[7] At the beginning of an execution cycle variables have specific values associated with them that satisfy the constraints. The values of one or more variables are perturbed, and the values of the variables are adjusted so that the constraints are again

to a local search method, that can effectively use previous graph CSP solutions, but biases local refinements without explicit statement, and because we are interested in a more principled approach with control over most of the declarative constraints. We use the Choco Constraint Programming System[8] as refinement model API. Choco provides a Java library for CSPs built on an event-based propagation mechanism with backtrackable structures.

Constraints Formulation. First we code the vertex positions into a suitable representation for constraint formulation. Since the Choco system does not support tuple structures we divided a vertex position into two constraint variables. The distance between two vertex positions P_1 and P_2 is formulated as constraints for distances between two x-coordinates (x_1, x_2) and two y-coordinates (y_1, y_2). The domain of possible discrete values for these variables is limited to the fixed space on the handheld reserved for the graph presentation (480x600 pixels). For all pairs of vertices we formulate a distance constraint C_1 to avoid the overlapping of two or more vertices. This minimum distance constraint prevents overlapping by setting a minimum separation distance value between all vertices. As a second distance constraint C_2 in the opposite direction, we formulate a maximal distance constraint to place vertices next to each other for all pairs of related vertices. The distance between two 2D positions is normally computed as Euclidean distance $distance = \sqrt{(x_1 - x_2)^2 + (y_1 - y_2)^2}$. Because neither power nor radical operators available in the Choco system due to complexity and performance reasons, we used distance approximations by elementary calculation types, like addition and subtraction. We experimented with a simple approximation derived from the Manhattan distance $(L_1 norm)$, where the distance is computed as:

$$distance = |x_1 - x_2| + |y_1 - y_2|$$

This constraint defines a rectangular distance between two 2D vertex positions. We use it to formulate an algebraic minimisation constraint C_3 for the distance *dist* on each axis:

$$(|x_1 - x_2| > dist) \vee (|y_1 - y_2| > dist)$$

If constraint C_3 holds for two positions P_1 and P_2, then the distance between both positions is at least *dist*. Since Choco provides no option to formulate an absolute value in a constraint and we reformulate $(|x_1 - x_2| > dist)$ as:

$$((x_1 - x_2) < -dist) \vee ((x_1 - x_2) > dist)$$

Aesthetic Criteria Constraint Formulation. The first aesthetic criterion we formulate as CSP is to distribute vertices uniformly, the second is to keep vertices conform with the user expectations.[9] Our implementation of the first criterion

---

satisfied. The perturbation model has been used in interactive graphics systems like Sketchpad[17], ThingLab 1[18], Magritte[19], and Juno[20], and user interface construction systems such as Garnet [21].

[8] *http://www.choco-solver.net*

[9] C_2 already avoids edge crossings and keeps edge lengths uniform.

is an optimisation that tries to spread the graph as much as possible over the available screen. The Choco system offers the possibility to search for a general solution and thereby to maximise or minimise a certain variable. We use this to implement a node spreading behaviour (C_4). The variable denoting the objective value is the total distance between all vertices that are not related to each other (figure 3(b)). Our implementation of the second criterion corresponds to avoiding the following behaviour: If the user changes a relation, the positions of the vertices also change to the optimal positions for all remaining vertices, whereby all inactive relations are closed. In conjunction with the uniform distribution constraint C_4, this can lead to a completely new graph layout. Since a considerable new graph layout contradicts the aesthetic criterion to keep vertices conform with user expectations (small changes are needed to avoid human cognition problems), we try to implement a behaviour that leaves as many vertex positions as possible untouched while optimising the next graphical layout. The idea behind that is to rank the vertex positions according to the importance for the user. The measurement we use for this is the time delay of the node clicks the user performs during navigation. Roughly speaking, all the vertices that have been clicked recently keep their position in the new graph layout, all the vertices that have not been clicked recently release their positions more easily (C_5).

3.2   Limits of the Constraint-Based Programming Method

Although we can avoid inconsistent layout constraints ($C_{1...5}$ are consistent), we cannot exclude cases where the CSP returns no result. Limits of the CSP include that a solution, which satisfies all existing constraints, does not exist. In our case this happens for example, if the number of vertices is to large to fit on the available space. To avoid empty screens, we automatically reduce the amount of displayed vertices thereby reducing the amount of constraints to satisfy. Whenever the automatic layout algorithm is not able to find a solution to satisfy all the constraints, the amount of active vertices will be reduced to only those vertices with an active input or output relation to the current focus vertex. Due to this reduction, the inconsistencies will be iteratively eliminated, until a solution layout can be found. We illustrate this procedure in figure 4, adding three artificial relations TEST1, TEST2, and TEST3, to our match example. In STEP1, showing the match focus instance and three active relations, all additional test relations are inactive and the CSP solution is available. In STEP2, the focus changes to the date instance *08.06.1958* and thereby activates the relation TEST1 between the date and the result instance *Ereignis* → (*incidence* : *redcard*). The activation of TEST1 leads to a new C_2-constraint of the distance between the date and the result vertex. The CSP solution correctly forces the connected vertices to stay closer together. In STEP3, the user clicks on the *Ergebnis* relation, which activates the instance *3:1*. The invisible inactive relations TEST2, TEST3 are shown for illustration. The new C_2-constraints now force the TEST2, TEST3 related vertices to stay closer together. STEP4, clicking on the result instance *3:1*, results in an unsolvable CSP, because not all activated TEST1, TEST2, TEST3 relations can be placed in the *3:1* instance ring at the same time, without

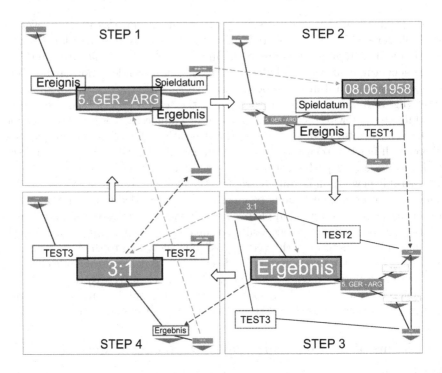

Fig. 4. Handling of inconsistent constraints

violating the central constraints $C_{1,2,3}$. On this account the layout algorithm is not able to find a solution to satisfy the central constraints of the six active relations and therefore the number of active relations/active nodes is reduced. As can be seen, only vertices with a direct connection to the current focus stay active and remain visible. This error recovery strategy resembles C_5 and turns out to be very robust.

4   Integration into a Distributed Dialogue System

A reaction and presentation component [22] is accountable for dialogue reaction and presentations in terms of the described semantic navigation structures. The added graph presentation planning capabilities includes (1) summarising multimodal result and finding an appropriate mapping toward a lower-level visual object and its attributes which we model in the interaction ontology, (2) finding out visual pattern interrelationships, (3) automating the visualisation of useful multimodal information which complements NLP generation output, and (4) provide consecutive information displays communicated from the server to the client. The semantic graph visualisation component is embedded into the reaction and presentation module. In SMARTWEB, the graph-based user interface on the client is connected to the graph layout module that resists on the server. All

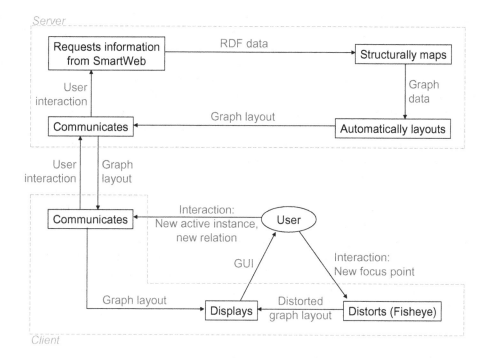

Fig. 5. Semantic graph visualisation data flow. Graph layouts for arbitrary RDF graph data are calculated on the server. On the client changing an active instance is communicated to the server.

data transfer between server and client is organised by special XML structures transmitted over socket connections in both directions. We extended this XML structure by an new *dynamic graph* environment, for the graph structure data to be exchanged, the graph node layout positions, and the user interactions. The data flow is shown in figure 5.

The graph structure and layout are sent from the server to the client to be displayed on the handheld. If the user exchanges an ontology instance by choosing a new active instance in the focus point menu with new relations, a new football match instance for example, the interaction is sent back to the server to get a new layout. The possible user interaction, to change the focus point, is handled locally on the handheld client and needs no communication with the server.

5 Related Work

In the mobile question answering (QA) context we put coherence between consecutive questions and answer turns, as well as coherence between different answer views, such as the main GUI and navigations in focus. Work on representation recall ability and affordance mapping [23], automatic generation of personalised multimedia and hypermedia presentations for desktops [24], multimodal fission and media design for symmetric multimodality for the mobile travel companion Smartkom [25] heavily influenced our current design. DFKI's RDFSViz[10] provides a visualisation service for RDF data and uses the Graphviz[11] graph visualisation software to visualise ontology instances. Navigation is supported by retrieving relationships to other instances of the ontology. Desktop-based ontology navigation tools such as Ontoviz plug-in[12] for Protege support the retrieval of term definitions, or the drawing of complete ontology trees. Displaying RDF data in a user-friendly manner is a problem addressed by various types of applications using different representation paradigms [26]. At least the following types can be identified: keyword search, e.g. Swoogle[13], faceted browsing [27], explicit queries, e.g. Sesame[14], and graph visualisations. IsaViz[15] represent RDF models as vertex-edge diagrams, explicitly showing their graph structure. More advanced navigation tools and other interaction capabilities are provided by MoSeNa [28] and Haystack [29]. Generally limited display resources are addressed in [30]. Our account for data transformations of basically non-spatial or non-numerical relational information into graph visualisations is straightforward for configurations, where the graph size is incremental but does not present too many concurrent nodes and constraints. A constraint-based method for user interfaces, the DeltaBlue Algorithm, has been discussed in [31]; we used a simpler method with satisfiable time performance for interactive graph display.

6   Conclusion and Future Work

We explored how to map highly structured RDF data as result structures in a QA scenario into a graph structure, and how the resulting graph can be visualised on a PDA, as an example of how to visualise Semantic Web data structures. We also discussed how multimodal interaction in dialogues can be established by additional graphical user interface capabilities—arranging RDF content through automatic layouting in a way that users better understand semantic relations of the contents by graph-based semantic navigation. The restrictions of the handheld device have also been taken into consideration, e.g., the small screen size, and we found a way to deal with the restricted computing power; the distributed SMARTWEB system turned out to be very useful for this purpose to compute automatic graph layouts on the server in order to solve the performance bottleneck on the handheld client. On the server, we used discretisation of the solution space to narrow down the search for a graph layout CSP to increase the reaction time of our interactive system. Using an automatic layout algorithm to arrange the vertices of the graph, our graph presentation system is able to deal with

---

[10] http://www.dfki.uni-kl.de/frodo/RDFSViz/
[11] http://www.graphviz.org/
[12] http://protege.cim3.net/cgi-bin/wiki.pl?OntoViz/
[13] http://swoogle.umbc.edu/
[14] http://www.openrdf.org/
[15] http://www.w3.org/2001/11/IsaViz/

arbitrary RDF graph data. We finally proposed a way to handle situation where no solution exists to satisfy all layout constraints.

During the development of our constraint system we conducted two evaluation phases that involved 20 users to get feedback at an early stage (cf. [32]). We formulated four evaluation objectives: (a) *the possibilities to interact with the graph are easily to understand*, (b) *it is possible to extract information from the graph structure and the node labels*, (c) *the user gets aware of the difference between an instance and relation nodes*, (d) *the user realises the dependencies between related active instances whose labels appear*. A first evaluation showed that users had problems with changing layouts and extracting information from the distorted graph. By refining the fisheye distortion (from distance-based to topology-based distortion), text of non-focussed nodes could be displayed in a bigger font on the screen. Significantly, the formulation of the four aestetic criteria of section 3.1 also facilitates to understand the graph structure—as shows the second evaluation. 85% describe the graph interaction possibilities as easy to understand (after an initial demonstration), 95% easily understand the difference between instance nodes and relation nodes. The figures suggest users can get a more precise understanding of the presented information in its whole complexity. These feedbacks are useful sources of suggestions for the further improvement of our graph presentation system, and show additionally, that graph visualisations and interactions are generally welcomed alternatives for highly structured result data in QA scenarios. Further evaluations should focus on the questions

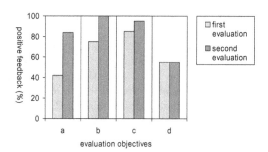

if users are able to reach and select the information they are most interested in. Extensions are editing and navigation functions via concurrent pen and voice to provide symmetric multimodal query fusion and query correction. We expect the concurrent use of pen and voice to reduce the initial learning phase toward intuitive use of multimodal mobile Semantic Web interfaces.

Acknowledgements

The research presented here is sponsored by the German Ministry of Research and Technology (BMBF) under grant 01IMD01A (SmartWeb). We thank our project partners, our reseach assistants, and the evaluators. The responsibility for this papers lies with the authors.

References

1. Wahlster, W.: SmartWeb: Mobile Applications of the Semantic Web. In: Dadam, P., Reichert, M. (eds.) GI Jahrestagung 2004, vol. 3238, pp. 50–51. Springer, Heidelberg (2004)
2. Sonntag, D., Engel, R., Herzog, G., Pfalzgraf, A., Pfleger, N., Romanelli, M., Reithinger, N.: Smartweb handheld - multimodal interaction with ontological knowledge bases and semantic web services. In: Huang, T.S., Nijholt, A., Pantic, M., Pentland, A. (eds.) Artifical Intelligence for Human Computing. LNCS, vol. 4451, pp. 272–295. Springer, Heidelberg (2007)
3. Sonntag, D.: Interaction Design and Implementation for Multimodal Mobile Semantic Web Interfaces. In: Smith, M.J., Salvendy, G. (eds.) Human Interface and the Management of Information. LNCS, vol. 4558, pp. 645–654. Springer, Heidelberg (2007)
4. Wahlster, W. (ed.): VERBMOBIL: Foundations of Speech-to-Speech Translation. Springer, Heidelberg (2000)
5. Wahlster, W.: SmartKom: Symmetric Multimodality in an Adaptive and Reusable Dialogue Shell. In: Krahl, R., Günther, D. (eds.) Proc. of the Human Computer Interaction Status Conference 2003, Berlin, Germany, DLR, pp. 47–62 (2003)
6. Reithinger, N., Fedeler, D., Kumar, A., Lauer, C., Pecourt, E., Romary, L.: MIAMM - A Multimodal Dialogue System Using Haptics. In: van Kuppevelt, J., Dybkjaer, L., Bernsen, N.O. (eds.) Advances in Natural Multimodal Dialogue Systems, Springer, Heidelberg (2005)
7. Wahlster, W.: SmartKom: Foundations of Multimodal Dialogue Systems (Cognitive Technologies). Springer-Verlag New York, Inc., Secaucus, NJ, USA (2006)
8. Oviatt, S.: Ten myths of multimodal interaction. Communications of the ACM 42(11), 74–81 (1999)
9. Reithinger, N., Bergweiler, S., Engel, R., Herzog, G., Pfleger, N., Romanelli, M., Sonntag, D.: A Look Under the Hood Design and Development of the First SmartWeb System Demonstrator. In: ICMI 2005. Proceedings of 7th International Conference on Multimodal Interfaces, Trento, Italy (October 04-06, 2005)
10. Oberle, D., Ankolekar, A., Hitzler, P., Cimiano, P., Sintek, M., Kiesel, M., Mougouie, B., Vembu, S., Baumann, S., Romanelli, M., Buitelaar, P., Engel, R., Sonntag, D., Reithinger, N., Loos, B., Porzel, R., Zorn, H.P., Micelli, V., Schmidt, C., Weiten, M., Burkhardt, F., Zhou, J.: DOLCE ergo SUMO: On foundational and domain models in SmartWeb Integrated Ontology (SWIntO). Journal of Web Semantics: Sci. Services Agents World Wide Web (2007)
11. Myers, D.G.: Psychology. Worth Publishers (2004)
12. Fensel, D., Hendler, J.A., Lieberman, H., Wahlster, W.: Spinning the Semantic Web: Bringing the World Wide Web to Its Full Potential. MIT Press, Cambridge (2005)
13. Boselli, R., Paoli, F.D.: Semantic navigation through multiple topic ontologies. In: SWAP 2005. Semantic Web Applications and Perspectives (2005)
14. Tsang, E.: Foundations of constraint satisfaction. Academic Press, New York (1993)
15. Battista, G.D., Eades, P., Tamassia, R., Tollis, I.G.: Graph Drawing: Algorithms for the Visualization of Graphs. Prentice-Hall, Englewood Cliffs (1999)
16. Graf, W.: Constraint-based graphical layout of multimodal presentations. In: Advanced Visual Interfaces, pp. 356–387 (1992)

17. Sutherland, I.E.: Sketchpad: A man-machine graphical communication system. In: Spring Joint Computer Conference, pp. 329–345 (1963)
18. Borning, A.: The Programming Language Aspects of ThingLab, a Constraint-Oriented Simulation Laboratory. ACM Trans. Program. Lang. Syst., 353–387 (1981)
19. Gosling, J.A.: Algebraic Constraints. PhD thesis, Carnegie-Mellon University (1983)
20. Nelson, G.: Juno, a constraint-based graphics system. In: SIGGRAPH, pp. 235–243 (1985)
21. Myers, B., et al.: The Garnet Toolkit Reference Manuals: Support for Highly-Interactive, Graphical User Interface in Lisp. Technical report, Carnegie Mellon University, Computer Science Department (1990)
22. Sonntag, D.: Towards combining finite-state, ontologies, and data driven approaches to dialogue management for multimodal question answering. In: Proceedings of the 5th Slovenian First International Language Technology Conference (IS-LTC) (2006)
23. Faraday, P., Sutcliffe, A.: Designing effective multimedia presentations. In: CHI 1997. Proceedings of the SIGCHI conference on Human factors in computing systems, pp. 272–278. ACM Press, New York, NY, USA (1997)
24. Andre, E., Mueller, J., Rist, T.: WIP/PPP: automatic generation of personalized multimedia presentations. In: MULTIMEDIA 1996. Proceedings of the fourth ACM international conference on Multimedia, pp. 407–408. ACM Press, New York, NY, USA (1996)
25. Wahlster, W. (ed.): SmartKom: Foundations of Multimodal Dialogue Systems. Springer, Berlin (2006)
26. Pietriga, E., Bizer, C., Karger, D., Lee, R.: Fresnel: A browser-independent presentation vocabulary for RDF. In: International Semantic Web Conference, pp. 158–171 (2006)
27. Yee, K.P., Swearingen, K., Li, K., Hearst, M.: Faceted metadata for image search and browsing. In: CHI 2003. Proceedings of the SIGCHI conference on Human Factors in Computing Systems, pp. 401–408. ACM Press, New York, NY, USA (2003)
28. Becker, J., Brelage, C., Klose, K., Thygs, M.: Conceptual Modeling of Semantic Navigation Structures: The MoSeNa-Approach. In: Proceedings of the Fifth ACM International Workshop on Web Information and Data Management, pp. 118–125 (2003)
29. Quan, D., Huynh, D., Karger, D.R.: Haystack: A platform for authoring end user semantic web applications. In: International Semantic Web Conference, pp. 738–753 (2003)
30. Gerstmann, D.: Advanced visual interfaces for hierarchical structures. In: Human Computer Interaction (March 2001)
31. Sannella, M., Maloney, J., Freeman-Benson, B.N., Borning, A.: Multi-way versus one-way constraints in user interfaces: Experience with the deltablue algorithm. Software - Practice and Experience 23(5), 529–566 (1993)
32. Heim, P.: Graph-Based Visualization of RDF Soccer Data and Interaction Possibilities on a Handheld. Diploma thesis, Computational Visualistics, University of Koblenz/Germany (2007)

Personalization of Content in Virtual Exhibitions

Bill Bonis[1], John Stamos[1], Spyros Vosinakis[2], Ioannis Andreou[3],
and Themis Panayiotopoulos[1]

[1] Department of Informatics, University of Piraeus, Piraeus, Greece
{bonisb,jstamos,themisp}@unipi.gr
[2] Department of Product and Systems Design Engineering, University of the Aegean,
Syros, Greece
spyrosv@syros.aegean.gr
[3] Department of Technology Education and Digital Systems, University of Piraeus,
Piraeus, Greece
gandreou@unipi.gr

Abstract. Presentation of content is an important aspect of today's virtual reality applications, especially in domains such as virtual exhibitions. The large amount and variety of exhibits in such applications raise a need for adaptation and personalization of the environment. This paper presents a content personalization framework for Virtual Exhibitions, which is based on a semantic description of content and on information implicitly collected about the users through their interaction. The proposed framework uses stereotypes to initialize user models, adapts user profiles dynamically and clusters users into interest groups. A 3D virtual museum has been implemented as a case study, and an evaluation has been conducted.

Keywords: Virtual Reality, Virtual Museums, User Modeling, Personalization.

1 Introduction

This advent of Virtual Reality in recent years has enabled the development of novel interactive and immersive applications that emphasize on the presentation of content and satisfying user experience. Examples of such applications are 3D Virtual Exhibitions, i.e. interactive 3D environments containing a large collection of media objects. While the graphical representation of the exhibits is an important aspect of Virtual Exhibitions, the large amount and variety of these elements presents another challenge to the designers of such environments, namely to adapt the presentation of a collection according to user interests, i.e. content personalization.

A number of Virtual Exhibition applications are currently available and run standalone [1], or over the Internet [2,3,4], representing real or fictional museums and exhibitions. It can be noted that most of today's exhibition environments emphasize on the content and aim to the deeper understanding of entities and concepts through user navigation and interaction in 3D. Most of these applications contain, nevertheless, static collections arranged in predefined positions, and the design of the virtual space and its contents is based entirely on the author's categorization. Therefore, the user's role is limited to a passive observer and the presentation of large

B. Falcidieno et al. (Eds.): SAMT 2007, LNCS 4816, pp. 172–184, 2007.

collections may fall into the obstacle of navigational difficulties in 3D environments [5], which eventually leads to the reduction of user interest and to the disability to explore and search for the desired content [6].

The problem of presenting and categorizing large quantities of content has been effectively addressed in Web [7,8,9,10] and multimedia applications [11], where user modeling has been used to personalize content presentation based on the users' own interests. The authors claim that virtual environments could also benefit from user modeling and adaptation techniques that make assumptions about user interests and intentions concerning the application, and construct the virtual space accordingly. Such a personalized space may reduce the navigational burden while it still retains the metaphor of being immersed in a 3D environment. Although the concept of personalizing content is not new, there have been few approaches towards user modeling in 3D environments. In analogy to the theory of content adaptation on the Web [12], content personalization in 3D could be supported by a) a process of recognizing user interaction patterns in the 3D environment [13], b) a mechanism that makes assumptions about user preferences based on these interactions, and c) respective modifications of the environment that reflect user needs and increase her satisfaction.

As a first step towards this theory, Chittaro and Ranon have proposed AWE3D [14], an architecture for presenting personalized 3D content on the Web. In this approach, the system monitors and records user interactions with the environment and applies a rule set on the recorded interactions in order to modify the user model, and personalizes the environment based on assumptions concerning user preferences. They present a 3D E-Commerce application as a case study. The work of dos Santos and Osorio [15] also focuses on content adaptation in 3D environments. They use a rule based approach with certainty factors to modify the content and structure of the environment according to user interests. A distance learning environment has been created as a prototype. Both approaches adopt a rule based mechanism to generate and update the user models, and the responsibility of defining and validating these rules lies entirely on the designer. Any modification of the content will require respective changes in the rules.

Celentano and Pittarello [13] propose an approach to facilitate adaptive interaction with the virtual environment, which is based on the following: a structured design of the 3D interaction space, the distinction between a basic virtual world layer and an interaction layer, and the recording of the environment's usage by the user in order to find interaction patterns. The aim is to facilitate the system's usage by monitoring user behavior and predicting future needs for interaction purposes. When the system recognizes the initial state of an interaction pattern, it executes the final state without letting the user engage in the intermediate ones.

The authors propose a novel approach towards adaptive virtual environments, in which objects are dynamically distributed among rooms and users may experience personalized presentations based on their previous interaction with the system. The proposed framework is supported by a semantic graph, defined by the designer, which describes the nature of the exhibits by hierarchically categorizing the content, and drives the user modeling process. Additionally, the authors support the claim that users' experience in a virtual exhibition is enhanced through communication and collaboration with other users in a shared environment [16]. In this context, the proposed framework supports dynamic clustering of user groups based on the similarity of user models, which may lead to the formation of e-societies with similar interests. The authors have implemented a science fiction virtual museum as a case study of the proposed framework and have performed a user evaluation of the application. Evaluation results indicate the impact of the user modeling process in adapting the presentation of content to the user preferences and a significant degree of user satisfaction from the system in general.

The rest of the paper is structured as follows: section 2 presents the proposed framework for content personalization in virtual exhibitions, while section 3 presents a science fiction virtual museum as a case study and section 4 presents an evaluation of the implemented system. Finally, section 5 draws the conclusions and states the future work.

2   A Framework for Content Personalization

The proposed framework for designing and implementing virtual exhibitions with content personalization is based on four methods that enhance a static 3D environment with dynamic characteristics: user model generation, content selection and presentation, user model update and clustering. In the next paragraph a full user interaction session with an application designed under this framework is presented.

When a new user enters the environment for the first time, a user model is assigned to her based on stereotypes [17] that correspond to her selection of an avatar, i.e. her graphical representation in the 3D environment. While the user is browsing the environment, her navigation and interaction with content are monitored and the recorded behavior is utilized to make assumptions about her interests and preferences, which are then incorporated into the user's profile. At any time, the user can ask to be transported to a personalized environment, which reflects her assumed preferences and recommends new content that might be of her interest. The user can also join communities with similar preferences, visit other personalized environments, and exchange opinions about the content. User interest groups are proposed by the environment through an automated clustering process.

From the designer's point of view, the framework can be employed to construct new dynamic virtual exhibitions without having to define explicit rules for content personalization and adaptation. The designer has to provide the 3D content, i.e. the rooms and objects of the environment, the semantic graph, i.e. an ontological description of the content, and the user stereotypes that contain templates of initial user preferences concerning the content. A presentation process then creates the exhibition rooms and distributes the exhibits dynamically based on the above data. The personalized environments also depend on the interaction history of the respective users. Furthermore, exhibitions generated using the proposed framework can be easily adapted or enhanced by altering or inserting new 3D content and making appropriate changes in the semantic graph and the user stereotypes.

The proposed framework is based on a thin client/server architecture (Fig.1), in which the users interact with the environment on the client side and user modeling takes place on the server side. This architecture follows the paradigm of decentralized user modeling architectures [18], and allows clustering of users into groups with similar preferences, whilst it is also necessary for the support of multi-user environments and for immediate adaptation and expansion of exhibition data by the designer.

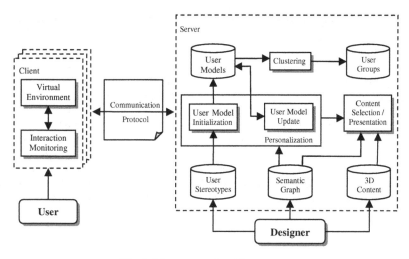

Fig. 1. The proposed architecture

The rest of this section describes the proposed framework and the related architect-tural components in detail.

2.1 The Semantic Graph

A vast number of applications that utilize user modeling methodologies try to address the user's need for quick and efficient access to a subset of information that meets her interests and preferences, without having to search through a larger set of objects. A widely used term in the literature for describing these applications is recommender systems [7,10,11,17,19,20,21]. A distinctive characteristic of these systems compared to information retrieval and filtering systems and to search engines is the output of individualized information based on a priori knowledge about the content and assumptions about user preferences.

A thematically uniform set of objects can be grouped together and categorized based on a number of criteria; relations can be determined between objects and categories or between categories themselves, e.g. relations of affinity and inheritance. For example, in an art exhibition, exhibits can be grouped with respect to their creators, the epoch or the style. These categories can be generalized into broader categories or specialized into subcategories. A categorical hierarchy of this type forms a tree with nodes being the categories and edges being the relations between them. The entirety of the categories can thus be represented as a forest (a set of distinct trees). Nodes at higher levels in each tree imply categories with broader meaning and relations between nodes can be viewed as inheritance relations. Nodes of different trees can be connected implying a semantic *union* relation. Because of the union relations this categorization scheme is used as a graph instead of a forest in the context of determining related categories. The actual objects are attached to the categorization trees using connection(s) with one or more lowest-level category nodes (the leafs of the categorization trees) via an instance relation. Thus, the resulting structured hierarchical semantic taxonomy forms a directional graph, the Semantic Graph (SG), in which nodes (distributed into levels) stand for objects and concepts, edges represent the relations between them [19,22,23] and the levels represent the degree of generalization.

Nodes are divided in two categories, the *object nodes*, which represent the actual elements of the environment, and the *categorization nodes*, which represent object categories. The latter are distributed hierarchically based on generalization relations; a broader term lies at a higher level than a more specialized one.

Figure 2 presents a sample part of a semantic graph that is used for categorizing cars. The dashed line is a union relation that connects two categories belonging to different trees. In the example the node Type is divided into two subcategories, the nodes Professional and Racing that are also divided into two subcategories each. The EVO 5 has been used as a WRC racing car and it was manufactured by Mitsubishi, thus it can be connected with the nodes WRC and Mitsubishi. The union relation connects the respective nodes to imply the association between Ferrari manufacturer and F1.

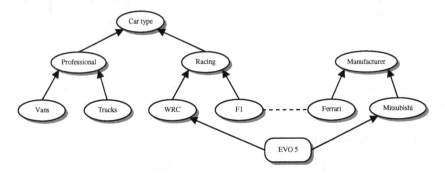

Fig. 2. A part of a semantic graph for categorizing cars

Let A be the node at the lower end of a categorization edge and B the node at the upper end. All edges (which represent the aforementioned relations) have a numerical weight in the range of (0, 1], which is the degree of membership of the object or concept that node A represents to the set that the node B portrays. The use of the degree of membership is analogous to the respective term from the Fuzzy Set Theory [23]. For example, the Terminator II movie can be said to be a well known sci-fi movie of the 90's, so the degree of membership of this movie in the Sci-fi and 90's categories is significant. In the case of union edges, the weights represent the degree of association between the connected nodes. The choice of weights can be made by using expert knowledge or by using machine learning techniques. The degrees of membership and the degrees of association are used during the execution of the personalization-recommendation algorithm.

In the proposed framework, the SG is the core element, because it drives most computational procedures:

- modeling user preferences in the environment,
- providing recommendations to users,
- clustering users with similar preferences to groups, and
- dynamically distributing the content into separate rooms and determining the connections between them.

The authors argue that the SG facilitates access to the exhibition's content repository, by reflecting a natural content interpretation, effectively serving the users' informational needs and preferences.

2.2 Dynamic Content Generation and Rendering

The hierarchical structure of the SG is used for the creation of the virtual exhibition's spatial structure. Each categorization node can be represented by a set of conceptually relevant rooms that are connected via doors, following the graph's edges and the relations of kinship between semantic nodes. This approach provides a multidimensional navigation paradigm, in which rooms are connected based on their semantic similarity. A user can navigate inside the exhibition environment and browse the content by following paths that correspond to any of the concepts that can characterize it. E.g. a room with objects from Steven Spielberg movies is connected with the room of US directors (at a higher level). By dynamically structuring the environment, the effort of designing and creating every room in the exhibition is reduced to a minimum.

The designer provides a number of template rooms that are used by the client application to construct the actual exhibition rooms. The system requires a default template room and, optionally, any number of thematic rooms related to existing categories of the semantic graph. Each template room is a 3D model of a section of the exhibition environment, in which two types of objects are inserted: *doors* and *exhibit containers*. When the client application has to construct a thematic exhibition room it searches for a predefined template that matches the respective category. In the case that no such template exists, the default room is used. Then, the system dynamically links the room with related ones using the existing doors, each labelled by the name of the room it leads to. Finally, the exhibits are dynamically placed in the containers defined by the designer. A default exhibition is constructed by creating an entrance room that is connected to all the general (top-level) category rooms of the semantic graph.

The collection of objects that will populate each room is generated by traversing the graph from the respective categorization node to the set of object nodes. The number of objects can be significantly large, especially when browsing the content of higher-level nodes. In that case, objects are distributed into a set of interconnected rooms.

2.3 Personalization

All available objects in the proposed framework are categorized by the SG in a semantic taxonomy. When and if a user is interested in a certain object, it can be assumed that she is also interested in one or more categories to which the object belongs (she likes the movie series Star Wars because she likes sci-fi movies). If this belief is reinforced during the interaction with the system, a recommendation set with members originating from this particular category would probably be a preferable choice. The *user model* is an instance of the SG, with all nodes being given a numerical value (positive or negative), which represents the degree of interest that the user is assumed to have for the term or object that the node portrays. The process of calculating these degrees of interest is analytically explained in the next paragraphs.

User Stereotypes. As stated in [24], the explicit creation of a user profile may annoy users that are unwilling to state their interests and lead to user models that do not actually reflect the user preferences. Therefore, an indirect method for generating an initial user model has been utilized, in which a set of stereotypes [17] is used to initialize the model of a new user. At registration time, the user selects an avatar from a provided library, and the system creates an initial user model and thus makes an assumption about her preferences and interests based on her selection. Each avatar is related with assumptions about the user, which are lexical values of properties defined by the designer. A possible set of properties can be age, sex, education or anything that is considered to characterize the users and the respective values can be young, old, female, male, high, low etc. The stereotypes are rules that relate each value of each property with estimated degrees of interest for a set of nodes in the SG.

The degree of interest in the categorization nodes is calculated as follows. Initially all the categorization nodes have a zero value. For each value of each property, the degree of interest declared in a stereotype (if any) is added to the respective categorization node. This initial user model is used for the formation of a recommendation set (which will be explained in the following paragraphs) prior to the user interaction with the 3D environment. This approach deals with the new user problem [20], with a reduced accuracy nevertheless. This is an initial estimation, however, and as the user interacts with the system, information is accumulated in the user model, updating it and thereby increasing its accuracy.

The aforementioned methodology is based on researches [25,26] that state the possible relation between the choice of an avatar by the user and the users' intrinsic characteristics and personality trades. Expert knowledge and/or user segmentation researches can be used to create the set of stereotypes.

Data Acquisition and User Model Updating. The algorithm that updates the user model based on her interactions uses a forward propagation mechanism. Initially, it collects the summing degree of interest for each object node of the SG that the user has interacted with. This is the result of a monitoring process that takes place on the client side. A variety of acquisition methods can be used, such as measuring the time spent by the user observing an object, taking into account the type of interaction, and providing a rating system that lets users express their preferences. The framework supports various types of acquisition methods and interpretations. At the end of this step, all object nodes have a degree of interest ranging from negative values (dislike for the particular object) to positive values (positive interest).

For algorithmic uniformity, each union relation in the SG is substituted with a dummy node (union node) placed at one level higher than the maximum level of the categorization nodes it connects. These nodes are connected to the union node with dummy relations that have weights equal to w, where w^2 equals to the degree of association of the union relation. For the rest of the paper we refer to both categorization and union nodes with the term *semantic nodes*.

Let SN be a semantic node and $DI(SN)_t$ the degree of interest in node SN at time instance t. Let $CSN = \{ CSN_1, CSN_2, ..., CSN_N \}$ be the set of children nodes of node SN, with $DI(CSN_i)_t$, $i \in [1,N]$ being the respective degrees of interest. Let also W_i be the weight of the edge that connects CSN_i to SN. Then, the degree of interest at time instance $t+1$ for node SN is calculated by:

$$DI(SN)_{t+1} = \sum_{i=1}^{N}(DI(CSN_i)_t \cdot W_i) + DI(SN)_t$$

Calculations begin at level 0 (the object nodes level) and they proceed level by level until the maximum level of the SG. Object nodes maintain the values from the initial step. The resulting degrees of interest comprise the updated user model.

Adaptation of the Environment. For the creation of a personalized room, a set of objects is chosen to be recommended to the user by populating the room with them. This task is realized using a backward propagation mechanism based on the user model. This mechanism is computationally equivalent to the forward propagation mechanism described before, having only the opposite direction. Thus, instead of the CSN, the $ASN = \{ASN_1, ASN_2, ..., ASN_K\}$ set is used which represents the set of ancestor nodes of node SN and N is replaced by K. During the backward propagation procedure an estimated degree of interest is assigned to the object nodes. The choice of objects that will be part of the user's personalized room depends on the ratings of the respective object nodes. Consequently, the simplest approach is to choose the M top rated nodes of level zero (the level with the object nodes), where M is the capacity of the room, whilst variations of the proposed room can be produced by employing methods, such as to insert a few random elements to increase the diversity, or to avoid selecting objects that the user has already interacted with.

Besides the selection of the recommendation set, a personal room should also have connections to other rooms of the virtual environment, allowing the user to further explore the exhibition. The set of rooms that are connected to the personalized room should also reflect the user's preferences. As mentioned earlier, every categorization node is represented as a set of rooms. So, to connect the personal room with the rest of the environment, L doors, where L is the door capacity of the personal room, are dynamically created. These doors are connected to a single room of each room set from the L top rated categorization nodes of the user's model.

2.4   Clustering: Formation of User Communities

In internet based applications, added grouping capabilities can promote the formulation of e-communities, thus increasing the sense of immersion in the virtual environment, enhancing communication opportunities, and satisfying the need for social interaction and awareness. To create a group of users with similar interests and preferences in the proposed framework, the user models must be compared. By assigning a unique index to every node of the SG, a user model can be taken as a numerical vector with each item being the degree of interest in the corresponding node. The dissimilarity between a pair of vectors can be computed using multidimensional Euclidean distance, and thus a group can be created based on distances smaller than a given threshold. This technique can provide satisfactory results in finding users with similar interests and presenting the respective personalized rooms to promote user communication. The user can choose a personalized room from a subset of recommended rooms, the owners of which have been clustered by the system, based on their interests.

3 Case Study: A Science Fiction Museum

In order to assess the proposed framework, a Virtual Museum concerning science-fiction movies, has been developed. It lets users navigate in thematically different rooms (e.g. containing characters, vehicles etc.), enter a personalized room with various exhibits that the user might enjoy, according to her interests, or enter online exhibitions of users with similar interests. The exhibition's content (3D models) has been created using external modeling tools and imported to the environment, while an authoring tool has been implemented to facilitate the process of construction, manipulation and maintenance of the SG and the system's database management. The authoring tool can be used by the designer of the virtual exhibition, effectively reducing the effort of creating all the needed elements. The Virtual Museum contains 87 exhibits with a SG consisting of 34 nodes distributed over 3 levels.

As the users navigate inside the virtual museum, they look at exhibits that fall inside their field of view. In order to perceive an exhibit, a user must be oriented towards it and be within a predefined distance. Furthermore, the users have the option of manipulating an exhibit, i.e. rotating it in order to gain a full perspective. Presentation of additional information concerning the exhibit is achieved through description pages in a provided panel. The users are also able to provide feedback about an exhibit by rating it and entering their personal message/comment. At the same time, users can view comments written by others, chat with them, observe their actions in the virtual space (via their avatars) and enter their personalized room.

The initial assumptions concerning user preferences are based on stereotypes related to the avatar gallery, and consequent interaction in the exhibition's space provides the appropriate degrees of interest for the profile update. Three types of interaction are monitored (ordered by the degree of influence upon the rating of an object, ascending):

- *viewing time*: the time spent looking at an exhibit. It is considered that the more time the user looks at an exhibit, bounded by an upper limit, the more interested she is in it.
- *manipulation of an exhibit*: interaction with an exhibit has a stronger impact than viewing time, as the user expresses her preferences more explicitly.
- *commenting and rating exhibits*: users can write comments that others can read and rate the exhibits accordingly. There are three available rating values: positive, negative and indifferent, which can reveal the user's opinion directly.

User interactions produce degrees of interest that are used by the server to assembly the personalized room and to distribute its contents dynamically upon user request. Users have the added ability to be instantly transported to any exhibition room (including the personalized room), in order to access rapidly a desired category/exhibit room.

The museum's 3D content is modeled in VRML and is loaded/rendered by the client application, which is implemented in Java, using the Java3D API. At the server side, the exhibition and user data were stored into an SQL Server 2000 database and

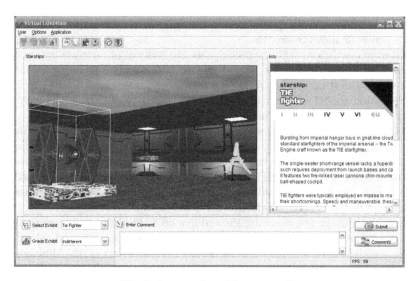

Fig. 3. A screenshot of the case study

their retrieval/manipulation is coordinated by the server application, implemented in Delphi programming language. The client-server communication is achieved with the use of TCP/IP sockets. Figure 3 presents a screenshot of the virtual museum application.

4   User Evaluation

In order to provide insight on the effectiveness of the proposed user modeling framework, a user evaluation has been conducted, using the aforementioned implemented system. A group of 20 users participated in the evaluation process, all of which were undergraduate students in the Computer Science department of the University of Piraeus. The users had to interact with the system and then fill out a questionnaire that was given to them. The degree of familiarity with 3D environments was almost uniformly distributed, with 11 participants stating that they had a fairly good experience in such environments, while 9 of them had almost never used a 3D application.

In the first step of the experiment process the user had to choose an avatar from the provided library, register into the system and state her opinion about the degree of representation by the avatar. After logging into the system, the user had to complete the main process of the evaluation, which included navigating in the personalized room, interacting with the recommended objects and finally navigating through the rest of the exhibition for a short period of time, interacting with other objects of interest. Every object in the personalized room had to be marked in the questionnaire as known or unknown to the user and to be rated in a scale of -3 (total dissatisfaction) to 3 (total satisfaction). This evaluation process was repeated three times, while the system was adapting her user model, based on the user interaction with objects. At the end of the evaluation, the participant was asked to rate her experience with the system, and write down comments.

Following discussions with the participants and analysis of their written comments indicated that the quality of the rendered graphics and models is an important factor in object rating and therefore it affects total user experience, i.e. a visually appealing, high-resolution model is more interesting than a low-resolution model with poor quality textures. Additionally, the quality of navigation (smooth movement, collision detection etc.) and the interaction capabilities of the objects significantly influenced user opinion. Several users stated that if the aforementioned factors were improved, the total system rating, which was formulated in a scale of 0 to 5 at an average of 3,65, would probably be higher. Overall, the experience gained during the conducted user evaluation, indicates that intrinsic characteristics of dynamic 3D exhibitions have an important impact on user satisfaction about the set of the recommended objects.

The analysis of the questionnaire showed that 66,85 % of the recommended object ratings were positive, 18,82 % were negative and 14,33 % were neutral, whereas the average positive rating was 2,12 and the average negative was -1,85. Although these are preliminary results, it can be inferred that the contents of the personalized room have matched user preferences to a satisfying degree. Additionally, user satisfaction by the object recommendations has improved from the first round of the evaluation to the last by an average factor of 10,47%. This improvement, although small, was considered satisfactory by the authors, as the duration of the interaction periods was small and object recommendations are implicitly influenced by the initial collection of objects contained in the personalized room, during early user interactions with these objects.

5   Conclusions and Future Work

This paper presented a user-oriented framework for designing and implementing virtual exhibition environments. Implicit generation and adjustment of user models allows applications that are based on this framework to dynamically adapt content presentation to user interests and preferences, based on user stereotypes and prior interaction of users with exhibits. A semantic graph is used as a basis for both content categorization and user modeling. This semantic representation of content enhances presentation capabilities and simplifies the alteration and/or extension of existing environments. Additionally, it is possible to detect similarities among user models, leading to the formulation of user interest communities. A case study i.e. a virtual museum application has been implemented, in order to gain insight about the effectiveness of this approach. A user evaluation of this case study produced favorable feedback towards the application of this framework in virtual exhibitions and virtual content presentation systems in general.

In the future, the authors are planning to fine grain the proposed framework, through the development of more complex case studies, in order to better evaluate the user modeling process. Another objective for the research team would be to provide an open-ended platform for virtual reality content presentations thus promoting researcher, developer and designer cooperation in the field. Finally, an ongoing extension of this work is to enrich the framework by allowing users to customize their personal space, thus providing stronger feedback on the quality of their user model.

References

1. Lepouras, G., Charitos, D., Vassilakis, C., Charissi, A., Halatsi, L.: Building a VR-Museum in a Museum. In: Proc. of VRIC Virtual Reality International Conference (2001)
2. Su, C.J., Yen, B., Zhang, X.: An Internet Based Virtual Exhibition System: Conceptual Design and Infrastructure. Comp. and Ind. Engineering 35(3-4), 615–618 (1998)
3. Ciabatti, E., Cicnoni, P., Montani, C., Scopigno, R.: Towards a Distributed 3D Virtual Museum. In: Proc. of Working Conference on Advanced Visual Interfaces, Italy, pp. 264–266 (1998)
4. Patel, M., White, M., Walczak, K., Sayd, P.: Digitization to Presentation – Building Virtual Museum Exhibitions. In: Proc. of Vision Video and Graphics, pp. 1–8 (2003)
5. Vinson, N.G.: Design Guidelines for Landmarks to Support Navigation in Virtual Environments. In: Proc. Of CHI 1999, pp. 278–285 (1999)
6. Modjeska, D., Waterworth, J.: Effects of Desktop 3D World Design on User Navigation and Search Performance. In: Proc. of IEEE Information Visualization, pp. 215–220 (2000)
7. Mukherjee, R., Sajja, E., Sen, S.: A Movie Recommendation System—An Application of Voting Theory in User Modeling. User Modeling and User-Adapted Interaction 13(1-2), 5–33 (2003)
8. Mobasher, B., Cooley, R., Srivastava, J.: Automatic personalization based on web usage mining. Communications of the ACM 43(8) (2000)
9. Eirinaki, M., Vazirgiannis, M.: Web Mining for web personalization. ACM Transactions on Internet technology 3(1), 1–27 (2003)
10. Schafer, J.B., Konstan, J.A., Riedl, J.: E-Commerce Recommendation Applications. Data Mining and Knowledge Discovery 5, 115–153 (2001)
11. Raskutti, B., Beitz, A., Ward, B.: A Feature-based Approach to Recommending Selections based on Past Preferences. User Modeling and User-Adapted Interaction 7, 179–218 (1997)
12. Brusilovsky, P.: Adaptive Hypermedia. User Modeling and User-Adapted Interaction 11, 87–110 (2001)
13. Celentano, A., Pittarello, F.: Observing and adapting user behavior in navigational 3D interfaces. In: Proc. of AVI 2004, pp. 275–282 (2004)
14. Chittaro, L., Ranon, R.: Dynamic Generation of Personalized VRML Content: a General Approach and its Application to 3D E-Commerce. In: Proc. of the seventh international conference on 3D Web technology, pp. 145–154 (2002)
15. dos Santos, C., Osorio, F.: An intelligent and adaptive virtual environment and its application in distance learning. In: Proc. of the working conference on Advanced visual interfaces, pp. 362–365 (2004)
16. Ko, D., Sumi, Y., Choi, Y., Mase, K.: Personalized Virtual Exhibition Tour (PVET): an experiment for Internet collaboration. In: Proc. of Systems, Man, and Cybernetics IEEE SMC 1999, pp. 25–29 (1999)
17. Rich, E.: User Modeling via Stereotypes. Cognitive Science 3, 329–354 (1979)
18. Fink, J., Kobsa, A.: A review and analysis of commercial user modeling servers for personalization on the world wide web. User Modeling and User-Adapted Interaction 10, 209–249 (2000)

19. Kim, W., Kerschberg, L., Scime, A.: Learning for automatic personalization in a semantic taxonomy-based meta-search agent. Electronic Commerce Research and Applications 1, 150–173 (2002)
20. Burke, R.: Hybrid Recommender Systems: Survey and Experiments. User Modeling and User-Adapted Interaction 12, 331–370 (2002)
21. Perugini, S., Goncalves, M.A., Fox, E.A.: Recommender Systems Research: A Connection-Centric Survey. Journal of Intelligent Information Systems 23(2), 107–143 (2004)
22. Minsky, M.: A framework for representing knowledge. The Psychology of Computer Vision, pp. 211–277. McGraw-Hill, New York (1975)
23. Itzkovich, I., Hawkes, L.W.: Fuzzy extension of inheritance hierarchies. Fuzzy Sets and Systems 62, 143–153 (1994)
24. Rongen, P.H.H., Schroder, J., Dignum, F.P.M., Moorman, J.: A Multi Agent Approach to Interest Profiling of Users. In: Proc. of Multi-Agent Systems and Applications IV, pp. 326–355 (2005)
25. Kang, H., Yang, H.: The Effect of Anonymity on the Usage of Avatar: Comparison of Internet Relay Chat and Instant Messenger. In: Proc. of the Tenth Americas Conference on Information Systems, New York, August 2004, pp. 2734–2743 (2004)
26. Nowak, K.L., Rauh, C.: The Influence of the Avatar on Online Perceptions of Anthropomorphism, Androgyny, Credibility, Homophily, and Attraction. Journal of Computer-Mediated Communication 11(1), 153–178 (2005)

The Concept of Interactive Music: The New Standard IEEE P1599 / MX

Denis Baggi[1] and Goffredo Haus[2]

[1] Scuola Universitaria Professionale della Svizzera Italiana
Galleria 2, CH-6928 Manno, Switzerland
denis.baggi@supsi.ch
[2] Dipartimento di Informatica e Comunicazione
Università degli Studi di Milano
Via Comelico, 39/41, I-20135 Milan, Italy
haus@dico.unimi.it

Abstract. Music is much more than passive listening to a binary file. It can easily become a *complete experience*, the *adventure of entering a new world*, of *understanding a narration* and *seeing images*. Music has always incorporated the newest technology of a given epoch, and the marriage with computer science is at least as old as the early attempts by *Mozart* and *Haydn*. For at least forty centuries in all cultures, music has used *symbols* to represent its contents and give hints for its performance, thus this standard is the continuation of this tradition with its use of *human and machine readable symbols using the XML language*. This article illustrates the possibilities offered by the new standard IEEE P1599, locally known as project MX, through a few *applications* meant to show its flexibility and its role as *enabling technology*.

Keywords: music standards, symbolic music, XML, music layers, synchronisation.

1 Introduction

True listening to a piece of music is much more than consuming a binary, MP3 file. Music lovers see it as an *adventure* akin to enter the complex world of a great novel (*War and Peace* comes to mind) with fictional characters that live not unlike real ones. A whole world unfolds like an adventure, with narration, emotions, images, relations, and connections to multimedia events of many kinds. And enthusiasts who want to know more want to investigate the *way that music is built*, its score if it exists, other notations, historical anecdotes, and especially *musicological analysis,* as it has been performed for the last two centuries.

All this is taken for granted because, for at least forty centuries in most cultures, music has been written with *symbols*, to allow *annotation, reproduction* and *instructions to render its expression.* It is only for about a century that in the West a method of *music packaging* has negated all other attributes of music. And this is especially true for MP3 and WAV files, because the LP had at least a jacket with various kinds of indication.

B. Falcidieno et al. (Eds.): SAMT 2007, LNCS 4816, pp. 185–195, 2007.

The new standard IEEE P1599, a project internally known as MX, proposes a *complete vision* of the music experience thanks to the representation of all music events as symbols in XML, and the synchronisation of all different *layers* [1].

The article start with some short historic considerations, proceeds with the description of the standard, and gives examples of some applications that have been proposed and realised to show its power.

2   The Standard IEEE P1599 – MX

2.1   Past and Present Standards for Symbolic Music

A tribute ought to be given to all those efforts that have made this standard possible. The idea of representing music with symbols is not new and, for computer applications, has gone on for several decades, as shown by the *Plaine-And-Easie Code* [2] and *DARMS* [3]. Closer to us, attempts have been made to use the new technology brought about by SGML, a subset of which has been defined for music, namely, SMDL, or *Standard Music Description Language* [4]. It had been well defined, but it failed to attract much attention because of lack of applications.

Presently, there are some de-facto standards using XML, two of which are important for this work. *MusicXML* is a proprietary standard by company Recordare [5] and is used in dozens of existing applications on the market, including popular *Finale*. This standard has been in existence for many years, and new versions are constantly being made available. The *Music Encoding Initiative,* or *MEI* [6], is a project by the Digital Library of the University of Virginia and it has been used in a few projects of music encoding.

2.2   The New Standard IEEE P1599

A brief description will be given here on the *structure* of new standard P1599, since details on its inner working are given elsewhere in this series of articles. Its main features are based on the observation that music contents involve *multimedia information*, such as *text* and *symbolic contents* – such as *scores* and *librettos* – *still images* like score scans and photos, *audio*, and *video* – concert recordings, video clips, an the like. This richness is taken into account thanks to the identification of *six* different levels of music description: *General, Logic, Structural, Notational, Performance,* and *Audio layers,* as shown in Figure 1, each of which represents one aspect of the music. Briefly, the *General Layer* describes *catalogue information* about the piece, the *Logic Layer* is about *music semantics,* the *Structural Layer* about *music objects,* the *Notational Layer* is about *graphical representations of the score*, the *Performance Layer* about *execution* of music, and, finally, the *Audio Layer* deals with recordings of the piece. All of them, as stated, are described in greater details elsewhere. The *spine* is a fundamental structure to synchronise all layers, as shown in the examples of below.

The existence, or shape, of layers depends on the type of music: a *score* would make no sense in e.g. African drum music, while a *harmonic grid* is senseless in

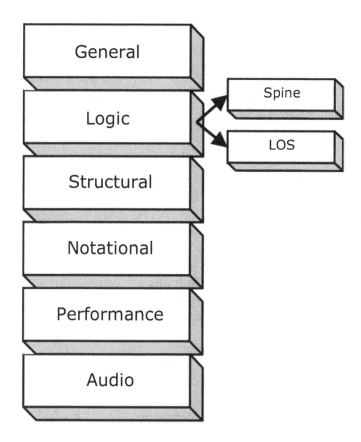

Fig. 1. The layers defined of P1599

Gregorian chant or Indian music. Thus, not all layers need be present. The spine exhibits links among different representations, allowing navigation among layers. Applications could, in theory, reconstruct some layers from others (e.g., in some cases, build a score from the audio).

In the hierarchical XML structure, each layer is represented in a sub-tree of the global representation. The use of synchronised *layers*, and the total independence from the standard to reproduce audio or video, opens up new possibilities for P1599, for which the only limitation is fantasy, as shown in the next paragraphs.

3 Proposals for Applications with P1599 to Significantly Increase Music Enjoyment

These are examples of applications of the standard, which could easily be only realised thanks to the *symbolic expression of music* and the presence of *synchronised layers*, this opening up a new way for music fruition.

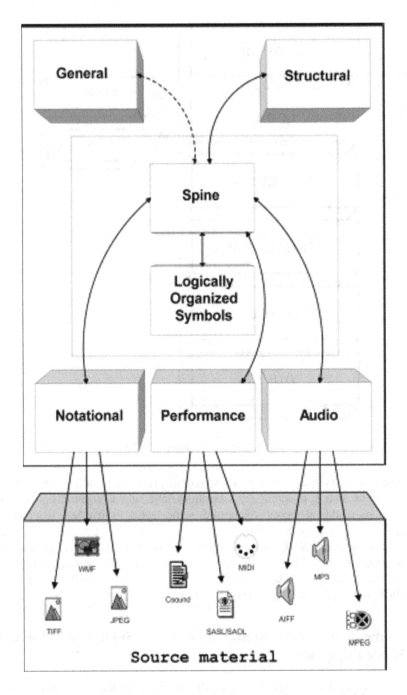

Fig. 2. Relationships among layers and links towards source material

1. **A piece by a jazz Big Band.** The *harmonic grid* is displayed and the *name of the soloist* pops up at the beginning of each solo.
2. **An opera.** A DVD of an opera allows the user to: *see the play* on the screen, *hear the music, see the score* (including *manuscripts*); *read the libretto*; choose *excerpts of alternative renditions.*
3. **Music with a "program" or story.** E.g., Vivaldi's *Four Seasons* come with poems that refer to segments of the music.
4. **Music with no apparent meaning.** For instance, free jazz of the 60's -70's is perceived by many as a random collection of meaningless sounds, while an associated video, generated anew each time, may help understand what is meant.
5. **A fugue.** The *theme* is highlighted as it gets passed among the different voices.
6. **A piece of Indian classical music.** The *scale* of the raga is shown and the melodic development is highlighted.
7. **A piece of several drums**, as in *African Drumming*, to show how the hits do not fall together.
8. **Preservation of the music heritage from the past.** To store documents in any media [7].
9. **Musicological study.** Ease of queries – for example, all pieces utilising the lowest note of a grand piano; questions as to why a certain note is used in a given harmonic context.

4 Examples of Realised Applications Using P1599

The following examples are meant to demonstrate the new features of P1599:

1. *King Porter Stomp*, piano music with score, audio and video, curiosity pictures
2. *Crazy Rhythm*, jazz jam session, Paris, with harmonic grid and different solos
3. *Tosca*, opera with the original manuscript by Puccini, audio and video, original pictures
4. *La Caccia*, from *L'Autunno, Quattro stagioni*, music with a program and correlation between the music and its descriptive sonnet
5. *Proposal to illustrate a piece of Free Jazz.*

The scre enshot will be shown and a brief description will be given for each example.

4.1 King Porter Stomp

This is an application built, like all others, at the *Laboratory for Musical Informatics* of the University of Milan, Italy. The screenshot of Figure 3 contains different windows, of which those with the extra caption *real time* operate in synchronism while the music is being played.

The user starts with the *piece selection* window: in this case, there are two choices, *King Porter Stomp 1924,* and *King Porter Stomp 1939*. They refer to two published scores of that piece, once famous, by American composer and pianist *Jelly Roll Morton*, or *Ferdinand Joseph La Motte*, 1889-1941.

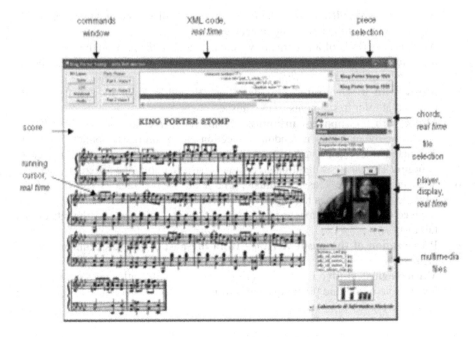

Fig. 3. The screenshot for the *King Porter Stomp* music browser

In the *file selection* window, the user can choose among alternate multimedia files, in this case a recording from 1926 in MP3 format, a MIDI rendition of the 1924 score encoded in MP3, and an excerpt from Louis Malle's movie *Pretty Baby* of 1977 – a character patterned after Morton is heard composing the piece in the background. The latter, i.e. the movie, is the one shown here in the window *player, display*, which for plain music shows instead a common music player.

Upon that selection, several synchronised activities start and execute in real time.The music starts playing, in this case the movie segment starts, with its sound. On the *score*, the *running cursor* indicates what is being played, here the beginning of the 7th bar. The user can move the red cursor with the mouse and initiate playing from another point in the score while the other real time windows adjust synchronously, and of course move the player cursor.

The *XML code* window shows the encoded events, in this case those of the LOS, Logical Organized Symbols of Figure 1, scrolling with the music. In the *command window*, the user can select which XML code is displayed: *spine, LOS, notation* and *audio*, again those of Figure 1, and in the same window he can choose the *voice* the running cursor will follow – there are three voices in this case.

The *chords* window displays the elements of the music harmony of the piece, again synchronously with the playing, and the window for the *multimedia files* allows selection of pictures or other, portraits of Morton, of his band, and curiosities, including a map of Storyville at the turn of the 20th century, destroyed in the 1930's.

4.2 Crazy Rhythm

Figure 4 is the screenshot for jazz piece *Crazy Rhythm*. Instead of a score, it displays the *harmonic grid*, pointed to by the running cursor, and lets picture and name of each soloist pop up at the appropriate moment. There are four saxophonists taking solos: André Ekyan on alto and Alix Combelle on tenor, both from France, followed by Afro-Americans Bennie Carter on alto and Coleman Hawkins on tenor, both in Europe at the time, top jazz specialists of their time. The rhythm section consisted of violinist Stéphane Grappelli on the piano, Django Reinhardt on guitar, Eugène d'Hellemmes on bass and Tommy Benford on drums. Though entirely improvised, the recording, made on April 27, 1937, counts among the best of jazz history, and everybody plays extremely well.

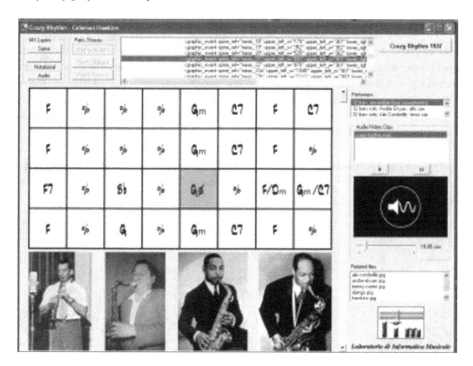

Fig. 4. The browser for *CrazyRhythm*

The ensemble exposes the theme once, in four voices, hence 32 bars. At each solo, image and name of soloist appears: thus, it is possible to compare styles by jumping on the image of another soloist, and even to compare the sound of the alto with that of the tenor saxophone. Bars and grid are of course synchronised. Each soloist takes 32 bars, though Hawkins, in the middle of the development of sentences that keep building up, takes another one, after the encouraging shout by Django "Go on, Go on", which is automatically displayed at the 31st bar. It is details like this that would be totally lost to a casual listener, and which instead constitute the essence of improvised music. The ensemble takes over at the 30th bar of Hawkins's second solo.

In this application, the standard and its browser are meant to teach a would-be jazz expert what to look for: distinguish among soloists, instruments, follow the improvisation at each bar, understand what everybody, from horn to drums passing through the whole rhythm section, is doing. This is a tool to learn jazz appreciation.

4.3 Tosca

Figure 5 is the screen to demonstrate the standard with the opera *Tosca,* with music and original manuscript by Giacomo Pucini. It has been presented at the Exhibition *Tema con variazioni: musica e innovazione tecnologica,* Parco della Musica, Rome, December 2005 – January 2006. The manuscript has been made available by publishing house *Ricordi,* which has the manuscripts of practically all great Italian composers of the last two centuries.

The usual synchronisation of events applies here, as in the other examples. The user can select between the original manuscript and a printed score with the words, also between clarinet and the music sung by the tenor, and among three performers, including a video. There are also various images of the time, of Rome, Castel Sant'Angelo, posters and the like.

It shows how easy it is to gain knowledge of a complex piece like an opera, including libretto and different performances that differ considerably.

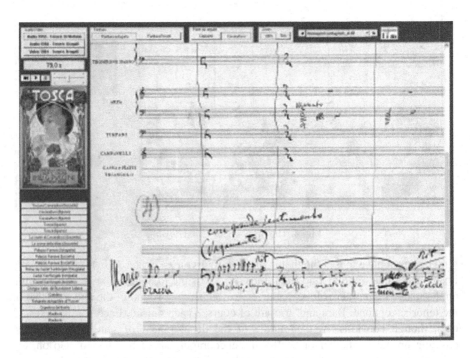

Fig. 5. Application for *Tosca* with the original manuscript by Giacomo Puccini (courtesy of Ricordi)

4.4 La Caccia

La caccia, or *The Hunt,* is the third movement of the concert *L'autunno,* or *Autumn,* from the sequence *Quattro Stagioni,* or *Four Seasons,* by Antonio Vivaldi. This a collection of several scenes about hunting a fox.

The whole music score comes with sonnets, each for each season, which describe what is happening and link to marks in the score. They are rumoured to have been written by Vivaldi himself. For La Caccia, it is possible to identify music excerpts describing the following situation: the hunters ride their horses, they blow their hunting horns, they see the fox, dogs bark and the hunters shoot, the fox is hit but keeps running, the fox is exhausted, the fox finally dies, and the hunters go away with the prey.

Again the browser allows the usual synchronisation, this time with the sonnet. That may seem trivial, but most people are not aware that the whole piece was conceived in this way by the composer.

Fig. 6. Music with a program, *La Caccia* by Antonio Vivaldi

4.5 Piece of Free Jazz

Free jazz is a musical current popular in the 60's and 70's that broke with many traditions. Among others, it rejected tonality, the diatonic scale, the regular tempo and meter, used sounds forced outside of the range of the instrument, and any device that was not part of a learned repertoire. It was, therefore, hard even for seasoned jazz lovers to make sense of such music, which sounded to many like a set or random sounds.

However, experience shows that, given that the public is accustomed to be told the *meaning* of something, true or supposed, often just a simple gesture, image or explanation opens up something akin a to *revelation*. Figure 7 shows a proposal for a "creative" browser that automatically displays varying images to convey the mood of the piece – in this case, the segment *Aum* in the suite containing also *Venus* and *Capricorn Rising* from record *Tauhid,* by Pharoah Sanders, on Impulse, 1969. While the *music*, represented by the central (blue) stripe, scrolls with *musical symbols* over an imaginary landscape, pictures appear that represent *ancient Egypt* – pride of the ancient past heritage claimed by the composer – rising *cosmological events*, *astrological charts*, and the *fallacy of a modern urban landscape*, shaking while it hides social injustice. A generative grammar in XML can be used to that end, and it ensures that the user would approach the system with renewed curiosity, to discover new aspects of the music, since the whole would look different whenever accessed.

Fig. 7. A browser to explore record *Tauhid* by tenor saxophonist Pharoah Sanders, Impulse, 1969, LP and CD

5 Conclusions

There is nothing wrong with listening to binary, closed and unreadable files if one is merely interested in the sound of music. However, *music is more than sound*, and to show all those non-obvious features, one needs *symbolic representation* and *synchronised layers*. Displaying technology is available, in the form of portable players with a display, i-Pods, cellular phones, game players, that would run a multi-layered representation of music, beyond Web browsers and video screens. P1599 is a standard that allows this technology to be used to make music both an enjoyable experience and a *world to navigate*.

Acknowledgement

This work has been endorsed by the global program *Intelligent Manufacturing Systems*, www.ims.org, and financially supported by the Commission for Technological Innovation of the Swiss Federal Government. The IEEE CS Standards Activity Board has been the official sponsor for Standard IEEE P1599.

References

1. Haus, G., Longari, M.: A Multi-Layered Time-Based Music Description Approach based on XML. In: Computer Music Journal, Spring issue (2005)
2. Brook, B.S.: The Plaine and Easie Code. In: Brook, B.S. (ed.) Musicology and the Computer, pp. 53–56. City University of New York Press, New York (1970)
3. Erickson, R.F.: DARMS, Digital Alternate Realization of Musical Symbols. The Darms Project: A Status Report, Computers and the Humanities 9(6), 291–298 (1975)
4. Newcomb, S.R.: Standard Music Description Language complies with hypermedia standard. IEEE COMPUTER, 76–79 (July 1991)
5. http://www.recordare.com/xml.html
6. Roland, P.: The Music Encoding Initiative (MEI) DTD and the Online Chopin Variorum Edition, http://www.lib.virginia.edu/digital/resndev/mei/mei_ocve.pdf
7. Haus, G.: Rescuing La Scala's Audio Archives. IEEE COMPUTER 31(3), 88–89 (1988)
8. Baggi, D.: An IEEE Standard For Symbolic Music. IEEE COMPUTER, 100–102 (November 2005)
9. Baggi, D.: Capire il Jazz, le strutture dello Swing, Surveys of CIMSI/SUPSI, Scuola Universitaria Professionale della Svizzera Italiana, multimedia book with CD-ROM (2001) (in Italian), www.supsi.ch

Outline of the MX Standard

Luca A. Ludovico

LIM – DI*Co* - Università degli Studi di Milano
Via Comelico 39/41 – I-20135 Milano, Italia
luca.ludovico@dico.unimi.it

Abstract. MX is a new XML-based format to describe comprehensively heterogeneous music contents. In a single MX file, music symbols, printed scores, audio tracks, computer-driven performances, catalogue metadata, text and graphic contents related to a single music piece are linked and mutually synchronised within the same framework. Heterogeneous contents are organised in a multilayered structure that supports different encoding formats and a number of digital objects for each layer.

1 A New Standard for Music

MX, an acronym which stands for *Musical application using XML*, is the code-name for a new file format whose international standardisation is in progress. Its development follows the guidelines of IEEE P1599, *Recommended Practice Dealing with Applications and Representations of Symbolic Music Information Using the XML Language*.

MX is the result of research efforts at the *Laboratorio di Informatica Musicale*, or LIM, of the *Università degli Studi di Milano*. P1599, the IEEE Standard, is sponsored by the Computer Society Standards Activity Board and was launched by the Technical Committee on Computer Generated Music (IEEE CS TC on CGM) [1]. In 2002, a prototypal version of the format was released, originally known as *Musical Application using XML*, or MAX [2]. This format was discussed at IEEE *MAX 2002* international conference. The IEEE final evaluation process, known as balloting, is currently being performed, with the aim of making MX/P1599 an international standard.

The most recent version is *MX Release Candidate 1*, whose DTD and documentation can be downloaded from http://www.mx.dico.unimi.it. Tools for *music visualisation* [3], *content-based retrieval* [4], and *automatic segmentation* [5] are currently available.

2 Key Features of MX

MX is an XML-based format. There are many advantages in choosing XML to describe information in general, and music information in particular. For instance, an

B. Falcidieno et al. (Eds.): SAMT 2007, LNCS 4816, pp. 196–199, 2007.

XML-based language allows inherent readability, extensibility and durability. It is open, free, easy to read by humans and computers, and can be edited by common software applications. Moreover, it is strongly structured, it can be extended to support new notations and new music symbols, and it can thus become a means of interchange for music with software applications and over the Net. Most of these topics have been treated in [6] and [7].

A comprehensive description of music must support heterogeneous materials. Thanks to the intrinsic capability of XML to provide structures for information, such representations can be organised in an effective and efficient way. MX employs *six different layers* to represent information, as explained in [8] and shown in Figure 1:

- *General* – music-related metadata, i.e. catalogue information about the piece;
- *Logic* – the logical description of score symbols;
- *Structural* – identification of music objects and their mutual relationships;
- *Notational* – graphical representations of the score;
- *Performance* – computer-based descriptions and executions of music according to performance languages;
- *Audio* – digital or digitised recordings of the piece.

Not all layers must, or can, be present for a given music piece. Of course, the higher their number, the richer the musical description.

Richness has been mentioned in regard to the number of heterogeneous types of media description, namely symbolic, logic, audio, graphic, etc. But the philosophy of MX allows one extra step, namely that each layer can contain many digital instances. For example, the *Audio* layer could link to several audio tracks, and the *Structural* layer could provide many different analyses for the same piece. The concept of multi-layered description – as many different types of descriptions as possible, all correlated and synchronised – together with the concept of multi-instance support – as many different media objects as possible for each layer – provide rich and flexible means for encoding music in all its aspects.

It is possible to adopt some *ad hoc* encoding in addition to already existing formats to represent information. In fact, while a comprehensive format to represent music is not available, popular existing standards must be taken into account. This is a not a contradiction because of the two-sided approach of MX to music representation, which is: keep intrinsic music descriptions inside of the MX file – in XML format – and media objects outside of the MX file – in their original format. The symbols that belong to the score, such as chords and notes, are described in XML, in the *Logic* layer. On the contrary, MP3 files and other audio descriptions are not translated into XML format, rather they are linked and mapped inside the corresponding MX layer, the *Audio* layer.

It should be clear that the description provided by an MX file is flexible and rich, both in regard to the number and to the type of media involved. In fact, thanks to this approach, a single file can contain one or more descriptions of the same music piece in each layer. For example, in the case of an operatic aria, the MX file could house: the catalogue metadata about the piece, its author(s) and genre; the corresponding portion of the libretto; scans of the original manuscript and of a number of printed scores; several audio files containing different performances; related iconographic contents, such as sketches, on-stage photographs, and playbills. Thanks to the

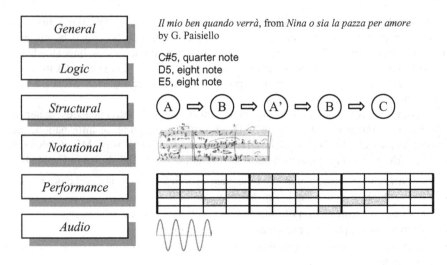

Fig. 1. The characteristic multi-layered structure of MX. In the right part of the figure, intuitive graphical examples are provided to illustrate the purpose of the layers.

comprehensive information provided by MX, software applications based on such a format allow an integrated enjoyment of music in all its aspects.

The *spine*, the second key concept of the MX format, consists of a sorted list of events, where the definition and granularity of events can be chosen by the author of the encoding. The spine provides both an abstraction level and the glue among layers, and represents an abstraction level, as the events identified in it do not have to correspond to score symbols, or audio samples, or anything else. It is the author who can decide, from time to time, what goes under the definition of music event, according to the needs. Since the spine simply lists events to provide a unique label for them, the mere presence of an event in the spine has no semantic meaning. As a consequence, what is listed in the spine structure must have a counterpart in some layer, otherwise the event would not be defined and its presence in the list (and in the MX file) would be absolutely useless. For example, in a piece made of *n* music events, the spine would list *n* entries without defining them from any point of view. If each event has a logic definition – e.g. it is a note or a rest – it can be graphically represented in many scores and rendered in a number of audio tracks. These aspects are treated in the *Logic*, *Notational*, and *Audio* layers respectively.

Music events are not only listed in the spine, but also marked by unique identifiers. These identifiers are referred to by all instances of the corresponding event representations in other layers. Thus, each spine event can be described:

- in 1 to *n* layers; e.g., in the *Logic*, *Notational*, and *Audio* layers;
- in 1 to *n* instances within the same layer; e.g., in three different audio clips mapped in the *Audio* layer;
- in 1 to *n* occurrences within the same instance; e.g., the notes in a song refrain that is performed 4 times (thus the same spine events are mapped 4 times in the *Audio* layer, at different timings).

Thanks to the spine, MX is not a simple container for heterogeneous media descriptions related to a unique music piece. It shows instead that those descriptions can also present a number of references to a common structure. This aspect creates synchronisation among instances within a layer (*intra-layer synchronisation*), and – when applied to a complex file – also synchronisation among contents disposed in many layers (*inter-layer synchronisation*).

3 Conclusions

The format proposed here has been designed to achieve a comprehensive description of music, content interoperability, and deliverability. On one side, this way of encoding music pays special attention to on-line accessibility, digitalisation of analogue material and preservation of artefacts from the past, independently of the cultural origin and language. On the other, the format is meant to provide an integrated and evolved fruition of music, thus representing a new approach to music delivery and enjoyment.

We hope that such an effort will open the way for a large number of new applications with increased power and flexibility, as well as new markets for these kinds of applications. The repercussions may have a wide effect on music education, media entertainment, enjoyment and development of music as a whole.

Specific applications of the MX format to music description and representation will be provided in other articles of this special session.

References

1. Baggi, D.L.: Technical Committee on Computer-Generated Music. Computer 28(11), 91–92 (1995)
2. Haus, G., Longari, M.: MAX 2002. Proceeding of the First International IEEE Conference on Musical Application using XML. IEEE Computer Society, Los Alamitos (2002)
3. Baggi, D.L., Baratè, A., Haus, G., Ludovico, L.A.: A computer tool to enjoy and understand music. In: Proceedings of EWIMT 2005 – Integration of Knowledge, Semantics and Digital Media Technology, pp. 213–217 (2005)
4. Baratè, A., Haus, G., Ludovico, L.A.: An XML-Based Format for Advanced Music Fruition. In: Proceedings of SMC 2006 – Sound and Music Computing 2006 (2006)
5. Haus, G., Ludovico, L.A.: Music Segmentation: An XML-oriented Approach. In: Wiil, U.K. (ed.) CMMR 2004. LNCS, vol. 3310, pp. 330–346. Springer, Heidelberg (2005)
6. Roland, P.: The Music Encoding Initiative (MEI). In: Proceedings of the first IEEE International Conference MAX 2002 – Musical Application using XML, pp. 55–59 (2002)
7. Steyn, J.: Framework for a music markup language. In: MAX 2002. Proceeding of the First International IEEE Conference on Musical Application using XML, pp. 22–29. IEEE Computer Society, Los Alamitos (2002)
8. Haus, G., Longari, M.: A Multi-Layered, Time-Based Music Description Approach Based on XML. Computer Music Journal 29(1), 70–85 (2005)

Automatic Synchronisation Between Audio and Score Musical Description Layers

Antonello D'Aguanno and Giancarlo Vercellesi

Laboratorio di Informatica Musicale (LIM)
Dipartimento di Informatica e Comunicazione (DICo)
Universitá degli Studi di Milano,
Via Comelico 39,
I-20135 Milano, Italy
{daguanno,vercellesi}@dico.unimi.it
http://www.lim.dico.unimi.it

Abstract. This work describes algorithms dedicated to score and audio alignment using the MX / IEEE P1599 format. The format allows description of the score, and management of synchronisation points, linking them with different versions of the performed music. An algorithm is proposed here that allows alignment of an MX score and its execution, coded in PCM format, which produces an output for the MX Spine that contains synchronisation between notes and audio signal. The proposed architecture is based on two different steps: the first deals with the audio level and the extraction of features like pitch, onset and the like, while the second determines the alignment between the features extracted in the first step and the notes present in the MX score. For each step, different algorithms are proposed and discussed, and analysis and comparison of synchronisation capabilities are provided.

Keywords: IEEE P1599, MX, synchronisation, PCM, MIDI.

1 Introduction

Contemporary digital music archives consist of huge collections of heterogeneous documents. For a music piece, an archive may contain corresponding scores in different versions, for example, voice and piano or orchestral, as well as several interpretations, e.g. played by different performers and recorded in diverse formats (CD recordings, MP3, FLAC and so on). The heterogeneity of music information makes retrieval hard to accomplish [1] and as a consequence many problems remain unsolved. One important problem that needs a solution is synchronisation, which requires the implementation of algorithms that automatically link different audio streams of the same piece to symbolic data formats representing the different scores.

In particular, for the Notational and Audio layers in MX, synchronisation means that some algorithms, for a given event in some representation of the music score, are capable of determining the timing of the corresponding audio

B. Falcidieno et al. (Eds.): SAMT 2007, LNCS 4816, pp. 200–210, 2007.

events within an audio representation in some format. After execution of a synchronisation process in MX it is possible, for a given audio event, to find the correct position in the score, in the *Notational* and the *Audio* layers.

As stated in [2]: " 'Such synchronisation algorithms have applications in many different scenarios: following some score-based music retrieval, linking structures can be used to access some suitable audio CD accurately to listen to the desired part of the interpretation." 'This possibility represents a useful tool for music students, who can listen to music audio and, at the same time, see the corresponding notes. The MX spine and MX Music Events are the linking structure described in [2]. Furthermore, the MX linking structure can be useful for musicologists who can use synchronising algorithms to link the interpretation layer and the score layer, and then use MX for the investigation of agogics and tempo. In addition, temporal linking of score and audio data can be useful for automatic tracking of score positions during a performance.

This article is organised as follows. Section 2 presents an overview of score following and synchronisation algorithms proposed in literature. Section 3 describes the ComSi algorithm. Section 4 describes the power of MX for representing audio synchronisation results. Section 5 provides the analysis of results and, finally, conclusions will be summarised in 6.

2   Related Works

Many algorithms have been proposed in literature that deal with synchronisation. The majority of them can be subdivided into two groups: in the first one, audio and score have to be analysed, while the second one requires the realisation of correct links between these two layers ([3][2][4][5][6][7][8]). The algorithms proposed in the literature use several different systems to implement audio analysis, with well-known tools from audio signal processing. For example, [7] uses a Short Time Fourier Transform, in [2] proposes an onset detection followed by pitch detection. In [6] the feature extraction procedure performs these operations: decomposition of the audio signal into spectral bands corresponding to the fundamental pitches and harmonics, followed by the computation of the positions of significant energy increases for each band - such positions are candidates for note onsets. The most popular solution proposed in the literature to select the correct links between audio and score is based on the *template matching technique* ([7][3][8]). Such algorithms build a MIDI score to obtain a template of the real execution, which is compared to the real audio using a DTW[1] programming technique [9]. The correct synchronisation is then obtained from the difference between the agogics of the real execution and that of MIDI. The algorithm described here uses a different approach based on recursive research, as explained in see section 3.3).

Actually, the unsolved problems of synchronisation are not limited to finding a suitable algorithm. There is at least one more open question: once the correct

---

[1] Dynamic Time Warping is a technique for aligning time series that has been well known in the speech recognition community since the 1970s.

synchronisation has been obtained, how can this result generalised to use it in other applications? The answers proposed in the literature are trivial and not sufficient to allow interoperability or future utilisation. For example, the last MIREX[2] task, focused on synchronisation, used a text file to bridge audio and score, by letting every line in its text file connect a MIDI event with an audio instant. While this solution meets experimental needs, it is not suitable for commercial applications. The algorithm proposed in section 3 uses the features of MX to solve this problem (See section 4).

3   The COMSI Algorithm

This algorithm proposed here is able to synchronise a MX music score with one or more PCM (Pulse Code Modulation) audio executions. It is recursive, and consists of three different phases (figure 1):

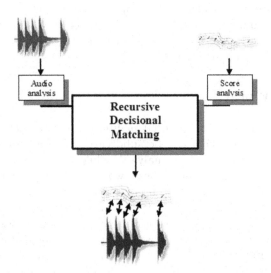

Fig. 1. The general overview of the COMSI algorithm

- in the first phase, the MX score is read in order to extract all relevant musical events. This algorithm is also able to read MIDI files.
- in the second phase, the PCM audio signal is analysed to identify all possible musical notes
- in the third and last phase, a decisional matching is performed to put relate the event at the score level with the same event at the audio level. The decisional method is based on the attack time of notes extracted at audio level.

Each phase is described in the following sections.

---

[2] Music Information Retrieval Evaluation eXchange is a community-based endeavor which involves many researchers active in MIR field.

3.1   The MX Score Analysis

In this phase, the score is read to extract the musical events that are related to the audio signal by means of decisional matching. For each measure, only the notes with strong accent are selected, because it is easier to recognise them at audio level, when the audio signal is analysed. For every strong accent, the verticalisation of the score at that point is computed. An example is shown in the table below (1) where fundamental frequencies are referred to notes in the score (i.e. 440 Hz corresponds to A):

Table 1. This table contains the time described as measures and events and the fundamental frequencies related to the note known at that time. The algorithm is able to perform these operations on MIDI scores as well.

Measures	Events	Fundamental Frequencies (Hz)
1	1	1047, 311, 415, 523, 1109, 988
1	2	784, 1175, 1568, 294, 440, 466, 698, 1397
2	1	523, 156, 208, 262, 554, 494
2	2	392, 587, 784, 147, 220, 233, 349, 698

3.2   The PCM Audio Analysis

A recursive, second-order notch filter bank is used to perform audio signal analysis. The aim of this phase is the detection of each attack-time note represented at the symbolic level in the score. For each verticalised chord extracted from the score (see table (1)), the audio signal is filtered with a notch filter centred on the frequency of the note examined, as shown in figure 2.

Consider note A at 440Hz. While (a) shows its waveform signal, this signal is filtered with the notch filter with center frequency 440Hz (b), which yields a filtered signal with high energy. Note that if the signal is filtered at different centre frequencies, the signal will contain less energy: see the cases for centre frequency at 220Hz (c) and 880Hz (d).

The bandwidth of the notch filter is 10 Hz. This value has been selected because it covers the whole frequency bandwidth, starting from the MI at 82,406889 Hz (first octave) and avoiding superposition between adjacent semitones (figure 3).

The audio signal in time-windows is split at 100ms. Thanks to the notch filter bank, the energy of each time-window can be computed. For each note, a possible attack-time is considered to be the audio segment which has an energy value above a convenient relative threshold, obtained from the average energy of the filtered signal.

When two or more notes are present in a single musical event, the energies of each single note that composes the analysed musical event are multiplied together. This allows the creation of a pseudo-score that contains, for each note extracted at the score level, the attack-time hypothesis at the audio level. This attack-time hypothesis gets defined as the execution event.

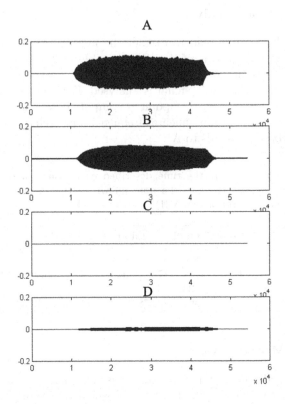

Fig. 2. An example of a 440Hz signal (a) filtered with a notch filter having its centre frequency at 440Hz (b), 220Hz (c) and 880Hz (d)

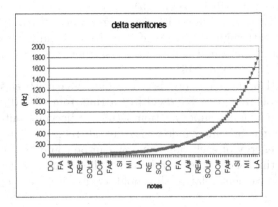

Fig. 3. Differences between two adjacent semitones

3.3   The Decisional Matching

In this phase, the musical event at score level is related to the same one found at audio level. In other words, the attack-time note for each note contained in the score has to be found. It is based on (figure 4):

– the expected attack-time note,
– the relative position into the symbolic score
– the duration of audio signal

  This decisional matching is performed in two steps:

1. recursive research of attack-time related to the musical event that corresponds to the first event of each measure in the score. The average duration of each measure is computed as the ratio between the whole duration of the audio signal and the number of measures at score level. Hence, for every event, the expected attack-time is computed and the nearest execution event at audio level is examined. The adjacent measure average duration (right and left intervals) is computed with respect to the expected attack-time. This attack-time is considered valid if the differences between the average time of the left and the right intervals are less than 30% of the expected average duration. Then, all the steps listed above are applied recursively to detect all attack-times.
2. Sequential research of each musical event contained in a measure. The algorithm is similar to the previous one, but in this case the algorithm is sequential. The average duration of quarter notes is examined, with a tolerance of 30% of every average duration, further reduced to 20%.

4   The Linking Structure in MX

In this section, the MX capabilities to represent the synchronisation anchors between audio and score are described (more information about MX can be found in [10][11][12]). In MX, synchronization anchors are saved in three different layers: *Audio, Notational, Logic*. The track_event tag in the *Audio* layer has two attributes: event_ref, which contains the identifier for the current event, and start_time, which contains the related audio time. In this example this means that in the audio file, event named vno1_001_01 is present after 30sec.

  Similarly, in the *Notational* layer, graphic_event tag is present and has five different attributes: event_ref the name of the event we considered and the other four attributes which represent the position of the event in a graphic image. The measurement unit of the position is specified by the measurement_unit tag (expressed in pixels in this example).

  The origin of the axis is in the upper left corner. Every event is represented by a bounding box, and in the Notational layer the co-ordinates of the upper-left corner and lower-right are annotated. Looking further at this example, it appears that the upper-left corner of the event *vno100101* is positioned 529 pixels to the

Fig. 4. The schema of the decisional matching phase

right and 109 pixels below the axis origin, while the lower right corner is 555 pixels to the right and 172 pixels below the axis origin. The element spine in the *Logic* layer links graphic event and track event, which have the same *eventref* attribute. Attributes timings and *hpos* have the following meaning:

– `timing` represents the temporal co-ordinate of the event. It describes the time offset from the occurrence of the preceding event. Time is expressed in *Virtual Time Units* (VTU). The measurement unit can be defined according to user needs, but all the numeric values must be integer. Since 0 means simultaneity in this relative system, each event is simultaneous to all those with 0 timing. Besides, if the event itself presents has a timing equal to 0, it is also simultaneous to all the preceding events up to the first event with a non-zero value, that event included.

– `hpos` refers to a virtual horizontal dimension. The value is expressed in a relative spatial unit called *VirtualPiXels* (VPX), which provides reference points for the vertical alignment of different symbols in different voices, parts or staves. The score is viewed in this case from an abstract vantage point (real scores are treated in the Notational layer). As a consequence, staff systems never generate new lines: there is only one staff system, including all symbols on a single line, so that only the horizontal axis is used in establishing vertical alignment.

Example of the MX potential to represent the results of the synchronisation algorithm - AudioLayer

```
<audio>
  <track file_name="nina.wav" encoding_format="audio_wav"
```

Fig. 5. The role of Spine in MX. In this image a score is shown with the related MIDI sequence and the performance file. The Spine links them all.

```
                              file_format="audio_wav">
     <track_indexing timing_type="seconds">
     ...
     <track_event start_time="0.30" event_ref="vno1_001_01"/>
     <track_event start_time="0.30" event_ref="vno1_001_02"/>
     ...
     <track_event start_time="12.00" event_ref="vno1_15_03"/>
     <track_event start_time="12.00" event_ref="vno2_15_03"/>
     ...
     </track_indexing>
   </track>
</audio>
```

Example of the MX potential to represent the results of the synchronisation algorithm - Notational Layer

```
<notational>
  <graphic_instance_group description="Manuscript">
    <graphic_instance file_name="M2_0004.jpg" file_format="image_jpeg"
                      encoding_format="image_jpeg" position_in_group="1"
                                       measurement_unit="pixels">
      ...
      <graphic_event event_ref="vno1_001_01" upper_left_x="529"
           upper_left_y="109" lower_right_x="555" lower_right_y="172"/>
      <graphic_event event_ref="vno1_001_02" upper_left_x="557"
           upper_left_y="105" lower_right_x="580" lower_right_y="165"/>
      ...
      <graphic_event event_ref="vno1_15_03" upper_left_x="627"
           upper_left_y="693" lower_right_x="656" lower_right_y="719"/>
      <graphic_event event_ref="vno2_15_03" upper_left_x="846"
           upper_left_y="689" lower_right_x="883" lower_right_y="713"/>
      ...
    </graphic_instance>
  </graphic_instance_group>
</notational>
```

Example of the MX potential to represent the results of the synchronisation algorithm - Spine

```
<logic>
  <spine>
    ...
    <event id="vno1_001_01" timing="12" hpos="12"/>
    <event id="vno1_001_02" timing="0" hpos="0"/>
    ...
    <event id="vno1_15_03" timing="48" hpos="48"/>
    <event id="vno2_15_03" timing="0" hpos="0"/>
    ...
  </spine>
</logic>
```

5   Analysis of Results

Different tests have been performed on the ComSi algorithm in regard to synchronisation of several polyphonic music tracks and MX scores. In these tests, scores have been considered with different time signatures. Synchronisation is considered correct when the ComSi-time of the analysed musical event lies between the real-time of the previous musical event and the next one, as in the formula:

$$previous_{real-time event} < ComSi_{synchtime} < next_{real-time event} \qquad (1)$$

Only the error related to the first event of each measure are reported. This approach has been chosen to simplify the analysis of the results.

The table 2 shows the results: the error columns report the percentage of measure for which the first event of the measure had been confused with the first events of another measure. This two error columns refer to a measure which has a distance of |1| or |2|, respectively from the considered measure. Presently, a new test infrastructure is being developed which will use the MX capabilities to host many different synchronisation methods between the Audio and the Notational layer. This new system will allow intensive tests for many different synchronisation algorithms and will take into account every musical event.

Table 2. Results of ComSi test

Track	Time Signature	Correct Measure	One Error Measure Event	Twice Error Measure Event
Chopin	3/4	66%	24%	10%
Beethoven	4/4	71%	18%	11%
Beethoven (faster)	4/4	69%	16%	15%
Fiore di Maggio	4/4	70%	17%	13%
Cavatina	6/8	43%	48%	9%
Aria	6/8	61%	31%	8%

6   Conclusions

In this paper we have described algorithms dedicated to score and audio alignment using the MX / IEEE P1599 format. This algorithm allows alignment of an MX score and its execution, coded in PCM format, and produces an output for the MX Spine that contains synchronisation between notes and audio signal. The analysis of results 5 show the reliability of our algorithm. Furthermore, the MX standard will allow intensive tests for many different synchronisation algorithms and will take into account every musical event.

References

1. Stephen, J.: Music information retrieval. In: Annual Review of Information Science and technology, ch. 7, Blaise Cronin, Medford, NJ, USA, vol. 37, pp. 295–340 (2003)
2. Arifi, V., Clausen, M., Kurth, F., Muller, M.: Automatic synchronization of music data in score-, midi-and pcm-format. In: ISMIR. Proc. of Intl. Symp. on Music Info. Retrieval (2003)
3. Dannenberg, R.B., Hu, N.: Polyphonic audio matching for score following and intelligent audio editors. In: Proceedings of the 2003 International Computer Music Conference, Singapore, October 2003, pp. 27–33 (2003)

4. Clausen, M., Kurth, F., Müller, M., Arifi, V.: Automatic Synchronization of Musical Data: A Mathematical Approach. MIT Press, Cambridge (2004)
5. Dixon, S., Widmer, G.: Match: A music alignement tool chest. In: 6th International Conference on Music Information Retrieval (2005)
6. Müller, M., Kurth, F., Röder, T.: Towards an efficient algorithm for automatic score-to-audio synchronization. In: ISMIR 2004. 5th International Conference on Music Information Retrieval, Barcellona, Spain (October 2004)
7. Soulez, F., Rodet, X., Schwarz, D.: Improving polyphonic and poly-instrumental music to score alignment. In: 4th International Conference on Music Information Retrieval, pp. 143–148 (2003)
8. Turetsky, R.J., Ellis, D.: Ground-truth transcriptions of real music from force-aligned midi syntheses. In: Proc. Int. Conf. on Music Info. Retrieval ISMIR, vol. 3 (2003)
9. Rabiner, L.R., Juang, B.H.: Fundamentals of Speech Recognition. Prentice, Englewood Cliffs, NJ (1993)
10. Haus, G., Longari, M.: A multi-layered, timebased music description approach based on xml. Computer Music Journal 29(1), 70–85 (2005)
11. Haus, G., Longari, M.: Towards a symbolic/time-based music language based on xml. In: MAX 2002. Proceedings of the First International IEEE Conference on Musical Applications Using XML (2002)
12. Baraté, A., Ludovico, L.A.: An xml-based synchronization of audio and graphical representations of music scores. In: WIAMIS 2007. Proceedings of the 8th International Workshop on Image Analysis for Multimedia Interactive Services, Santorini, Greece (2007)

Interacting at the Symbolic and Structural Levels

Adriano Baratè

Laboratorio di Informatica Musicale (LIM),
Dipartimento di Informatica e Comunicazione (DiCO),
Università degli Studi di Milano,
via Comelico, 39 - 20135 Milano, Italy
barate@dico.unimi.it

Abstract. This article presents an approach to music interaction and creativity at the *symbolic* and *structural levels*. While common music interaction occurs at the symbolic level, on a sequence of notes, it is the structural level, the focus of this writing, that allows interaction with music contents at a higher degree. This requires identification and creation of *fragments of music* in existing and new compositions, then the resulting set of music themes can be treated with Petri Nets, even in real time. Algorithms can be applied to modify the music synthesised or played, and, thanks to MX, this interaction is performed simultaneously at different levels. After an introduction of the main concepts, the article describes a case study based on "Peaches en Regalia", by Frank Zappa, presented at the *International Conference of Esemplastic Zappology, ice-Z*, which took place in Rome on 9-10 June 2006. Finally, advanced MX scenarios will be presented.

Keywords: Music, MX, Petri Nets, IEEE P1599.

1 Introduction

Interaction between *Petri Nets* and music can be investigated in two respects, namely, the *analytical* and the *compositional*. The first approach uses the formalism to achieve a description of an existing music piece, to let the music structures emerge. The latter has the purpose of creating new pieces according to procedures defined by Petri Nets. In brief, one approach uses Petri Nets as a description tool, while the other as a generative instrument.

In other articles, results of each use of Music Petri Nets have been presented, demonstrating in many cases their applicability both as a description and as a compositional tool.

In this article, the focus is on real-time interaction with Petri Nets that model compositions, using them as tools to create new experiences for the listener. In this perspective – by modifying parameters of Petri Nets on-the-fly – the composer can create different versions of the piece for music fruition, by working at a higher level of interaction, i.e. with music structures, and not with single notes or audio samples.

B. Falcidieno et al. (Eds.): SAMT 2007, LNCS 4816, pp. 211–221, 2007.

2 An Overview of Petri Nets

A *Petri Net* is an abstract and formal model to represent the dynamic behaviour of a system with asynchronous and concurrent activities ([9], [10], [11]).

Such a net can be represented as a bipartite graph in which nodes are called *places* and *transitions* – graphically represented, respectively, with circles and rectangles. Oriented *arcs* connect only nodes of different kinds, i.e. *places to transitions* and vice versa.

In Figure 1, an example of an elementary Petri Net is shown. P1, P2, P3, P4 are places, T1, T2, T3 are transitions, and the oriented lines represent arcs. The number sometimes associated to arcs is called *arc weight*, the meaning of which is explained below.

The key concept that lets Petri Nets evolve and self-modify is the idea of *marking*, realised by using *tokens*: at a given time, every place holds a non-negative number of tokens, indicated by the upper numerical value inside the circle. The lower value indicates the *capacity* of the place, i.e. the maximum number of tokens that can be housed.

Tokens can be transferred from place to place according to policies known as *firing rules*. The dynamic evolution of a Petri Net is determined by the following rules:

- A *transition* is enabled when all the incoming places of that transition present a number of tokens greater or equal to the weights of the corresponding incoming arcs, and – after the fire of the transition – the marking of all the output places will be less than or equal to their capacities
- When a transition is enabled, its firing subtracts from the incoming places a number of tokens equal to the weights of the incoming arcs, and adds to each outgoing place a number of tokens equal to the weights of the corresponding outgoing arc.

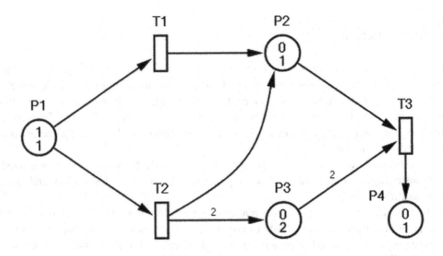

Fig. 1. Example of Petri Net

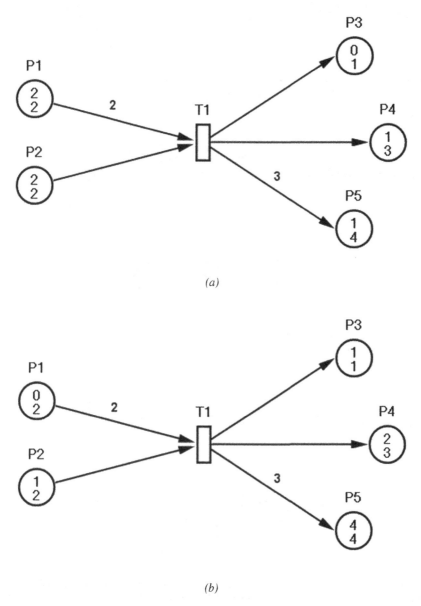

Fig. 2. Firing rules in a Petri Net: in (a) transition T1 is enabled to fire; in (b) the final situation after the fire

Finally, note that Petri Nets can be described using the concept of *refinement*, a simple morphism that describes complex Petri Nets in terms of simpler ones, as shown in Figure 3.

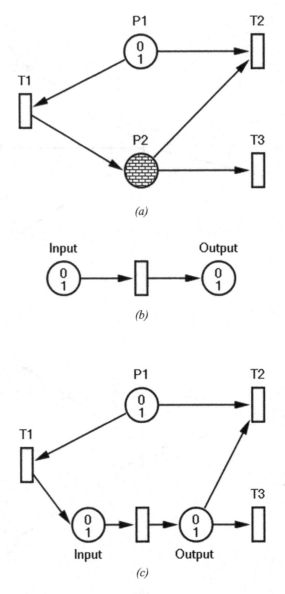

Fig. 3. Refinements and subnets: in *(a)* the original net, in *(b)* the subnet, and in *(c)* the expanded net

2.1 Petri Nets and Music

Petri Nets have been applied to various fields of information management and description. For music applications, a specific extension known as Music Petri Nets is available. In this implementation, music objects can be associated to places, and music operators to transitions. A *music object* may be anything that could have a musical meaning and could be thought as an entity, either simple or complex, abstract or detailed. A *music operator* provides a method to transform music objects, i.e. by transposing, time stretching, and generally by applying transformational algorithms. In the following figure, two simple examples of music objects are presented.

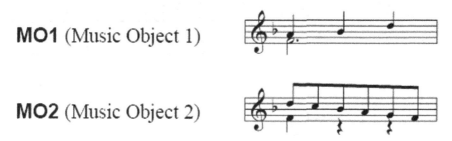

Fig. 4. Two examples of music objects

In this approach, when a place receives a token, the associated music object, if present, is played, and the token will remain locked for the whole duration of the fragment. When a transition fires, its music operator changes the music objects associated with the incoming places and transfers them to the outgoing places. Transitions without music operators are used for mere net evolution.

Fig. 5. A sequence structure

Figure 6 provides some examples of music structures, in order to illustrate respectively a *fusion* (two objects generating one object), a *split* (an object generating two objects), an *alternative* (a non-deterministic choice between two objects), and a *joint structure* (a logical connection between two objects).

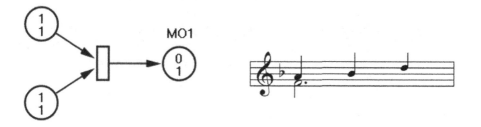

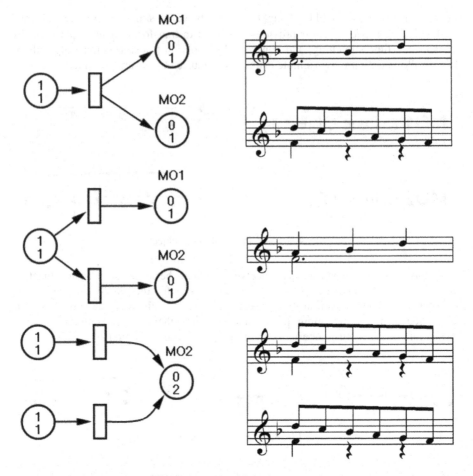

Fig. 6. Examples of music structures

At LIM,[1] Petri Nets have been applied to music since 1982. In particular, early papers such as [5] investigated the possibility of describing causality in music processes through the formal approach of Petri Nets, while latter studies focus on music creation.

Apparently, different applications of Petri Nets to music analysis do lead to contradictory results. Even though Ravel's *Bolero* has been described as in [8], some limitations of this approach have become evident when attempts were made to model a complex work, such as Stravinsky's *Rite of Spring* [6].

Two factors clearly influence the analytical power of the Petri Nets formalism:

1. The intrinsic characteristics of the piece to be described. A Petri Net model is suitable for music works that consist of few music objects, or with a very simple

---

[1] Laboratorio di Informatica Musicale, Università degli Studi di Milano.

structure. For instance, the music form known as *canon*, based on the literal repetition of the same music objects in different voices at different instants, can be represented in a very efficient and compact way with Petri Nets. But such characteristics are not peculiar of Baroque music or dead forms: for instance, Ravel's *Bolero* is characterised by an intentionally simple and repetitive scheme. Of course, on the other hand, it would be very difficult to describe a romantic piece or a jazz improvisation with Petri Nets.

2. The restriction of investigation to those music objects which prove to be efficient from the point of view of Petri Nets. In fact, Petri Nets is a description tool which is more or less effective depending on how clearly the identification of music objects, and simple relationships among them, can be expressed. Here, the concept of music object is deliberately vague and can include whole episodes of a music work, as well as atomic musical events. For this context, *segmentation* will be defined as the activity of isolating music objects and discovering their mutual relationships.

Another important parameter to determine the adequacy of Petri Nets as description tool is the degree of detail the analysis wants to reach. This statement appears to justify the contradictory results obtained when considering different music pieces. The test case presented at CMMR 2005 [4] has taken into consideration the first movement of a sonata by W.A. Mozart. Intentionally, this case study represents an intermediate situation between a strongly structured piece and a completely unstructured one. At the same time, its score can not be described as a sequence of repetitive music objects (like a canon or a fugue); however musicologists agree about the existence of characteristic grouping structures (principal theme, secondary theme, middle themes) and about the macroscopic segmentation in the sonata (exposition, development, repeat, and coda). Thus, for the test case discussed in [4], Petri Nets have proved to be:

- a very synthetic instrument to describe the macro-structure of sonata form, but limited to a very small amount of information.
- an efficient description instrument at a medium degree of abstraction (repetition of rhythmic and melodic patterns, transformation of themes in order to generate new music material).
- a completely inadequate tool to depict the status and the evolution of atomic music elements (chords, notes, rests).

3 Petri Nets and Interaction

After the investigation of aspects of music analysis with Petri Nets, the use of formalism to interact with music in real-time will be described here.

If a Petri Net is used as a model for the structure of a music piece, its simple graphic representation is sufficient to understand it. With some practice, a music analyst is able to recognise, in a Music Petri Net, repetitions of themes, loops, transformations, and so on.

Thanks to the MX format, we at LIM have implemented an evolution of this formalism, in which music objects – which can be, as stated, associated to places – are expressed in its XML dialect. This means that a Petri Net can contain places with associated MX fragments that express only the *Logic* part, or have a linked MIDI file, or have a linked video clip. When such a model is executed, the *mixing process* involves all MX fragments, even those with their associated various kinds of music representations, thus creating a global MX file as output.

The adoption of the MX format allows the composer of Music Petri Nets to concentrate on the structure of the music piece he wants to obtain, without minding lower level material involved in the mixing process, such as. the file formats of the linked objects. The final result automatically generates synchronisation of various kinds of music representation, permitting a new type of musical experience [1] [2] [3].

3.1 Real-Time Interaction

A real-time application of a Petri Net can be realised thanks to modifications of some peculiarities of the net at execution time. To this end, there are two possibilities: modifications of the net structure (dynamic addition of places, transitions and arcs), and modifications of the net parameters (number of tokens, arc weights). Since the first approach can be viewed as an evolution of the second, a workflow divided in two phases has been considered as a first simple step, thanks to:

- definition of a fixed structure of the music piece, and its subsequent implementation with a Petri Net model.
- interaction with the produced model in terms of addition-subtraction of tokens present in places, during the net execution.

With this approach in mind, the composer has two possibilities:

- by performing both steps, he creates several different compositions with the execution of its model, by modifying the net parameters during the execution.
- by performing only the first step, he creates a dynamically changing composition, for which only its main structure is stable and defined, and allows performers (or co-composers) to fix the precise form of the piece during a specific execution of the net.

Thus, creation of the Petri Net can be accomplished by analysing an existent music piece, and a performer has the possibility of interacting with such a model.

4 A Case Study: "Peaches and Regalia" by Frank Zappa

Using the concepts described above, an application based on "Peaches en Regalia" by Frank Zappa was presented at the ice-Z, International Conference of Esemplastic Zappology, held in Rome on 9-10 June 2006.

Figure 7 presents the main music objects identified by a very simple analysis of the music piece and implemented in the Petri Net model. After a short drum introduction, fragment *(a)* is repeated four times, then fragment *(b)* is repeated two times, and finally fragment *(c)* leads to a series of improvised solos.

Fig. 7. Music objects used in the Petri Nets model of "Peaches and Regalia"

The Petri Net shown in Figure 8 models the original version of the music piece as recorded by Frank Zappa. This means that, if the net is executed without changing the number of tokens in real-time, the original version is reproduced.

In this Petri Nets model, interaction is limited to addition/subtraction of tokens in the places represented with a white background, in order to maintain a similarity to the original result. Addition and subtraction of tokens has the following results:

- the number of tokens in the **Start** place controls how many times the entire music piece is repeated.
- the number of tokens in **Repeat A** and **Repeat B** places controls how many times fragments *(a)* and *(b)* are repeated.
- the presence of one or more tokens in the white input places of the two transitions **Transpose +** and **Transpose -** causes the transposition of fragment *(a)* by a major third higher or lower respectively. From Petri Nets execution rules, it derives that if both this places have more than zero tokens, a transition is chosen at random.

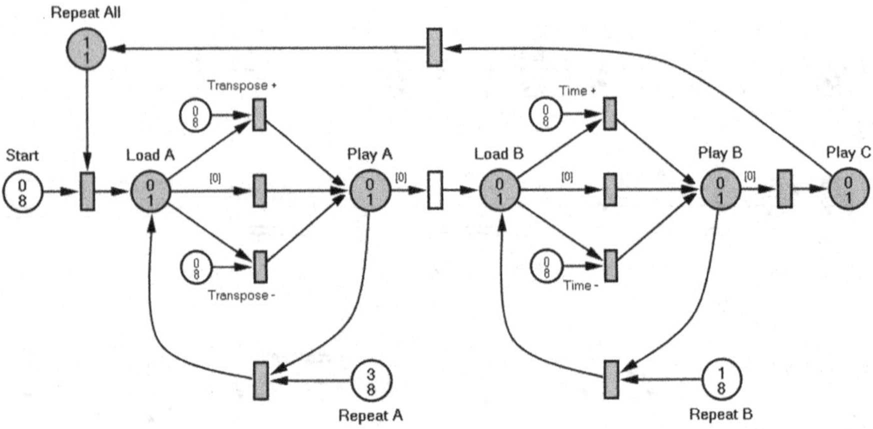

Fig. 8. The Petri Nets model of "Peaches and Regalia"

- the presence of one or more tokens in the white input places of the two transitions **Time +** and **Time -** causes time shrinking of fragment *(b)* to double or half theoriginal time respectively. From Petri Nets execution rules, it clearly derives that if both this places have more than zero tokens, a transition is selected at random.

Even though this model is very simple, the application demonstrates how an original music structure can be formalised, and how a performer can create different versions of the piece every time the net is executed.

Every MX fragment associated to places **Start** (the drum introduction), **Load A**, **Load B** and **Play C** has, aside from the *Logic* layer, links to a graphical representation of the *score* and to an *audio* clip of the fragment. Because of these links, during the execution of the net, the application plays the audio of that current fragment and displays on the screen the score being played, in a synchronised way. In addition to that, and thanks again to the MX format, an entire execution of the net mixes the MX fragments, produces a complete score, as well as a complete wave file of the music piece performed.

5 Related and Future Works

Since 1980, we at LIM have been using Petri Nets as the basic tool for music description and processing, creating applications to support our approach. The systems used to design, execute and debug the Petri Net described in this article has been called *ScoreSynth*, and was entirely developed in our laboratory [7].

In regard to future works, we propose to do more in the field of real-time transformations of Petri Net models, to allow changes of arc weights and even dynamic modification of the entire structure of the model. Extensive tests will have to be made with different types of data linked to MX fragments, to produce rich and heterogeneous outputs. These possibilities will be explored in the future.

6 Conclusions

In this article, a new perspective for *music creation* and *manipulation* has been described, to show how an existing music piece can be analysed, formalised with Petri Nets, and eventually modified in real-time. The originality of this approach is that music manipulation is performed at a high level. The composer and performer can concentrate on modifications of the internal structure of the piece, while the computer automatically treats lower level integration. With the adoption of the MX format, music information is automatically handled in a synchronised and heterogeneous way, allowing real-time visualisation of different music aspects, and the production of several representations of the output at the same time.

References

1. Baggi, D., Baratè, A., Haus, G., Ludovico, L.A.: A Computer Tool to Enjoy and Understand Music. In: EWIMT 2005. Proceedings of the 2nd European Workshop on the Integration of Knowledge, Semantics and Digital Media Technology, London, UK, pp. 213–217 (2005)
2. Baratè, A., Ludovico, L.A.: An XML-based Synchronisation of Audio and Graphical Representations of Music Scores. In: WIAMIS 2007. Proceedings of the 8th International Workshop on Image Analysis for Multimedia Interactive Services, Santorini, Greece (2007)
3. Baratè, A., Haus, G., Ludovico, L.A.: MX Navigator: An Application for Advanced Music Fruition. In: Proceedings of AXMEDIS 2006. 2nd International Conference on Automated Production of Cross Media Content for Multi-channel Distribution, Leeds, UK, pp. 299–305 (2006)
4. Baratè, A., Haus, G., Ludovico, L.A.: Music Analysis and Modelling through Petri Nets. In: Kronland-Martinet, R., Voinier, T., Ystad, S. (eds.) CMMR 2005. LNCS, vol. 3902, Springer, Heidelberg (2006)
5. Degli Antoni, G., Haus, G.: Music and Causality. In: Proceedings of 1982 International Computer Music Conference, La Biennale, Venezia, pp. 279–296. Computer Music Association Publ., San Francisco (1983)
6. De Matteis, A., Haus, G.: Formalisation of Generative Structures within Stravinsky's "Rite of Spring". Journal of New Music Research 25(1), 47–76 (1996)
7. Haus, G. (Scientific Direction): The Intelligent Music Workstation (IMW) CD-ROM, mixed mode CD-ROM (Macintosh HFS + CD-DA). IEEE Computer Society Press (1994)
8. Haus, G., Rodriguez, A.: Formal Music Representation; a Case Study: the Model of Ravel's Bolero by Petri Nets, Music Processing, Madison. Computer Music and Digital Audio Series, A-R edn., pp. 165–232 (1993)
9. Haus, G., Sametti, A.: Modelling and Generating Musical Scores by Petri Nets. Languages of Design 2(1), 7–24 (1994)
10. Petri, C.A.: General Net Theory. In: Proceedings of the Joint IBM & Newcastle upon Tyne Seminar on Computer Systems Design (1976)
11. Peterson, J.L.: Petri Net Theory and the Modelling of Systems. Prentice Hall, New Jersey (1981)

Interacting with Csound Timbral Textures

Elisa Russo

Laboratorio di Informatica Musicale (LIM)
Dipartimento di Informatica e Comunicazione (DICo)
Università degli Studi di Milano
Via Comelico 39/41 – I- 20135, Milano, Italia
elisa.russo@dico.unimi.it

Abstract. The possibility of interconnecting *Csound* with *MX* allows creation and manipulation of timbres in an exhaustive way. Thanks to Csound features, it is possible to let MX files sound with any synthesis algorithm. This approach supports all the connections to current files through MIDI, real-time controls, analysis and synthesis thanks to the direct interaction with files linked in MX.

1 Introduction

MX is an effective instrument for the multimedia representation of music, from many points of view. The multimedia representation of music for synthesis, especially in regard to Csound, will be examined in this context. Csound is a software for digital synthesis, created for the simulation of every kind of sound. Therefore, it is also a powerful instrument for the generation of timbres for MX files. Moreover, the integrated multimedia description of MX allows realisation of complete Csound timbres. The data needed for the synthesis representation have, among others, characters representing the *symbolic nature* of the music such as chords and rests, its *performance* as in MIDI and Csound, and its *audio* with WAV and MP3. The XML format can describe all these objects in detail ([1], [2], [3]).

2 Csound as a Performance Language for MX

The kernel for communication between Csound codes and MX layers is the structure csound_instance, a sub-layer of *Performance* layer.

The corresponding XML element in MX to a Csound score is csound_score. The following is an example of its use:

```
<csound_score file_name="example.sco">
<csound_spine_event line_number="8" spine_ref="p1v1_1"/>
</csound_score>
```

It can be seen that, for every score, the correspondence between all Csound notes and related MX events is coded in MX. The score lines to be linked in Csound are only those in which the first char is an "i", because, in this case, only the

B. Falcidieno et al. (Eds.): SAMT 2007, LNCS 4816, pp. 222–225, 2007.
© Springer-Verlag Berlin Heidelberg 2007

synchronisation between sounds and events is of interest, and not between Csound function tables and events. Therefore, all information about the function tables should be obtained from the instruments that use them – whose correspondence is coded in `csound_orchestra`.

The following is an example of the last MX element mentioned above:

```
<csound_orchestra file_name="example.orc">
  <csound_instrument_mapping instrument_number="1"
                        start_line="10" end_line="20">
    <csound_part_ref  part_ref="1"/>
    <csound_spine_ref  spine_ref="p1v1_1"/>
  </csound_instrument_mapping >
</csound_orchestra>
```

In this case, every instrument of the orchestra is univocally characterised by the identification number of attribute `instrument_number`. To clarify the references, `start_line` and `end_line` specify the exact lines of the instrument in the orchestra. This way every instrument can be related to particular MX contents – univocally defined through `part_ref` and `spine_ref`.

The *Performance* layer provides another sub-layer to enlarge the timbre potentialities of Csound, namely, `midi_instance`. Here is an example of how a MIDI file can be linked to MX files:

```
<midi_instance file_name="example.mid" format="0">
  ( ... )
  <midi_event timing="1" spine_ref="p1_v1_1"/>
  <midi_event timing="1" spine_ref="p1_v1_2"/>
  ( ... )
</midi_instance>
```

This code is useful when Csound contents have to be connected to MIDI for sound generation. All this is managed through MX. The most useful utilities of the interconnection between MIDI and Csound are access of Csound contents to MIDI, and real time interaction between Csound and MIDI - as explained in more details in [4] and [5] - which in this case happens under the supervision of MX, with the possibility of interconnecting among different kinds of contents.

3   Music Symbols Through MX and Csound

The possibility of importing notes in Csound from MX information in the *los* is described in detail in [6]. This section shows that it is intuitively possible to obtain all symbols of a Csound score from the symbolic contents of MX. One simple example will be shown to demonstrate how an algorithm of conversion from musical MX symbols into Csound works. Clearly, when all symbols are translated in Csound, together with the Csound instruments to play the score, the generation of audio is immediate.

The generation of a score from symbolic information depends basically on the single annotated sounds, such as chords, notes, rests, etc. The difficulty lies in the translation of every symbolic event so that no loss of information occurs.

In Csound syntax, with the exception of function declarations, every line of a score file represents a single sound. In particular, for every sound to be produced, the essential features are: the instruments to use, the start time, the duration, the amplitude and the frequency. Other parameters can be present, which for the sake of simplicity will not be considered here. Time can be expressed in seconds, amplitude as an absolute value, and pitch according to octave-point-pitch-class notation.

Figure 1 shows a C-Major chord and its corresponding MX and Csound code.

♪ (music notation)	`<chord event_ref="v1_e1">`
	  `<duration num="1" den="1"/>`
	  `<notehead>`
	    `<pitch step="C" octave="5"/>`
	  `</notehead>`
	  `<notehead>`
	    `<pitch step="E" octave="5"/>`
	  `</notehead>`
	  `<notehead>`
	    `<pitch step="G" octave="5"/>`
	  `</notehead>`
	  `<notehead>`
	    `<pitch step="C" octave="6"/>`
	  `</notehead>`
	`</chord>`
		`i1  0  4    5000 8.00`
		`i1  0  4    5000 8.04`
		`i1  0  4    5000 8.07`
		`i1  0  4    5000 9.00`

Fig. 1. A chord encoded in MX and in Csound

In MX, a chord is a single element that contains a number of sub-elements encoding each single note. On the other hand, in Csound every line represents a single sound event. Usually, in a performance-oriented language, only sounds are coded, while rests are derived from the absence of sound. The way of coding of rests in MX and in Csound is a typical example of the different approaches for symbolic and performance languages. It appears therefore that symbolic information allows a trivial conversion in terms of start time and duration time. However, a detailed management of more complex information, such as different kinds of articulation, tie symbols and irregular groups, requires a deeper approach.

4 MX and Audio Instances

From a "Csound point of view", the *Audio* layer can be used to connect Csound files to audio files. This interconnection allows manipulation and enrichment of audio information with some well-known techniques – for example, analysis-based. During the last few years, many approaches have been studied and developed in order to increase the possibilities of Csound in respect to manipulation of audio samples, which represents a precious resource for the creation of interesting and different kinds of sounds. Techniques widely used in research are FOF synthesis and analysis-based-techniques, Phase Vocoder, Heterodyne filter analysis, Linear predictive analysis, and so on. In this context, Csound has the advantage of being able to interact directly with various audio representations of the MX file.

The following excerpt of MX code shows how every event of a wave file can be represented in correspondence with other musical information.

```
<audio>
  <track file_name="example.wav" ... >
    ( ... )
    <track_event ... spine_ref="p1v1_1" />
    <track_event ... spine_ref="p1v1_2" />
    ( ... )
  </track>
</audio>
```

It can be seen that every event is interfaced with all the information of the MX file thanks to the reference in the spine. Therefore, the interaction between Csound and audio is immediate and, at the same time, easy to use and manipulate.

5 Conclusions

The MX structure has been analysed in regard to its usefulness for interacting with Csound. It has been shown that the advantages of Csound and its possibility of interfacing with MX are many and various. Suffice to say that, when Csound accesses MX, it is capable of adding every MX feature to its synthesis instruments. It is hoped that this contribution inspires further development of applications and frameworks that exploit all the potentialities of the MX-Csound interaction.

References

1. Haus, G., Longari, M.: A Multi-Layered, Time-Based Music Description Approach Based on XML. Computer Music Journal 29(1), 70–85 (2005)
2. Baratè, A., Haus, G., Ludovico, L.A.: An XML-Based Format for Advanced Music Fruition, Sound and Music Computing 2006 (2006)
3. Baggi, D.L., Baratè, A., Ludovico, L.A., Haus, G.: A Computer Tool to Enjoy and Understand Music. In: EWIMT 2005. The 2nd European Workshop on the Integration of Knowledge, Semantics and Digital Media Technology, pp. 213–217 (2005)
4. Selfridge-Field, E.: Beyond MIDI: the Handbook of Musical Codes. MIT Press, Cambridge (1997)
5. Boulanger, R.: The Csound book. MIT Press, Cambridge (2000)
6. Haus, G., Ludovico, L.A., Russo, E.: From Music Symbolic Information to Sound Synthesis: an XML-Based Approach. In: SMC 2007. Sound and Music Computing Conference, Lefkada, Greece (2007)

A Semantic Web Environment for Digital Shapes Understanding

Leila De Floriani[1,2], Annie Hui[2], Laura Papaleo[1], May Huang[2], and James Hendler[3]

[1] Department of Computer and Information Science – University of Genova, Italy
{deflo,papaleo}@disi.unige.it
[2] Department of Computer Science University of Maryland, College Park, USA
{huiannie,mayhuang}@cs.umd.edu
[3] Rensselaer Polytechnic Institute, Troy, NY, USA
hendler@cs.rpi.edu

Abstract. In the last few years, the volume of multimedia content available on the Web significantly increased. This led to the need for techniques to handle such data. In this context, we see a growing interest in considering the Semantic Web in action and in the definition of tools capable of analyzing and organizing digital shape models. In this paper, we present a Semantic Web environment, *be-SMART*, for inspecting 3D shapes and for structuring and annotating such shapes according to ontology-driven metadata. Specifically, we describe in details the first module of *be-SMART*, the *Geometry and Topology Analyzer,* and the algorithms we have developed for extracting geometrical and topological information from 3D shapes. We also describe the second module, the *Topological Decomposer,* which produces a graph-based representation of the decomposition of the shape into manifold components. This is successively modified by the third and the fourth modules, which perform the automatic and manual segmentation of the manifold parts.

Keywords: Semantic web, digital shapes, shape understanding, shape analysis, semantic annotation.

1 Introduction

In the last few years, there has been a general trend within different research and industrial communities to organize multimedia data (e.g. images, 3D models, videos) into digital libraries, and a lot of attention towards archiving 3D objects in an intelligent way [1,2,3]. This requires a strong integration of knowledge management technologies in different research fields, e.g. computer graphics and scientific visualization. The main objective is to make the knowledge embedded in 3D data explicit and sharable and to see the Semantic Web in action [4,5].

In particular, modeling and reasoning on 3D shapes requires digital representations that integrate geometry, topology, and semantics [2]. A shape is described as a collection of elementary cells, such as triangles, edges and vertices, which captures its geometry. A structural representation is a more concise description of a shape in which geometric details are abstracted and only important features remain. Thus, it is

B. Falcidieno et al. (Eds.): SAMT 2007, LNCS 4816, pp. 226–239, 2007.

a suitable basis for semantic annotation and reasoning. Examples of structural representations are skeleton-based descriptions, or part-based decompositions. Most of the existing representations are actually for manifold shapes. On the other hand, non-manifold shapes are common in many contexts, and they are often the result of an abstraction process applied to the manifold ones. This happens, for instance, in the *idealization* process which finite element meshes generated from CAD models undergo to meet simulation requirements [6].

Several steps are required to identify and associate semantics with a shape. Some of such steps are necessarily context-dependent. This means that semantics can have different levels of abstraction and the knowledge associated with a shape is generally not linear, but structured in a multifaceted mode. In summary, geometric and topological information is a useful resource for reasoning on shapes and a structural representations guided by topological information is essential for inferring semantic properties. In the near future, we foresee the need for systems capable of exploring, organizing and understanding digital representations. The purpose is also to populate repositories with semantically enriched digital models, improving in this way scientific communication and supporting the generation of new knowledge, specifically about digital models with complex geometric and topological structure.

In this paper, we present part of our research activity which leads to the development of *be-SMART (BEyond Shape Modeling for understAnding Real world representations)*, a system for inspection, structuring and semantic annotation of digital shapes. *be-SMART* is a complex system, modular and extensible which has been designed to support researchers in reasoning on digital shapes and in improving their knowledge about multimedia data available on the net.

In what follows, we describe the general architecture of the system (Section 2), presenting in details its first module, the *Geometry and Topology Analyzer (GTA)* (Section 3) and its second module, *Topological Decomposer (TD)* (Section 4). The *GTA* extracts geometric and topological information from a 3D model of a shape. Here, we provide the description of the algorithms we have developed for analyzing complex non-manifold shapes, and we discuss how a given model is annotated using ontology-driven metadata, specifically those defined in [8]. The *TD* module decomposes a non-manifold shape into manifold parts. The decomposition is represented as a graph (*decomposition graph*). We also briefly introduce the third and fourth modules (Section 5) which perform the segmentation of the manifold parts, and we show how these modules interact with the decomposition graph. Concluding remarks and future work are discussed in Section 6.

2 be-SMART: The Vision

With the advent of the Semantic Web [9], there has been focus in both industry and academia on developing techniques to annotate multimedia content, using Web ontology languages such as RDFS [10] or OWL [11]. The main reason is that, given semantically rich metadata, collections of heterogeneous multimedia (e.g., images, 3D models) can be more accurately searched and browsed by using new knowledge derived from existing annotations [12].

be-SMART is designed to be a system for geometric-topological inspection and semantic annotation and structuring of 3D shapes [13]. It relies on the idea to extract information about features and regions of interest and to provide an intuitive interface for reasoning on digital models. *be-SMART* aims at computing quantitatively replicable data, and at generating ontology-driven metadata about an object. This is achieved by *(i) extracting* (automatically) geometric and topological information from the model and by maintaining them using ontology-driven metadata; by *(ii) segmenting* the model (both manually and/or automatically) using editing technologies and context-dependent segmentation techniques and by *(iii) structuring* and *idealizing* (automatically) the shape in order to create a structural multi-level representation of the model guided by the associated semantic. The system can be coupled with a semantic web portal, which provides metadata management and interaction functionalities, such as [3,14].

be-SMART has been designed to be a Java-based system, which uses pre-loaded ontologies to provide the expressiveness required to assert the content of digital data, as well as information about the digital media itself (e.g. date of creation, upload). Extending the idea of PhotoStuff [15], which deals with images, *be-SMART* allows a user to annotate parts of a 3D shape with respect to concepts in any ontology specified in RDFS or OWL. As it works in 3D, *be-SMART* uses libraries for 3D modelling (Xj3D, X3D) [16,17]. An X3D file, representing the initial model, is successively modified and transformed (through ad-hoc XSLT transformations) according to the information extracted as well as to the segmentations performed. In the transformation process, new tags are added to the X3D file so that, at the final stage, all information is maintained, and the final model preserves the knowledge acquired and/or extracted during the different processing steps. Such information is also maintained in separate RDF files suitably linked with the initial X3D file.

2.1 be-SMART: The Modules

In *be-SMART* different modules act as a team in generating the final enriched shape representation, annotated at different levels of abstraction. We list here the modules we have defined (the architecture of the system is depicted in Fig. 1):

1. *Geometry and Topology Analyzer (GTA):* it analyses the input shape model and extracts geometrical/topological information which is maintained in the enriched shape model and as instance values of a given ontology.
2. *Topological Decomposer (TD):* starting from the information extracted by the *GTA* module, this module produces a graph-based representation (*decomposition graph*) of the shape model into nearly manifold components.
3. *Manual Segmentation module (MS):* This module offers both simple and advanced editing functionalities allowing a user to select portions of the model. The segmentation is maintained in the *decomposition graph*.
4. *Automatic Segmentation module (AS):* This module offers the possibility of applying automatic segmentation algorithms for decomposing the manifold components into meaningful parts (according to context-dependent criteria). The segmentation is maintained in the *decomposition graph*.

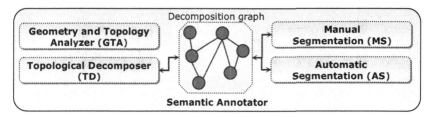

Fig. 1. The general architecture of *be-SMART*, with its five constituent modules. The *GTA* module exchanges information with all the other modules. Both the *GTA* and the *TD* modules are application-independent. The other modules are context-dependent.

5. *Semantic Annotator (SA):* This module offers the possibility of associating specific metadata values to specific portions of the decomposed model according to pre-loaded ontologies. Basically, it associates metadata with nodes of the *decomposition graph* which describe the decomposed model.

In the following two Sections, we present the first two modules by describing their functionalities. We show some examples of results and the mapping of the extracted information on the metadata defined in the Common Shape Ontology [2,8].

3 The Geometry and Topology Analyzer (GTA)

A 3D shape is most commonly described through a discretization of its boundary into a mesh consisting of triangles, bounded by edges and vertices, and by dangling edges, which do not bound any triangle, that we call *wire-edges*.

The *Geometry and Topology Analyzer (GTA)* addresses the problem of extracting topological characteristics from non-manifold 3D shapes containing parts of different dimensions. Informally, a *manifold* (with boundary) is a compact and connected subset Σ of the Euclidean space such that the neighbourhood of each point of Σ is homeomorphic to an open ball or to an open half-ball. Objects, which do not fulfil this property at one or more points, are called *non-manifold objects*, and if they also contain parts of different dimensionalities, are called *non-regular*.

The *GTA* module is based on a representation of the underlying mesh as a *Triangle-Segment (TS)* data structure (see Subsection 3.1). We have identified the following characteristics as relevant for the topological analysis of non-manifold 3D shapes:

- *non-manifold singularities*, which can be non-manifold isolated points, or non-manifold curves (see the two examples in Fig. 2(b) and Fig. 2(c));
- 1-dimensional parts of the shape, such as the *spider-webs* (see Fig. 2(a));
- *connected components* of the shape;
- *maximal connected components* which do not contain non-manifold isolated points (such as the object in Fig. 2(b)). Note that the object in Fig. 2(c) is formed by two of such parts (the cone and the planar surface).

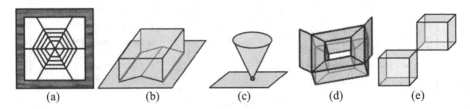

| (a) | (b) | (c) | (d) | (e) |

Fig. 2. (a) A spider-web on a window; (b) a block that touches a plane at two straight-lines; (c) a cone touching a plane at a single point; (d) an object that encloses one void, that is it has a 2-cycle (β_2=1), and contains two 1-cycles (β_1=2), which define the handle; (e) an object with two 2-cycles, each of which is the interior of a cube.

A topological signature for a non-manifold shape can be defined on the numbers of its non-manifold singularities (isolated points and curves), the numbers of different components listed above, and its Betti numbers, β_0, β_1 and β_2, which are the number of connected components, the number of 1-cycles and of 2-cycles of the shape, respectively. The 1-cycles can be interpreted as independent tunnels or handles, and the 2-cycles as portions of the shape enclosing a void (see for example Fig. 2(d)). The Betti numbers are related to the number of vertices, edges and triangles through the Euler-Poincare's formula: v-e+f=β_0-β_1+β_2, where v, e and f denote the number of the vertices, edges and triangles, respectively. Note that β_2 needs to be computed by extracting the 2-cycles from the model (see Subsection 3.4), and β_1 can be then derived by the formula. The *GTA* module computes the different characteristics, including β_0 and β_2 and successively derives β_1. Going into details, the *GTA* module identifies the following features:

- non-manifold vertices and non-manifold edges, which correspond to the non-manifold singularities of the shape;
- wire-edges;
- connected components;
- maximal connected components formed by edges (i.e., the wire-edges). We call them *wire-webs*, since they describe 1-dimensional parts of a shape (the web in Fig. 2(a))
- 1-connected components, which describe the parts of the shape which do not contain non-manifold isolated points. Note that a 1-connected component is a component in which, for every pair of triangles, there exists a path composed of triangles and edges such that any edge in the path belongs to the boundary of the two triangles preceding and following it in the path.

The *GTA* module also computes the Betti numbers, by computing the portions of the mesh enclosing voids, which gives β_2 (the number of 2-cycles). Note that β_0 is the number of connected components, and β_1 is derived from the Euler-Poincare's formula knowing β_0 and β_2. All information extracted by the *GTA* module are then attached to the initial model using ontology-driven metadata. Table 1 illustrates the mapping of the properties extracted by the *GTA* into the metadata identified for the concept *NonManifoldMesh* in the common shape ontology designed within [2] and described in [8].

Table 1. The mapping between the geometric and topological characteristics extracted by the *GTA* module and the metadata described in [8] for the concept *NonManifoldMesh* in the Common Shape Ontology

Information extracted by the *GTA*	*NonManifoldMesh* property
number of triangles	hasNumberOfCell
number of vertices	hasNumberOfVertice
number of edges	hasNumberOfEdge
number of wire-edges	hasNumberOfWireEdges
number of non-manifold vertices	hasNumberOfNonManifoldVertex
number of non-manifold edges	hasNumberOfNonManifoldEdge
number of connected components made of 1-dimensional simplexes	hasNumberOfConnectedSimplexes-OfDim1
number of connected components made of 2-dimensional simplexes	hasNumberOfConnectedSimplexes-OfDim2
β_0 (number of connected components)	hasNumberOfConnectedComponents
number of 1-connected components	hasNumberOf1ConnectedComponents
β_1	hasNumberOf1Cycles
β_2 (number of 2-cycles)	hasNumberOf2Cycles

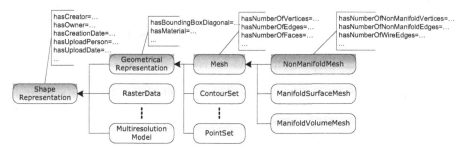

Fig. 3. Zoom on the concept *ShapeRepresentation* defined in the ontology described in [8]. The taxonomy of the concept is depicted illustrating the position of *NonManifoldMesh* and the relative metadata. The concept *NonManifoldMesh* has peculiar metadata and inherits the metadata from all its super-concepts.

In Fig. 3 a zoom on the concept *ShapeRepresentation* of the same ontology and the taxonomy of its sub-concepts, plus a list of useful metadata are shown. Using the *GTA* module, instances of the concept *NonManifoldMesh* can be automatically created with all the necessary metadata. Fig. 4 shows a model and the information extracted by the *GTA* module. In the following subsections, we present the *TS* data structure and we discuss how the *GTA* module finds, starting from a shape model encoded in the *TS*, all the information listed above. For each feature, we also discuss the mapping with the properties of the *NonManifoldMesh* concept in the Common Shape Ontology [2,8].

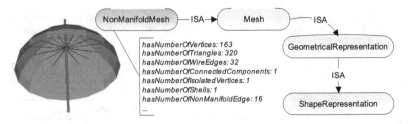

Fig. 4. A non-manifold mesh representing an idealised umbrella and the ontology-driven metadata extracted using the *Geometry and Topology Analyzer (GTA)*

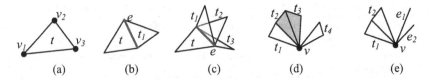

Fig. 5. Topological information encoded in the TS for each triangle and at each vertex v: (a) the three vertices for t; (b) if e is an interior manifold edge it is both the predecessor and successor at e of t; (c) non-manifold case, triangles t_1 and t_3 are predecessor and successor of t at e. (d) two 1-connected components incident at v. The vertex-triangle relation at v consists of an arbitrary triangle t_1 of the first component and of t_4 from the second. (e) The *vertex-wire-edge* relation at v encodes the wire-edge e.

3.1   Shape Representation Through the Triangle Segment Data Structure

Here we describe the data structure we use to encode the model and that we pass to the *Geometry and Topology Analyzer (GTA)*. The choice of the *Triangle-Segment* data structure is motivated by its compactness and by its efficient support in navigation [18]. The *Triangle-Segment* data structure (*TS*, in short) generalizes to the non-manifold case the *Indexed structure with Adjacencies (IA)* data structure [19] which describes triangle meshes embedded in the 3D Euclidean space. The *IA* data structure encodes only the vertices and the triangles of the mesh and, for each triangle, the references to its adjacent triangles. In a similar way, the *TS* encodes all vertices, and top simplexes, i.e., triangles and edges not bounding triangles, that we call wire edges. The wire-edges and triangles are defined by their bounding vertices, namely by the relations *wire-edge-vertex* and *triangle-vertex*. For each wire edge e, the *wire-edge-vertex* relation associates with e its two extreme vertices, while, for each triangle t, the *triangle-vertex* relations associates with t, its three vertices (see Fig. 5(a) as example).

For each edge e of a triangle t, the *TS* stores the *triangle-triangle-at-edge* relation, encoding the triangle(s) immediately preceding and succeeding t around e, when the triangles incident at e are sorted in counter-clockwise order around e. When there is no triangle adjacent to t along e, the predecessor and the successor in the *triangle-triangle-at-edge* relation are both t. When edge e is an interior manifold edge, the predecessor and the successor are identical and consist of the only other triangle t_0 which share e with t (see Fig. 5(b)). Note that, when e is a *non-manifold edge*, the predecessor and the successor of t around e are two different triangles (see Fig. 5(c)).

Moreover, to simplify the task of navigating around a vertex, the *TS* encodes, for each vertex *v*, the *vertex-triangle* relation which associates with *v* one triangle for each 1-connected component incident at *v* (Fig. 5(d)) and the *vertex-wire-edge* relation which associates with *v* all the wire edges incident at *v* (Fig. 5(e)).

It has been shown that the *TS* provides a very compact representation of the initial model [18]. A relevant property of the *TS* is its scalability to the manifold case and, also, it supports topological navigation in optimal time.

3.2 Extracting Non-manifold Singularities and Wire-Edges

Here we describe the algorithm that the *GTA* module applies for counting and extracting non-manifold vertices, non-manifold edges and wire edges[1]. Wire-edges are explicitly encoded in the *TS* and thus their retrieval and their counting is trivial.

A *non-manifold* vertex *v* is a vertex such that its *link* has at least one connected component. Recall that, the *star* of a vertex *v* is the set of edges and triangles in the model that are incident in *v* and that the *link* of *v* is the set of edges and vertices in its *star* not incident in *v*. The existence of at least one connected component in the link of *v* is identified by considering the *vertex-triangle* and *vertex-wire-edge* relations at *v* in the *TS*. Thus, *v* is manifold if only one of the two relations is not empty, and

- if the *vertex-triangle* relation is not empty, and it consists only one triangle;
- if the *vertex-wire-edge* relation is not empty, and it consists of either one or two wire-edges.

Otherwise the vertex *v* is non-manifold. A *non-manifold edge e* is detected by considering a triangle *t* incident at *e* and checking whether *t* has a predecessor and successor in the *triangle-triangle-at-edge* relation of *t* at *e*, which are different.

3.3 Extracting Connected, 1-Connected Components and Wire-Webs

While non-manifold vertices and edges are detected locally by examining their stars, in order to detect parts with certain connectivity properties, we need to perform a traversal of the entire model. Specifically, connected components are identified by applying a connected component labeling approach to the *1-skeleton* of the model (i.e., to the complex formed only by the vertices and edges). This is performed as a breadth-first traversal in the *TS* by using a queue *Q* as an auxiliary data structure. The traversal of each connected component starts at an arbitrary unvisited vertex *v* of the mesh, and visits all triangles in the *vertex-triangle* relation, and all the wire-edges in the *vertex-wire-edge* relation at *v*. Then, all the vertices which are bounding such triangles and wire-edges are considered, and if they are not visited, they are inserted in *Q*. The traversal continues by extracting the first vertex in the queue and considering it as current vertex. A connected component is completely traversed when the queue is empty. By repeating this traversal process till no unvisited vertex is left, the algorithm retrieves all the connected components. β_0 is computed as the total number of the connected components (*hasNumberOfConnectedComponents*).

---

[1] Their numbers can be maintained in the concept *NonManifoldMesh* in the Common Shape Ontology [8] via the properties *hasNumberOfNonManifoldVertices*, *hasNumberOfNon ManifoldEdges* and *hasNumberOfWireEdges*, respectively.

Wire-webs are connected components formed only of wire-edges. To detect and count such components (*hasNumberOfConnectedSimplexes-OfDim1*), a similar traversal (as described above) is performed, but, at each vertex v, only the incident wire-edges are considered.

The 1-connected components are computed by considering the sub-mesh Σ' obtained from the original one eliminating all the wire-webs. The 1-connected components will be used for the extraction and the counting of the 2-cycles presented in the following subsection. The 1-connected components correspond to connected components of the dual graph representing Σ'. In this graph, the nodes correspond to the triangles of Σ' and the arcs to the edges shared by two or more triangles[2]. These components are extracted from the *TS* starting from an arbitrary triangle t. For each edge e of t, all the triangles incident at e are extracted, marked as visited and successively inserted in an auxiliary queue Q. Then, the first triangle in Q is extracted and the traversal is repeated from such triangle. A 1-connected component is completely traversed when Q is empty. By repeating this traversal process till no unvisited triangle is left, all the 1-connected components are found. This gives a count to the number of 2D connected components (*hasNumberOf1ConnectedComponents*).

3.4 Computing the 2-Cycles

In this Subsection, we describe how the *GTA* module computes the number β_2 of 2-cycles in the shape model. The 2-cycles correspond to maximal 1-connected sub-meshes enclosing voids and not containing triangles dangling at non-manifold edges. They are computed starting from the 1-connected components identified at the previous step (Subsection 3.3) and by cutting maximal 1-connected sub-meshes of triangles which do not bound a void. Once β_2 is computed, β_1 can be derived by the Euler's formula. In the algorithm we restrict our consideration only to representations describing orientable surfaces. The algorithm is based on the concepts of *triangle sides* and *oriented surfaces*. A triangle t has two *triangle sides* (with opposite normals) defined by the two possible orderings of its vertices. Each triangle side induces an orientation on each of its edge. In Fig. 6(a) an example is shown where the edge connecting vertices a and b has orientation $a{\rightarrow}b$ on one triangle side, and $b{\rightarrow}a$

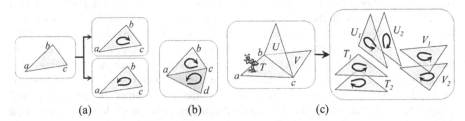

(a)   (b)   (c)

Fig. 6. (a) Example of the two possible *triangle sides* for a triangle t, (b) example of adjacency between two triangle sides. (c) example for an oriented surface. An *oriented surface* is basically the surface formed by all the triangle sides that are reachable (via adjacency) by a walking ant.

---

[2] Note that the arcs of the dual graph are actually hyperarcs since an edge in Σ' can be on the boundary of more than two triangles.

on the opposite side. Two triangle sides are *adjacent* at their common edge e if one is reachable from the other by crossing e. Note that two triangle sides adjacent at edge e induce opposite orientations on e itself (Fig. 6(b)).

For each 1-connected component Γ in the model, we extract connected collections of triangle sides for the triangles in Γ, that we call *oriented surfaces*. An *oriented surface* is a maximal 1-connected set of triangles sides, such that, for each pair of triangle sides s', s'' in the set, sharing an edge e, s' is the successor of s'' according to one orientation of e, and s'' is the successor of s' according to the opposite orientation of e. In other words, the two triangle sides that belong to two different triangles, which share a common edge, induce an opposite orientation on their common edge. To clarify the definition of oriented surface, consider the example shown in Fig. 6(c) and an ant walking on the upper side of triangle t (that is T_1). It reaches the triangle side U_1 by crossing edge (a,b). It reaches the opposite side of t (T_2) by crossing edge (a,c), because (a,c) is a boundary edge. An oriented surface is basically the surface formed by all the triangle sides that are reachable by a walking ant.

An oriented surface may enclose a void. An oriented surface S with boundary, such that for each triangle side s that belongs to S the other side s'' of the same triangle also belongs to S, cannot include any void. We call such oriented surfaces *folded surfaces*. Note that a 1-connected component Γ may consist of only one oriented surface, and in this case it must be a folded surface. Γ may consist of two surfaces, as it is always the case if it contains only manifold edges. For instance, if Γ is a triangulated sphere, Γ consists of two oriented surfaces, each of which corresponding to an orientation of the surface of the sphere. If Γ consists of two triangulated cubes sharing a face, then there are three oriented surfaces associated with Γ, namely the two cubes with normals pointing towards the inside and the oriented surface consisting of the union of the two cubes with the normals pointing towards the external empty space.

It can be easily seen that, if we consider Γ, there exists exactly one oriented surface in the set of oriented surfaces associated with Γ, that we denote S_{out}, such that the normals to the triangle sides forming it point towards the outside space. S_{out} defines a cycle of triangles which can be expressed as the linear combination of the other oriented surfaces associated with Γ. Thus, unlike the other oriented surfaces in Γ, S_{out} does not define a 2-cycle, since it contains the remaining oriented surfaces. Note that, if Γ consists of one folded surface, this latter does not define a 2-cycle as well. Since there is exactly one oriented surface for each 1-connected component which does not define a 2-cycle, we have that the number of 2-cycles in the model is equal to the total number of oriented surface minus the number of 1-connected components.

The oriented surfaces are computed through a traversal of the mesh by taking into account the triangle sides, thus simulating the walk of an ant on a surface.

4   The Topological Decomposer

In this Section, we present the *Topological Decomposer* (*TD*). The *TD* module is based on the algorithm described in [19] and it is mainly focused on the construction of a graph-based representation, which encodes a semantic-oriented decomposition of the input model Σ.

The decomposition of Σ into wire-webs and oriented surfaces naturally leads to a semantic-oriented decomposition consisting of three types of components, namely:

- *wire-webs*
- *sheets*, which are maximal 1-connected components of oriented surfaces with boundary such that, for each triangle side *s* that belongs to the surface, the other side *s″* of the same triangle also belongs to it.
- *shells*, which are maximal 1-connected components of oriented surfaces which contain only one side for each triangle.

We compute the semantic-oriented decomposition based on the results produced by the algorithms presented in the previous section. Wire-webs are extracted as described in Subsection 3.3. The first step is to identify the one oriented surface for each   1-connected component Γ which has the normals of the triangle sides forming it poin- ting towards the exterior of the complex. If Γ consists of only one oriented surface, then this latter is a *sheet*. In all the other cases, this *external* oriented surface is eliminated. If, after eliminating the external surface, we are left with only one oriented surface, then this latter must include a void, and thus it forms a *shell*. In all the other cases, so when Γ contains non-manifold edges, we need to cut oriented surfaces to generate shells and sheets. We consider an oriented surface *S* and identify the triangles in *S* such that both sides of such triangles belong to surface *S*. These triangles are detached from *S*, which thus becomes a shell. All the detached triangles are joined together at their shared edge and they form a sheet.

The semantics-oriented decomposition can be described as a hypergraph that we call the *decomposition graph*, which captures the connectivity among *wire-webs, sheets* and *shells*. The decomposition graph supports the extraction of global topological features of the decomposed shape.

The *decomposition graph* is a hypergraph $H=<N,A>$ in which the nodes in N correspond to the components of the decomposition (namely wire-webs, sheets and shells), and the hyperarcs in A capture the structure of the connectivity among these components. The hyperarcs are defined as follows: any k components $C_1, C_2, ..., C_k$ ($k>1$) in the decomposition such that the intersection of all of them is not empty defines one or more hyperarcs with extreme nodes in $C_1, C_2, ..., C_k$. The intersection of components $C_1, C_2, ..., C_k$ consists of isolated non-manifold vertices, maximal 0-connected 1-complexes formed by non-manifold edges, or maximal 1-connected mesh portions formed by triangles. A hyperarc is a connected component of such intersection.

Thus, hyperarcs are classified as follows: (*i*) *0-hyperarcs*, which consist only of one non-manifold vertex; (*ii*) *1-hyperarcs*, which are maximal 0-connected mesh portions formed by non-manifold edges; (*iii*) *2-hyperarcs*, which are maximal 1-connected mesh portions formed by triangles. A 2-hyperarc may connect only two nodes, because a connected set of triangles may belong to at most two shells. Note that there may exist one or more hyperarcs connecting the same set of nodes.

All hyperarcs which connect the same k components can be grouped into a macro-hyperarc, which defines the structure of the intersection of the k intersecting components. Fig. 7(a) shows a non-manifold model of a chair and the related *decomposition graph* with three *shells* and two *wire-webs*.

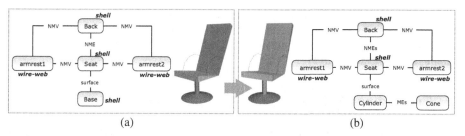

Fig. 7. (a) A non-manifold model for a chair with the Decomposition Graph. 3 shells (back, seat and base) and 2 wire-webs (armrests) are identified. (b) the same chair successively segmented via the MS module and the updated Decomposition Graph. NMV, NMEs and MEs indicates non-manifold vertex, non-manifold edges and manifold edges, respectively.

5 Manual and Automatic Segmentation Modules

The third and the fourth modules of *be-SMART* are the *Manual* and the *Automatic Segmentation Modules* (*MS* and *AS*). We have shown that the *TD* module produces a semantics-oriented decomposition of a non-manifold shape into manifold parts guided only by the extracted context-independent topological information. On the contrary, the *MS* and *AS* modules perform segmentations of a shape on the basis of specific heuristics. If the initial model is non-manifold, the two modules start from the decomposition graph produced by the *TD* module and they perform segmentation only on the manifold parts.

Here we briefly describe our on-going work in the development of the *MS* and *AS* modules. Both the *MS* and the *AS* modules update the decomposition graph which is passed to the *Semantic Annotator* module so that the user can associate semantic meaning to each portion according to the expert knowledge, possibly maintained in specific context-dependent ontologies. In this sense, we are going to concentrate on CAD/CAM related applications, and we will use existing or newly defined ontologies and knowledge bases. This will be a proof of our system in terms of both usability and efficacy of the results.

Within the *MS* module, we are working on providing simple and intuitive tools with which the user can specify a meaningful portion of a model. The *MS* module will allow the user to draw a screen-space split line (*lasso*) and then it will partition the mesh vertices according to their screen space projections. Also, the user will be able to select a sequence of vertices on the mesh and cut along the shortest paths between them. Finally, we will implement the method of *intelligent scissoring* defined in [22] which allows the user to paint *strokes* on the model for specifying where cuts should be made. Each stroke will have a user-specified width representing a value of uncertainty within which the *MS* should construct the cut.

As far as the *AS* module is concerned, we are concentrating on CAD/CAM applications and we are integrating the segmentation method presented in [23] with the merging heuristics described in [24]. Additionally, since we are also interested in determining features like *handles* and *tunnels*, we are currently considering the approach presented in [25] in which handle and tunnels are extracted from manifold shapes. Finally, *be-SMART* will allow the *MS* and the *AS* modules to interact in order to iteratively improve the quality of the segmentation. Note that, in case of the segmentations   performed   by the *MS* and the *AS* modules, only *1*-hyperarcs, composed by closed chains of manifold edges will be created in the decomposition graph (see for example in Fig. 7(b)).

6 Concluding Remarks

We have presented the design and part of the development of *be-SMART*, a system for geometric-topological inspection and semantic structuring and annotation of 3D shapes. We described the architecture of the system going into details on its first module, the *Geometry and Topology Analyzer* and describing the characteristics it extract and the algorithms we have implemented for computing these information. We also introduced the second module, the *Topological Decomposer*, which decomposes non-manifold models into manifold parts. Finally, we briefly described our on-going work in the implementation of the *Manual* and *Automatic Segmentation* modules.

Note that the decomposition obtained by the application of the *Topological Decomposer* could be further combined with descriptions of the manifold parts, in terms of a skeleton, thus forming the basis for a two-level shape recognition process and for the creation of an iconic representation of the initial model. In this sense, we are investigating the possibility to add a new module to *be-SMART* (the *Iconizator*) which will provide an high-level description of the model on the basis of the topological information which can be used for effective searching and retrieval.

be-SMART is at step two of five of its development: some functionalities have to be implemented ex-novo, some others can be inherited by existing tools. In particular, *be-SMART* has been inspired by PhotoStuff [15] and we are working on the integration of its functionalities into *be-SMART*. We believe that *be-SMART* will be an important step towards an actual use of the Semantic Web in multimedia content.

Acknowledgments. This work has been partially supported by the European Network of Excellence AIM@SHAPE under contract number 506766, by the National Science Foundation under grant CCF-0541032, by the MIUR-FIRB project SHALOM under contract number RBIN04HWR8 and by the MIUR-PRIN project on "Multi-resolution modeling of scalar fields and digital shapes".

References

1. Razdan, A., Rowe, J., Tocheri, M., Sweitzer, W.: Adding Semantics to 3D Digital Libraries. In: Proc. of the 5th International Conference on Digital Libraries, Singapore (December 11-14, 2002)
2. The European Network of Excellence AIM@SHAPE contract number 506766, see: www.aimatshape.net
3. MINDSWAP, Maryland Information and Network Dynamics Lab Semantic Web Agents Project. see: http://www.mindswap.org/
4. Davies, J., Fensel, D., van Harmelen, F. (eds.): Towards the Semantic Web: Ontology-driven Knowledge Management. Wiley Press, Chichester (2002)
5. Maedche, A.: Ontology Learning for the Semantic Web. The International Series in Engineering and Computer Science, vol. 665 (2002) SBN: 978-0-7923-7656-9

6. Leon, J-C., Fine, L.: A new approach to the preparation of models for FE analyses Intern. Journal of Computer Applications in Technology 23(2/3/4), 166–184 (2005)
7. Albertoni, R., Papaleo, L., Pitikakis, M., Robbiano, F., Spagnuolo, M., Vasilakis, G.: Ontology-based Searching Framework for Digital Shapes. In: Gil, Y., Motta, E., Benjamins, V.R., Musen, M.A. (eds.) ISWC 2005. LNCS, vol. 3729, pp. 896–905. Springer, Heidelberg (2005)
8. Attene, M., Moccozet, L., Hassner, T., Leon, J.C., Sayegh, R., Tal, A., Papaleo, L., Robbiano, F., Gutierrez, M., Andersenn, O., Marini, S., Biasotti, S., Catalano, C., Cheutet, V., Albertoni, R., Belayev, A., Hammann, S., Alliez, P., Cignoni, P., Pitikakis, M.: Metadata for digital shape models (July 2005). IST-NoE 506766 AIM@SHAPE
9. Berners-Lee, T., Hendler, J., Lassila, O.: The Semantic Web, Journal Scientific American (May 2001)
10. Brickley, D., Guha, R.V.: Resource description framework (RDF) schema specification, Candidate recommendation, W3C Consortium (March 27, 2000)
11. Bechhofer, S., van Harmelen, F., Hendler, J., Horrocks, I., McGuinness, D.L, Patel-Schneider, P.F, Stein, L.A.: OWL Web Ontology Language Reference, W3C Candidate Recommendation (2003)
12. Halaschek-Wiener, C., Golbeck, J., Schain, A., Grove, M., Parsia, B., Hendler, J.: Annotation and provenance tracking in semantic web photo libraries. In: Moreau, L., Foster, I. (eds.) IPAW 2006. LNCS, vol. 4145, Springer, Heidelberg (2006)
13. Papaleo, L.: Towards a Semantic Web System for Understanding and Reasoning on Digital Shapes, Tech. Report #DISI-TR-06-09, University of Genova (May 2006)
14. The AIM@SHAPE Digital Shape Workbench, www.aimatshape.net/resources
15. Halaschek-Wiener, C., Schain, A., Golbeck, J., Grove, M., Parsia, B., Hendler, J.: A flexible approach for managing digital images on the semantic web. In: Proc. Inter. Workshop on Knowledge Markup and Semantic Annotation, Galway (2005)
16. The Xj3D Project, http://www.xj3d.org/
17. X3D specification, http://www.web3d.org/
18. De Floriani, L., Magillo, P., Puppo, E., Sobrero, D.: A multi-resolution topological representation for non-manifold meshes. Computer-Aided Design Journal 36(2), 141–159 (2004)
19. Paoluzzi, A., Bernardini, F., Cattani, C., Ferrucci, V.: Dimension-Independent Modeling with Simplicial Complexes. ACM Transactions on Graphics 12(1), 56–102 (1993)
20. De Floriani, L., Hui, A.: A Two-level Topological Decomposition for Non-manifold Simplicial Shapes. In: Proc. of Solid and Physical Modeling Symposium, ACM, Beijing, China (2007)
21. De Floriani, L., Huang, M., Hui, A.: TopMesh: Extracting topological characteristics from simplicial non-manifold shapes Technical Report CS-TR-4798, CS Department, University of Maryland, College Park (2006)
22. Funkhouser, T., Kazhdan, M., Shilane, P., Min, P., Kiefer, W., Tal, A., Rusinkiewicz, S., Dobkin, D.: Modeling by Example. In: ACM Transactions on Graphics (SIGGRAPH 2004), Los Angeles (2004)
23. Cohen-Steiner, D., Alliez, P., Desbrun, M.: Variational Shape Approximation. In: ACM Transactions on Graphics (SIGGRAPH 2004), Los Angeles (2004)
24. Sheffer, A.: Model simplification for meshing using face clustering. Computer-Aided Design 33(13), 925–934 (2001)
25. Dey, T.K., Li, K., Sun, J.: On computing handle and tunnel loops, IEEE Proc. NASAGEM07 (to appear). Tech Report OSU-CISRC-06/07-TR48 (June 2007)

Modeling Linguistic Facets of Multimedia Content for Semantic Annotation

Massimo Romanelli[1], Paul Buitelaar[2], and Michael Sintek[3]

[1] DFKI IUI, Saarbrücken, Germany
firstname.lastname@dfki.de
[2] DFKI LT, Saarbrücken, Germany
firstname.lastname@dfki.de
[3] DFKI KM, Kaiserslautern, Germany
firstname.lastname@dfki.de

Abstract. We provide an integrated ontological framework offering coverage for deep semantic content, including ontological representation of multimedia based on the MPEG-7 standard. We link the deep semantic level with the media-specific semantic level to operationalize multimedia information. Through the link between multimedia representation and the semantics of specific domains we approach the Semantic Gap. The focus of the paper is on the linguistic features of multimedia, the annotation of these features and their analysis.

1 Introduction

An important reason for the so-called 'semantic gap' (the difficulty in assigning high-level semantics to the results of low-level feature analysis) is the lack of alignment between different levels of semantics and levels of analysis for the different modalities. It is therefore important to develop an integrated model that aligns the foundational semantics level with the domain-specific semantics level, the semantics of the different modalities and the semantics of multimedia analysis. Additionally, in order to generalize this for all domains, the alignment of domain-specific and multimedia semantics should be organized on the foundational level.

In this paper we describe such an integrated model (working title 'Smart-MediaLing') that we developed in the context of the SmartWeb project[1] on mobile access to the Semantic Web [22]. In SmartWeb we were confronted with a number of different semantic analysis tasks (media annotation and presentation, multi-modal interaction, text analysis, etc.), each of which requiring a different level of representation, realized by a number of separate ontologies. In order to bring these different representation levels together we developed the SmartMediaLing integrated model as a common knowledge space that provides semantic interoperability between the different components of the SmartWeb system.

---

[1] http://www.smartweb-project.org

B. Falcidieno et al. (Eds.): SAMT 2007, LNCS 4816, pp. 240–251, 2007.

The SmartMediaLing approach described here uses the DOLCE foundational ontology for this purpose as it already provides patterns for defining so-called 'information objects', on top of which we were able to define the alignment of the different semantic levels mentioned above. In particular, we used the DOLCE D&S (Description and Situation) and OIO (Ontology of Information Objects) patterns to align the SmartMedia ontology for defining multimedia objects and the LingInfo ontology for defining linguistic (textual) objects with the DOLCE foundational model.

The paper is organized as follows: in Sec. 2 we describe the different levels of semantic representation that we consider. In Sec. 3 we discuss the constituent ontologies (DOLCE, SmartMedia, LingInfo) that are integrated into SmartMediaLing. In Sec. 4 we discuss the alignment strategy and present the SmartMediaLing ontology in more detail. In Sec. 5 we discuss the relation to other approaches.

2   Narrowing Down the Task: Different Semantic Levels

In order to reach the goal of appropriately represent and processing different information deriving from different analysis perspectives of the same object a complex approach to representation of contents becomes indispensable. Different perspectives on a complex object corresponds to different representation levels specifying features, properties and relationships on the different analysis points of view. In the definition of our task to proper represent semantics of multimedia we evidence basically four different level of representation that has to interact: foundational, domain specific, multimedia, linguistic.

Additional evidence for the definition of different representation levels comes from the semiotic investigation of communication.

Semiotics is "the study of the social production of meaning through signs" [19]. As a Kantian philosopher Peirce, key figure in the early development of semiotics, distinguishes between the "word" and the "sign" [17]. As defined in Peirce semiotic theory, communication takes place between three subjects: a *sign* (also called *representamen*), that denotes an object, an *object* from the world, to which this sign refers and the *interpretant*, the sense made of that sign. Peirce further distinguishes three types of sign depending on the type of relation existing between sign and object: *symbol*, based on a conventional relation (e.g. spoken language, language of gesture), *icon*, based on a similarity relation (e.g. a portrait), and *index* a contextual relation (e.g. smoke indicating the presence of fire). We identify the interpretant as being the concept in an ontological system, the symbol as depicting the linguistic level of representation, the icon as depicting the multimedia level, and the index as depicting the discourse level (see Fig. 1).

The Foundational Level. The definition of a complex semantic framework with different levels of representation needs the specification of a conceptual relational common ground offering appropriate instruments for linking together these levels. As soon as complexity increases, the usability reasons suggest applying modularization and distribution of knowledge in an

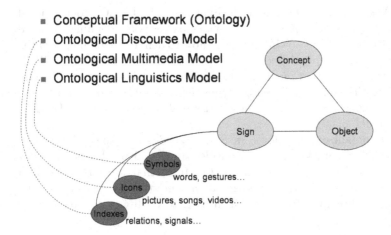

■ Conceptual Framework (Ontology)
■ Ontological Discourse Model
■ Ontological Multimedia Model
■ Ontological Linguistics Model

Fig. 1. Peirce semiotic triangle readapted and in context to different ontological levels

interoperating framework. To accomplish this task successfully a useful approach is to define a foundational level of representation from which every module of the framework can access basic ontological categories and relations. Foundational ontologies define these top level for the modularization and integration of meaning coming from different analysis sources.

The Domain-Specific Level. An ontology is said to be domain specific if it models the semantic of a specific domain. In our framework we define a different ontology for each domain and then align the all ontologies to the foundational one. Working this way we have the possibility to each time expand the world knowledge covered by the ontology by means of just adding new ontology branches without modifying basic relations in the framework.

The Multimedia Level. Videos, songs, pictures and so on are information objects with specific properties defining their realization in time (e.g. duration of a video) and space (e.g. number of pixels). On the other hand multimedia objects carry meaning that cannot be identified with the information object itself (e.g. an image depicting the Brazilian football team cannot be identified with the football team itself). We distinguish between a multimedia meta-data representation level, modeling characteristics of multimedia, and a domain specific level where concepts referred from media are completely specified.

The Linguistic Level. In order to ensure annotation for multilingual knowledge a rich representation of the linguistic symbols for the object classes that are defined by an ontology is needed. The linguistic level of information correspond to the symbol in Peirce triangle as depicted in Fig. 1. The purpose of such a semantic level is the definition of a grounding to the human cognitive and linguistic domain. Such domain is also important in the context of the interaction with multimedia objects where texts appear also in the perspective of a media object or as part of other media (e.g., the caption of a picture, the subtitles of a video).

The Analysis Level. Parallel to the already mentioned levels, that we can define as "static", we regard the "dynamic" dimension of multimedia as being the analysis level. We consider analysis, in both decomposition and annotation cases, as being a process activated by an agent, allayed to a multimedia object (domain) and resulting in his decomposition.

3   The Constituent Modules

Following the approach in [15] in order to define a framework with the features specified in Sec. 2 we have to first select a foundational ontology that matches the described requirements and enables us to reuse existing components. We decided to adopt the DOLCE ontology providing together a well defined formalization for basic relations and a number of modules, among others for the definition of contexts (D&S) and knowledge content (OIO). The second step is the specification of an adequate multimedia domain capable of describing annotation and decomposition of multimedia and a straight forwarded ontology description of linguistic feature.

In this section we shortly present the DOLCE ontology with the two modules *Descriptions & Situations* and *Ontology of Information Objects*. We then describe in two dedicated subsections the *SmartMedia* ontology for the coverage of the surface representation of the multimedia level. Finally we present the *LingInfo* ontology modeling facets of the linguistic domain.

3.1   DOLCE, D and S, OIO

DOLCE belongs to the WonderWeb library of foundational ontologies [12]. It is intended to act as a starting point for comparing and elucidating the relationships and assumptions underlying existing ontologies of the WonderWeb library. DOLCE (Descriptive Ontology for Linguistic and Cognitive Engineering) [7] is based on the fundamental distinction between enduring and perduring entities. An endurant is an entity that is wholly present, i.e., whose parts are all present, at any time at which it exists. A perdurant is an entity that enfolds in time, i.e., for any time at which it exists, some of its parts are not present. Meaning that *participation* is the main relation between *Endurants* (i.e., objects or substances) and *Perdurants* (i.e., events or processes): an *Endurant* exists in time by participating in a *Perdurant*. For example, a natural person, which is an *Endurant*, participates in his or her life, which is a *Perdurant*. DOLCE introduces *Qualities* as another category that can be seen as the basic entities we can perceive or measure: shapes, colors, sizes, sounds, smells, as well as weights, lengths or electrical charges. Spatial locations (i.e., a special kind of physical quality) and temporal qualities encode the spatio-temporal attributes of objects or events. Finally, *Abstracts* do not have spatial or temporal qualities and they are not qualities themselves. An example are *Regions* used to encode the measurement of qualities as conventionalized in some metric or conceptual space.

In DOLCE the module *Descriptions&Situations* (D&S) [8] has been defined to standardize a variety of reified contexts and states of affairs.

The DOLCE module **OIO** (**O**ntology of **I**nformation **O**bjects) provides a design pattern that allows us to concisely model the relationship between entities in an information system and the real world. As emphasized in [9] INFORMATION OBJECTS can be seen as NON-PHYSICAL-ENDURANTS participating in computational activities. Information-Objects correspond to the spatio-temporal entities of abstract information formalizing Shannon's communication theory [20].

3.2   SmartMedia

MPEG-7[2] is conceived for describing multimedia content data. MPEG-7 is used to store meta-data about multimedia in order to tag particular events. In the context of the SmartWeb project we defined an MPEG-7 based ontology (Smartmedia) following the approach in [2] and [11] restricting the number of the modeled concepts to those that fit well to the project.

Primarily the concepts mpeg7:MediaFormat for format and the coding parameters, mpeg7:MediaPro for coding schemes like resolution, compression, and mpeg7:SegmentDecomposition for decompositions of the audio, visual, textual segments in space, time, and frequency are imported into Smartmedia in order to offer a well defined background for the specification of meta-data level describing multimedia events like synchronization or decomposition of media.

3.3   Linginfo

Automatic multilingual knowledge markup requires a rich representation of the features of linguistic expressions (such as terms, synonyms and multilingual variants) for ontology classes and properties. Currently, such information is mostly missing or represented in impoverished ways, leaving the semantic information in an ontology without a grounding to the human cognitive and linguistic domain. Linguistic information for terms that express ontology classes and/or properties consists of lexical and context features, such as:

- *language-ID* - ISO-based unique identifier for the language of each term
- *part-of-speech* - representation of the part of speech of the head of the term
- *morpho-syntactic decomposition* - representation of the morphological and syntactic structure (segments, head, modifiers) of a term
- *statistical and/or grammatical context model* - representation of the linguistic context of a term in the form of N-grams, grammar rules or otherwise

To allow for a direct connection of this linguistic information for terms with corresponding classes and properties in the domain ontology, [4] developed a lexicon model (LingInfo) that enables a linguistically motivated definition of terms for

---

[2] 7http://www.chiariglione.org/mpeg/standards/mpeg-7/mpeg-7.html

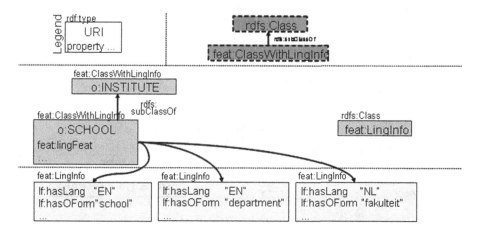

Fig. 2. LingInfo model with example domain ontology classes and LingInfo instances (simplified)

each class or property. The LingInfo model [3] is represented by use of the meta-class `ClassWithLingInfo` (and meta-property `PropertyWithLingInfo`), which allow for the representation of LingInfo instances with each class/property, where each LingInfo instance represents the linguistic features (`feat:lingInfo`) of a term for that particular class.

Figure 2 shows an overview of the model with example domain ontology classes and associated LingInfo instances. Figure 3 shows a sample application of the model with a LingInfo instance (and connected 'stem' instances) that represents the decomposition of the Dutch term "fakulteitsgebouw" ("department building"). The example shows a LingInfo instance (`Term-1` with `semantics "SCHOOL"`) that represents the word form "fakulteitsgebouw" (instance WordForm-1), which can be decomposed into "fakulteit" (`Term-2` , "fakulteit" with `semantics "SCHOOL"`) and "gebouw" (`Term-3` with `semantics "BUILDING"`).

4   Bringing It All Together

In Sec. 2 and 3 we presented respectively the different levels of representation and how we ontologically cover such levels in order to proper processing multimedia information. In this section we show how we connected these different levels.

This work were developed in the context of the SmartWeb project where the three ontologies introduced in Sec. 3 were all adopted as part of a comprehensive ontology named SWIntO (SmartWeb Integrated Ontology)[14]. The DOLCE ontology and the modules OIO and D&S were modified to meet the needs of the project and evolved respectively to SmartDOLCE, SmartOIO and SmartD&S. Basic functionality of Dolce remained unaffected. For more details on the use of DOLCE in SWIntO see also [6].

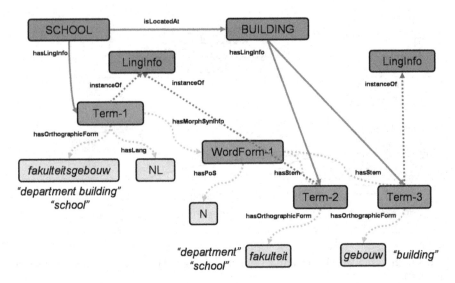

Fig. 3. LingInfo instance (partial) for the morpho-syntactic decomposition of the Dutch term "fakulteitsgebouw" ("department building")

In the context of the SmartWeb project we successfully used this framework for the disambiguation of cross-modal reference expressions and resolution of multi-modal expressions. This work enabled the system the use of multimedia in a multimodal context like in the case of mixed gesture and speech interpretation, where every object that is visible on the screen must have a comprehensive ontological representation in order to be identified and processed [18][21].

4.1 Alignment Strategy

The alignment process of a domain ontology to a core ontology is directly dependent from several factors [10]: intended use of the aligned ontology, intended form of the framework (modular, distributed), etc.

In our case we played a particular attention at following parameters:

- A modular reuse of the different component ontologies in other projects.
- Ontology alignment is different from equivalence because any element in the alignment depend on other elements and there will be degree of confidence between aligned elements.

In order to reach these means we decided to align ontologies to DOLCE as follows:

- non destructive: alignment happens without modifications for the core ontology.
- non reusing: properties are completely defined in the domain ontologies and then aligned as sub-properties to properties of DOLCE. No properties of DOLCE are directly reused in the ontology.

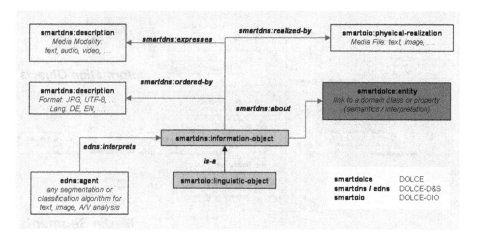

Fig. 4. Basic concepts and patterns of the DOLCE OIO module used in SWIntO for the integration of the foundational, multimedia and linguistic levels

4.2   The Integrated Model: 'SmartmediaLing'

In [9] information objects are introduced on the base of an example (Dante's Comedy). To align the smartmedia ontology to the DOLCE we applied this example to the world of multimedia. In Fig. 4 are depicted basic relations and concepts modeled in the OIO framework and adopted for the alignment.

In the case of the analysis of a picture of the Brazilian football team we identify the picture itself as an *information object* that is *about* an *entity*, a *particular* modeling the Brazilian football team in a DOLCE aligned domain specific ontology. The picture can be decomposed to different *segments*, each segment being about a different player from the same domain specific ontology. A *Segment-Decomposition* is an information object carrying the result of a segmentation process, a *perdurant* applied by some *agent*, some classification or segmentation algorithm that *interprets* the information object *using* visual descriptors. In Fig. 5 we give a graphical representation of such relations.

Exactly the same way we can see e.g., a semantic parser as an agent participating in a parsing process that is identified by an information object of type *textDecomposition*. The result of such a decomposition are again linguistic entities like sentences, words, morphemes and so on, as represented in the LingInfo ontology (See figure 6). Each linguistic information object is *about* an entity from a domain specific ontology and is itself a sort of *TextSegment* as specified from the Smartmedia ontology.

5   Related Work

A very interesting approach is that followed in [1] where the authors concentrate on the task of creating an ontological framework in the context of annotation

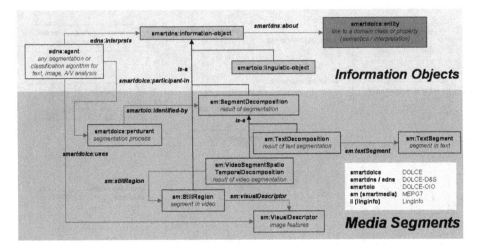

Fig. 5. The Smartmedia ontology integrated in the OIO module

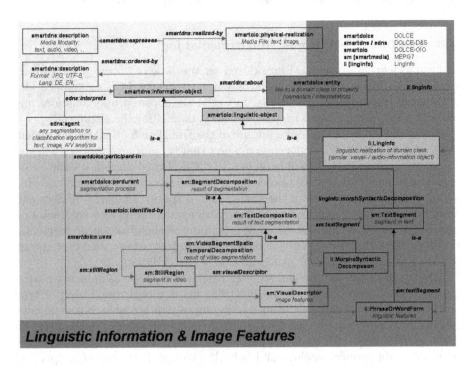

Fig. 6. The integrated framework for the representation of linguistic and multimedia objects as information objects

of multimedia objects. The approach is based on the DOLCE ontology and makes deep use of the *Descriptions and Situations* DOLCE module defining annotation, description and semantic patterns usable for the realization of an annotation tool. This work offers a well defined specification of interpretation and annotation processes. However deep analysis of relations between surface representation objects and deep semantic objects is not taken into account.

An other approach to semantic annotation for multimedia content similar to the one adopted in this work can be found in [16]. The work is based on the same ontological background and place emphasis mostly on the visual part of the ontology, the context analysis used for the visual analysis and a visual analysis algorithm.

6   Conclusion

In this paper we described an approach to the specification of a semantics for the annotation and use of multimedia objects in a comprehensive ontological framework. We analyzed the characteristics of multimedia objects and evidenced the necessity of specifying different levels of representation for covering the complexity of the task. On the other hand we established relations between the different level of analysis in order to ensure a proper treatment of processes of analysis and annotation of such multimedia objects. We stressed the necessity of a linguistic representation level in these framework and offered an ontological model of this level. Finally we showed how an alignment of different levels of semantic is possible in the context of a foundational ontology like DOLCE for a successful use in systems like SmartWeb.

Future work is needed to completely specify the processing part of the modeling. The approach in Sec. 5 will be taken into account for this purpose and actual work is concentrating on harmonizing the two approaches.

Acknowledgments

The research presented here is sponsored by the German Ministry of Research and Technology (BMBF) under grant 01IMD01A (SmartWeb). We thank our student assistants and the project partners. The responsibility for this papers lies with the authors.

References

1. Arndt, R., Staab, S., Troncy, R., Hardman, L.: Adding Formal Semantics to MPEG-7: Designing a Well-Founded Multimedia Ontology for the Web. Arbeitsberichte aus dem Fachbereich Informatik, Universit't Koblenz-Landau, ISSN (April 2007)
2. Benitez, A., Rising, H., Jorgensen, C., Leonardi, R., Bugatti, A., Hasida, K., Mehrotra, R., Tekalp, A., Ekin, A., Walker, T.: Semantics of Multimedia in MPEG-7. In: ICIP. Proc. of IEEE International Conference on Image Processing (2002)
3. Buitelaar, P., Declerck, T., Frank, A., Racioppa, S., Kiesel, M., Sintek, M., Romanelli, M., Engel, R., Sonntag, D., Loos, B., Micelli, V., Porzel, R., Cimiano, P.: LingInfo: A Model for the Integration of Linguistic Information in Ontologies.

In: OntoLex 2006. Proceedings of Workshop on Interfacing Ontologies and Lexical Resources for Semantic Web Technologies, Genova, Italy (May 27, 2006)

4. Buitelaar, P., Sintek, M., Kiesel, M.: A Lexicon Model for Multilingual/Multimedia Ontologies. In: Sure, Y., Domingue, J. (eds.) ESWC 2006. LNCS, vol. 4011, Springer, Heidelberg (2006)

5. Buitelaar, P., Cimiano, P., Racioppa, S., Siegel, M.: Ontology-based Information Extraction with SOBA. In: LREC 2006. Proc. of the 5th Conference on Language Resources and Evaluation (2006)

6. Cimiano, P., Eberhart, A., Hitzler, P., Oberle, D., Staab, S., Studer, S.: The SmartWeb Foundational Ontology. Technical report, Institute for Applied Informatics and Formal Description Methods (AIFB) University of Karlsruhe, SmartWeb Project, Karlsruhe, Germany (2004)

7. Gangemi, A., Guarino, N., Masolo, C., Oltramari, A., Schneider, L.: Sweetening Ontologies with DOLCE. In: Gómez-Pérez, A., Benjamins, V.R. (eds.) EKAW 2002. LNCS (LNAI), vol. 2473, Springer, Heidelberg (2002)

8. Gangemi, A., Mika, P.: Understanding the semantic web through descriptions and situations. In: Meersman, R., Tari, Z., Schmidt, D.C. (eds.) CoopIS 2003, DOA 2003, and ODBASE 2003. LNCS, vol. 2888, pp. 689–706. Springer, Heidelberg (2003)

9. Gangemi, A., Borgo, S., Catenacci, C., Lehmann, J.: Deliverable of the EU FP6 project Metokis (2005)

10. Hughes, T., Ashpole, B.C.: The Semantics of Ontology Alignment. In: I3CON. Information Interpretation and Integration Conference (August 2004)

11. Hunter, J.: Adding Multimedia to the Semantic Web - Building an MPEG-7 Ontology. In: SWWS. Proc. of the International Semantic Web Working Symposium (2001)

12. Masolo, C., Borgo, S., Gangemi, A., Guarino, N., Oltramari, A., Schneider, L.: The WonderWeb Library of Foundational Ontologies. WonderWeb Deliverable D17 (August 2002), http://wonderweb.semanticweb.org

13. Niles, I., Pease, A.: Towards a Standard Upper Ontology. In: Welty, C., Smith, B. (eds.) FOIS-2001. Proc. of the 2nd International Conference on Formal Ontology in Information Systems, Ogunquit, Maine (2001)

14. Oberle, D., Ankolekar, A., Hitzler, P., Cimiano, P., Schmidt, C., Weiten, M., Loos, B., Porzel, R., Zorn, H.-P., Micelli, V., Sintek, M., Kiesel, M., Mougouie, B., Vembu, S., Baumann, S., Romanelli, M., Buitelaar, P., Engel, R., Sonntag, D., Reithinger, N., Burkhardt, F., Zhou, J.: DOLCE ergo SUMO: On Foundational and Domain Models in SWIntO (SmartWeb Integrated Ontology). Journal of Web Semantics: Science, Services and Agents on the World Wide Web (2007)

15. Oberle, D., Lamparter, S., Grimm, S., Vrandecic, D., Staab, S., Gangemi, A.: Towards Ontologies for Formalizing Modularization and Communication. In: Large Software Systems. Applied Ontology, vol. 1(2), pp. 163–202. IOS Press, Amsterdam (2006)

16. Papadopoulos, G.T., Mezaris, V., Kompatsiaris, I., Strintzis, M.G.: Ontology-Driven Semantic Video Analysis Using Visual Information Objects. In: SAMT 2007. LNCS, vol. 4816, pp. 56–69. Springer, Heidelberg (2007)

17. Peirce, C.S., Burks, A.W.: Collected Papers of Charles Sanders Peirce. In: Hartshorne, C., Weiss, P. (eds.), vol. 1-6, vols. 7-8, pp. 1931–1935. Harvard University Press, Cambridge, MA (1958)

18. Romanelli, M., Sonntag, D., Reithinger, N.: Connecting Foundational Ontologies with MPEG-7 Ontologies for Multimodal QA. In: Avrithis, Y., Kompatsiaris, Y., Staab, S., O'Connor, N.E. (eds.) SAMT 2006. LNCS, vol. 4306, Springer, Heidelberg (2006)

19. Scollon, R., Scollon, S.W.: Discourses in Place - Language in the Material World. Routeledge, London (2003)

20. Shannon, C.: A Mathematical Theory of Communication. Bell System Technical Journal 27, 379–423, 623-656 (July, October 1948)

21. Sonntag, D., Romanelli, M.: A Multimodal Result Ontology for Integrated Semantic Web Dialogue Applications. In: LREC 2006. Proc. of the 5th Conference on Language Resources and Evaluation (2006)

22. Wahlster, W.: SmartWeb: Mobile applications of the semantic web. In: Dadam, P., Reichert, M. (eds.) GI Jahrestagung 2004, Springer, Heidelberg (2004)

Leveraging Ontologies, Context and Social Networks to Automate Photo Annotation

Fergal Monaghan and David O'Sullivan

Digital Enterprise Research Institute,
National University of Ireland, Galway,
IDA Business Park, Lower Dangan,
Galway, Ireland
fergal.monaghan@deri.org, david.osullivan@nuigalway.ie
http://www.deri.org

Abstract. This paper presents an approach to semi-automate photo annotation. Instead of using content-recognition techniques this approach leverages context information available at the scene of the photo such as time and location in combination with existing photo annotations to provide suggestions to the user. An algorithm exploits a number of technologies including Global Positioning System (GPS), Semantic Web, Web services and Online Social Networks, considering all information and making a best-effort attempt to suggest both people and places depicted in the photo. The user then selects which of the suggestions are correct to annotate the photo. This process accelerates the photo annotation process dramatically which in turn aids photo search for a wide range of query tools that currently trawl the millions of photos on the Web.

Keywords: Photo annotation, context-aware, Semantic Web.

1 Introduction

Finding photos is now a major activity for Web users, but they suffer from information overload as their attempts to find the photos they want are frustrated by the enormous and increasing number of photos [1][2]. The processing speed of computers can be leveraged to perform the hard work of finding photos for the user and to thereby alleviate the information overload. To enable the machine to retrieve photos for the user, we must first examine how users mentally recall photos themselves. Research indicates that users recall photos primarily by cues in the following categories, in descending order of importance: (i) who is in the photo; (ii) where the photo was taken; and (iii) what event the photo covers [3]. Searchable description metadata in these categories can be created for photos and search engines can then match user queries with these descriptions and present the best matches to the user. A key challenge is how to create this useful, searchable description metadata about photos. Manual annotation of photos is tedious and consumes large amounts of time. Automated content-based techniques such as face recognition rely on large training sets and are dependant

B. Falcidieno et al. (Eds.): SAMT 2007, LNCS 4816, pp. 252–255, 2007.
© Springer-Verlag Berlin Heidelberg 2007

on the illumination conditions at the scene of photo capture [4]. Complimentary context-based approaches provide a lightweight, scalable solution to support the abstract way in which users actually think about photos. Section 2 introduces the implementation of just such a context-based approach: the Annotation CReatiON for Your Media (ACRONYM) prototype[1].

Related Work. CONFOTO [5] is a semantic browsing and annotation service for conference photos. It combines the flexibility of the Resource Description Framework (RDF) with recent Web trends such as folksonomies, interactive user interfaces, and syndication of news feeds.

PhotoCompas [6] uses timestamps and co-ordinates captured by GPS-enabled cameras to lookup higher level contextual metadata about photos from existing Web services. Given metadata from previously annotated photos it suggests people that may be depicted in consequent photos.

ZoneTag[2] is a prototype for Nokia S60 smartphones that allows the user to upload images from the phone to the Flickr website. Zonetag leverages the context (e.g. location and time) captured by the smartphone to find a location tag and to suggest other Flickr tags based on tags previously entered by the user and their social network under a similiar context.

2 Annotation CReatiON for Your Media

ACRONYM is a Semantic Web-based photo annotation tool that can annotate any JPEG image on the Web with RDF. It focusses on the most important recall cues and takes advantage of RDF's powerful expressivity, interoperability and mobility while hiding its complexities from the user. ACRONYM makes use of the EXchangeable Image File (EXIF) format metadata that is created and stored inside JPEG photo files by off-the-shelf digital cameras. This commonly includes a timestamp of when the photo was captured, shutter speed, exposure time etc. but can also include the co-ordinates of the camera at the time of capture if the camera has been coupled with a GPS receiver. ACRONYM also makes frequent use of the GeoNames[3] geographical database, map, ontology and Web services.

The user logs in with their email address: this is hashed to provide a unique identifier and to link the user to any RDF metadata describing them. Once logged in the user can add people, places and import arbitrary RDF to the system with the click of a button. The user selects which JPEG photo to annotate by specifying its URL. The system displays the JPEG image and translates its EXIF metadata into RDF metadata formalised in a combination of the Dublin Core Terms, Friend-of-a-Friend (FOAF), World Geodetic System 1984 positioning and GeoNames ontologies. This RDF is then combined with similiar metadata from other photos to provide suggestions to the user for the creation of further metadata about the photo.

---

[1] http://acronym.deri.org

[2] http://zonetag.research.yahoo.com

[3] http://www.geonames.org

The user selects the people depicted in the photo from a list of suggested candidates that are described by FOAF social network metadata within the system. FOAF metadata about people and the relationships between them and the user can be created at the click of a button or imported from external sources. Instead of trying to identify faces in the content of the image, ACRONYM analyses the social context of the photo, ranking and ordering candidates in the list based on their social connection to the photo and the user as described by the ⟨foaf:knows⟩ relationship. A candidate receives one ranking point for each direction of a knows relationship between them and the user. Once at least one person has been selected as being depicted in the photo, an additional metric is used: one ranking point is added to each candidate per knows relationship between them and each person depicted. As the user selects which of the candidates are depicted, the list of candidates is updated: those people in the social circle of numerous depicted people float to the top of the list. This captures the social context of the photo and user and makes effective use of it by homing in on the most likely people to be annotated as depicted.

RDF metadata about places can be added via a full-text search field that queries a GeoNames Web service or can alternatively be imported from external sources. The user can also import from GeoNames all metadata on places within a specified kilometre radius of a selected place. Similiar to above the user then selects the places depicted in the photo from a list of suggested candidates in the system. If co-ordinates have been supplied by the EXIF metadata, these are used to lookup and import metadata about nearby places from GeoNames.

If (as is the common case) no co-ordinates have been supplied, ACRONYM again takes an analytical approach to estimate where the photo was captured. The algorithm takes the set of people already annotated as being depicted in the photo and estimates the location of each person at the time the photo was captured based on the co-ordinates of places they are co-depicted with in previously annotated photos. Firstly, each person is analysed to determine the temporally closest, previously annotated photo that depicts them. The mean of the co-ordinates of the places depicted in this other photo provides an estimate of where that person was at that time. The timestamped co-ordinates estimate for each person is then used to obtain weighted mean co-ordinates (weighted by temporal proximity to the photo being annotated) as a rough estimate of where all the people are co-depicted in this photo. This rough estimate is then used in place of hard data captured by a GPS receiver.

Each place in the system is then ranked according to its geographic proximity to where the photo was captured (or was estimated to have been captured) and this ranking is used to order the suggested places list for the user to select actual depicted places from. Once there is at least one place selected as being depicted, the mean co-ordinates of each depicted place are used in place of the rough estimate above to rank the suggested places. As the user selects which candidates are depicted, the list of candidates is updated: those places nearby numerous depicted places float to the top of the list. This captures the geographic context of the photo and makes effective use of it by homing in on the most likely places to be co-depicted with the given people and places at the given time.

3   Future Work

The main thrust of future work is to integrate event suggestion with that of people and places. Cluster analysis will be implemented on the temporal, geographic and social aspects of photos to detect abstract events which can be concretely named, suggested and annotated to photos. A key concept of future efforts will be two slider bars to tune precision and recall: only candidates with a ranking that meets the recall setting will be suggested and those suggested candidates that also meet the precision setting will be automatically annotated to the photo. The user will be able to quickly tune the automation level from fully manual to fully automatic. Furthermore, readily available face detection tools will be assessed to locate and count people depicted in photos.

4   Conclusions

ACRONYM's suggestion algorithm captures and makes use of key context cues. By looking up existing information and inferring higher level contextual knowledge the tool accelerates the photo annotation process. The end result is that from ground truth machine-readable metadata captured by cameras, such as time and co-ordinates, ACRONYM discovers human-readable fields such as placenames and people names that are actually useful to query engine users. This alleviates the information overload on users searching for photos on the Web.

Acknowledgments. This work is supported by Science Foundation Ireland (SFI) under the DERI-Líon project (SFI/02/CE1/1131).

References

1. Infotrends: Worldwide camera phone sales to reach nearly 150 million in 2004, capturing 29 billion digital images. Technical report (2004)
2. Infotrends: Worldwide consumer digital camera sales to reach nearly 53 million in 2004. Technical report (2003)
3. Naaman, M., Harada, S., Wangy, Q., Garcia-Molina, H.: Context data in georeferenced digital photo collections. In: MM 2004. International Conference on Multimedia, New York, NY (2004)
4. O'Toole, A.J., Phillips, P.J., Jiang, F., Ayyad, J., Penard, N., Abdi, H.: Face recognition algorithms surpass humans matching faces over changes in illumination. IEEE Transactions on Pattern Analysis and Machine Intelligence 29(9), 1642–1646 (2007)
5. Nowack, B.: Confoto: Browsing and annotating conference photos on the semantic web. Web Semantics: Science, Services, and Agents on the World Wide Web 4(4), 263–266 (2006)
6. Naaman, M.: Leveraging Geo-Referenced Digital Photographs. PhD thesis, Stanford University (2005)

Use Cases of Scalable Video Based Summarization and Adaptation Within MPEG-21 DIA

Luis Herranz and José M. Martínez

Grupo de Tratamiento de Imágenes, Escuela Politécnica Superior,
Universidad Autónoma de Madrid, E-28049 Madrid, Spain
{luis.herranz,josem.martinez}@uam.es

Abstract. Frame selection techniques based on semantic analysis are widely used to provide content-based summaries of video sequences. Video adaptation is also required to enable the consumption of these summaries through an increasing variety of devices and networks. Scalable video coding provides embedded coding of video where video adaptation is very simple and efficient. In this paper, we describe a model where both video summarization and adaptation are integrated within the MPEG-21 DIA framework, along with some use cases.

Keywords: MPEG-21, digital item adaptation, scalable video, video summarization, semantic video adaptation.

1 Introduction

Video summarization is a key technology required for efficient access and browsing to the huge amount of content available. In general, a summarized sequence is built selecting frames guided by some kind of semantic analysis of the content. Some examples of such video summarization techniques are static storyboards, fast playback for fast browsing and semantic frame dropping. Along with video summarization, content adaptation[1] is another key issue to bring effectively content from service providers to actual users, each one using their own terminals and networks, with their own capabilities and constraints. The MPEG-21 standard specifies a number of specific tools for content adaptation in MPEG-21 DIA[1]. As adaptation is usually a computationally expensive process, Scalable Video Coding[2] is becoming a useful technology to enable very efficient adaptation as the video bitstream is coded in a way that simplifies the adaptation process: a scalable video stream contains embedded versions of the source content that can be decoded at different resolutions, frame rates and qualities, simply selecting the required parts of the bitstream.

In most systems, summarization and adaptation are considered separately in two independent stages. In [3], both approaches are integrated in the same framework for efficient generation of adapted summaries. In this paper we describe the generation of adapted video summaries in that framework in several practical use cases.

B. Falcidieno et al. (Eds.): SAMT 2007, LNCS 4816, pp. 256–259, 2007.

2 Summarization Model of Scalable Video

In the integrated model of video summarization and adaptation of scalable video[3], the basic unit in the model is the Group of Pictures (GOP), as it is the basic coding and adaptation unit of scalable video. In a fully scalable video, the parts of the bitstream can be organized according to a three dimensional spatio-temporal-quality structure where each point (s, t, q) represents an adapted sequence. We will assume that the temporal level t corresponds to a GOP with 2^t frames.

The output of a video summarization algorithm is the set of frames that should be included in the adapted sequence. Different video summarization algorithms will differ on the way the frames are selected, but the final result for each frame is that it is either included or not. However, it is more convenient to express the resulting set of frames in terms of meaningful parameters for scalable video adaptation. For each GOP k, the parameter $skip_k$ indicates if it is skipped or included, and if so, the temporal level m_k to be selected. Some implicit quantization appears in the number of frames per GOP due to temporal scalability, but it is not important in most practical cases. Depending on the way this frame selection is achieved we will distinguish three types of summarization approaches (see section 3): semantic frame dropping, fast playback and static storyboard.

The last stage is the adaptation of the sequence itself to a given environment, generating the output sequence using a scalable video adaptation engine, additionally guided by the summarization parameters m_k and $skip_k$. The adaptation can be formulated as an optimization problem using the standard MPEG-21 DIA framework[4]. In the case of fully scalable video, the optimization problem usually consists of the selection of the version of the GOP (number of spatial, temporal and quality layers) that maximizes some measure of quality, satisfying the constraints of the usage environment (resolution, frame rate and network bitrate). The summarization is included in this MPEG-21 DIA framework modifying the optimization problem in order to include the results of analysis as a new *semantic* constraint.

3 Use Cases and Experimental Results

We focus on some applications of the proposed framework involving different cases of temporal summarization/adaptation, and show the adaptation results with a test sequence in a typical usage environment.

The test sequence is built by chaining three standard sequences: *akiyo, foreman* and s*tefan*. Each of these sequences has 300 frames with CIF resolution at 30 frames per second. This sequence was encoded with a wavelet scalable codec[5]. The sequence is adapted to be visualized in a constrained environment (e.g. a QCIF hand-held device with a limited bitrate network).

3.1 Semantic Frame Dropping and Fast Playback

We demonstrate the semantic adaptation engine using the activity based method proposed in [3]. The assumption is that in this case the skimming curve is

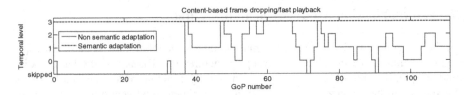

Fig. 1. Summarization parameters of the test sequence in different use cases

proportional to the activity of the GOP. The summarization temporal level resulting from the analysis is shown in Fig. 1 along with the curve for non semantic adaptation. It also shows which GOP are skipped.

Considering the case of constant time slot, the output sequence will drop many of the frames according to the skimming parameters. This is the case of semantic frame dropping. The same framework and analysis is used to generate a summarized sequence for fast playback of the video, using a playback frame rate of 30 frames per second. The only difference is that in this case the duration of each GOP is not constant, depending on the number of frames selected for each GOP. The duration is smaller than the duration of the source video, which is useful to provide an effective fast overview of the content.

3.2 Static Storyboard Summary

A set of few frames is a very useful representation, providing a fast overview of the sequence. In order to prove the utility of the framework in this case, a simple clustering method is used, adapted to the scalable video model. Each frame is represented by a YUV histogram computed over the lowest temporal and spatial resolution versions. The quality of the summaries is not almost affected[6] by these subsamplings while they are computed much faster. These feature vectors are clustered using K-means algorithm with the euclidean distance, resulting in K centroids. For each centroid, the closest frame of its cluster is selected as keyframe.

The resulting summary leads to the summarization parameters shown in Fig. 2. Note that only temporal level 0 is selected, as only 1 frame can be selected per GOP. If the GOP does not contain a keyframe, it is skipped. The resolution is constrained by the thumbnail resolution and the bitrate is constrained by the maximum size of each thumbnail (34 kB in the tests). The resulting summaries for K=5 keyframes are shown in Fig. 3, in the cases of 88x72 and 44x36 pixel thumbnails.

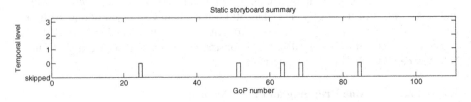

Fig. 2. Summarization parameters in the case of a static storyboard

88x72 thumbnails   44x36 thumbnails

Fig. 3. Summaries for different thumbnail sizes

4 Conclusions

In this paper we have described a method that integrates several types of summarization approaches with adaptation in the same framework. It uses scalable video adaptation and MPEG-21 DIA, with the advantage of very efficient adaptation described in a standard way. We have also described its application in the use cases of dynamic frame dropping, fast browsing and static storyboards and the model was tested with simple but useful analysis methods that provide very efficient solutions for effective browsing of video content.

Acknowledgements

Work partially supported by the Ministerio de Ciencia y Tecnología of the Spanish Government under project TIN2004-07860 (MEDUSA) and by the Comunidad de Madrid under project S-0505-TIC-0223 (PROMULTIDIS).

The authors would like to acknowledge Nikola Šprljan, Marta Mrak and Ebroul Izquierdo for their help and support with their scalable video codec.

References

1. Vetro, A.: MPEG-21 digital item adaptation: Enabling universal multimedia access. IEEE Multimedia 11, 84–87 (2004)
2. Ohm, J.R.: Advances in scalable video coding. Proceedings of the IEEE 93, 42–56 (2005)
3. Herranz, L.: Integrating semantic analysis and scalable video coding for efficient content-based adaptation. Multimedia Systems 13, 103–118 (2007)
4. Mukherjee, D., Delfosse, E., Kim, J.G., Wang, Y.: Optimal adaptation decision-taking for terminal and network quality-of-service. IEEE Transactions on Multimedia 7, 454–462 (2005)
5. Sprljan, N., Mrak, M., Abhayaratne, G.C.K., Izquierdo, E.: A scalable coding framework for efficient video adaptation. In: WIAMIS. Proc. Int. Work. on Image Analysis for Multimedia Interactive Services (2005)
6. Mundur, P., Rao, Y., Yesha, Y.: Keyframe-based video summarization using Delaunay clustering. International Journal of Digital Libraries 6, 219–232 (2006)

Ontology for Semantic Integration in a Cognitive Surveillance System

Carles Fernández[1] and Jordi Gonzàlez[2]

[1] Computer Vision Centre, Edifici O. Campus UAB, 08193, Bellaterra, Spain
[2] Institut de Robòtica i Informàtica Ind. UPC, 08028, Barcelona, Spain
{perno,poal}@cvc.uab.es

Abstract. The increasing interest in Cognitive Vision Systems (CVS) motivates the apparition of ad-hoc stages designed for the integration of multiple kinds of knowledge. This paper proposes a novel ontology to restrict and integrate high-level semantics for Human Sequence Evaluation (HSE), which targets multilingual capabilities and multipurpose end-user interfaces. The main contributions of this paper are the conception of a neutral semantic layer, which allows to link vision and linguistic domains; and the use of *situations* instead of verbs as basic elements for an ontological categorization of occurrences. In our approach, the domain has been restricted to outdoor surveilled scenarios, involving interactions among pedestrians, static objects, and vehicular traffic.

1 Introduction

There exists a growing involvement of researchers into Cognitive Vision Systems (CVS), which operate on different levels of abstraction and analysis related to human cognition. This is well accomplished by the conception of *Human Sequence Evaluation* (HSE), in which the interpretation of human behaviors in image sequences is performed by a modular architecture for user-oriented applications [1]. In such a framework, it is essential to develop proper criteria for high-level knowledge sharing and validation. Due to the broad spectrum of semantic representations, it is necessary to find mechanisms that clarify the structure of knowledge in given domains, for integration purposes. Towards this end, ontologies have been widely accepted as convenient tools.

This contribution addresses the use of ontologies as an integrative framework for knowledge representation, within a HSE system with multiple user interfaces and multilingual capabilities. The goal is to automatically extract behavioral descriptions from image sequences in restricted domains, in this case urban outdoor surveillance environments. We also discuss several criteria to model the semantic background of such ontologies, which are used to link the different representations at cognitive, high-level stages.

2 Representation Formalisms in HSE

Several formalisms are employed by a HSE system in order to represent semantic knowledge, which are conditioned to the application domain they address.

B. Falcidieno et al. (Eds.): SAMT 2007, LNCS 4816, pp. 260–263, 2007.

Table 1. Table of semantic representations in HSE

	STP	HLSP	LP
Type of semantics	Metric-temporal (basic relations)	Thematic roles (inferential role semantics)	Linguistic-oriented (NL semantics)
Models implied	Scene models, human motion models	Behavioral models (contextual and intentional)	Linguistic models (Syntax, morphology, alignment, etc.)
Benefits	Allows inference of higher-level predicates upon asserted facts	Linguistic-oriented, highest level of interpretation	Facilitates to convert between logic and NL
Limitations	Limited to metric-temporal reasoning	Domain-dependent and target-oriented	Language-dependent

Table 1 contains a summary of some remarkable features for the different semantic representation formalisms described.

- *Spatiotemporal Predicates* (STP) rely on Fuzzy Metric-Temporal Horn Logic (FMTHL), which facilitates a schematic representation of conceptual knowledge which is time-delimited and incorporates uncertainty [4]. We use it to represent and reason about spatiotemporal developments, by assigning fuzzy degrees-of-validity to quantitative values generated by the motion trackers.
- *High-Level Semantic Predicates* (HLSP) express semantic relations among entities, at a higher level than metric-temporal relations. They result from applying situational models over STP. These new constraints embed restrictions based upon *contextualization, integration,* and *interpretation* tasks. Hence, the set of HLSP reaches the highest account of semantics, in the cognitive sense that each one of them implies a perceived situation or behavior which is meaningful and remarkable by itself in the selected domain.
- *Linguistic Predicates* (LP) represent linguistic-oriented knowledge. They are incorporated using Discourse Representation Theory (DRT) [2]). They are used for NL generation and understanding. Each LP requires distinct thematic arguments depending on the language and situation. LP in different languages describing a single situation are related to a single HLSP.

We focus on HLSP for building the ontology, for them being language-independent and suitable for a neutral framework between vision and linguistics.

3 Ontologies for Integration of Knowledge

The main motivation for the use of ontologies is to *capture the knowledge involved in a certain domain of interest,* by specifying some conventions about the content implied by this domain. Ontologies are especially used in environments requiring to share, reuse, or interchange specific knowledge among entities involved in different levels of manipulation of the information.

There exist many approaches for the ontological categorization of visually perceived events. An extensive review of the most important ones is done in [3], from which we remark Case Grammar, Lexical Conceptual Structures, Thematic Proto-Roles, WordNet, Aspectual Classes, and Verb Classes. As an extension, our approach relates each situation from the ontology with a set of required entities, which are classified depending on the thematic role they develop. The main advantage of this approach in an independency of the particularities of verbs to a concrete natural language, thus facilitating the addition of multiple languages in the HSE system.

A taxonomy has been developed for the possible set of semantic entities in the described domain. The chosen list of entities include *agents* as those which can spontaneously act to change a situation, here pedestrians and vehicles; *objects* as static elements of the scene; *locations*; and also a set of abstract *descriptors* which permit to add fuzzy modifiers to the conditions related to the entities. Other roles such as experiencer, goal, location, or instrument are easily enclosed in the selected categories.

3.1   Ontological Categorization of Situations

The main target for the proposed ontology is to enumerate and correlate the instantiable situations which are detectable in the selected domain, using a proper cognitive-based semantic representation. Now that the possible semantic participants have been established and organized, the set of situations can be classified.

Talmy organizes conceptual material in a cognitive manner by analyzing what he considers most crucial parameters in conception: space/time, motion/location, causation/force interaction, and attention/viewpoint [5]. For him, semantic understanding involves the combination of these domains into an integrated whole. Our classification of situations agrees with these structuring domains: We organize semantics in a linear fashion, ranging from objective knowledge in vision processes (low-level) to uncertain, subjective knowledge based on attentional factors (high-level). It is structured as follows, see Table 2:

- The *Status* class contains metric-temporal knowledge, based on the information provided by the considered trackers: body, agent, and face. Its elements represent spatial configurations and analysis of agent trajectories.
- The *ContextualizedEvent* class involves semantics at a higher level, now considering interactions among semantic entities. This knowledge emerges after contextualizing different sources of information, what allows for anticipation of events and reasoning of causation.
- Finally, the *BehaviorInterpretation* class specifies event interpretations with the greatest level of uncertainty and the larger number of assumptions. Intentional and attentional factors are considered, here the detection of remarkable behaviors in urban outdoor scenarios for surveillance purposes.

Each of the described behaviors requires certain arguments, characterized by the mentioned entities. For instance, a *DangerOfRunover* situation involves at least two Agents, a Vehicle and a Pedestrian, and a *Theft* situation involves a minimum of two Pedestrians and an object of type *PickableObject*.

Table 2. Central part of the ontology: the taxonomy for a classification of situations

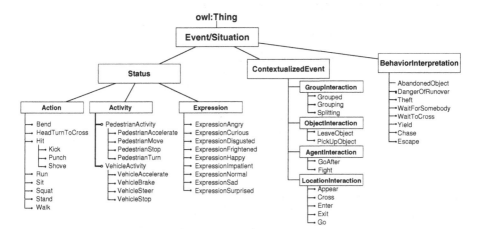

4   Conclusions and Future Work

An ontology has been designed to account and organize the universe of situations to be handled by a CVS for surveillance purposes. These situations are represented by HLSP, which hold a high level of semantics and are language-independent. The resulting ontology builds on a neutral framework between vision and linguistics. The proposed modeling is particularly useful for multilingual NL interfaces, making easier tasks of discourse categorization and disambiguation. It also restricts the domain of acceptance for semantic formalisms, facilitating prediction. One direct application is related to semantic indexation: The set of HLSP can be seen as the universe of high-level indexes in a domain, which facilitate further applications such as search engines and query-based retrieval of content. Several issues must be covered in next steps: proper communication between the semantic layer and the NL interface requires to relate the proposed ontology of situations to a linguistic-oriented one. In addition, the domains of application have to be enlarged.

References

1. Gonzàlez, J.: Human Sequence Evaluation: The Key-Frame Approach. PhD thesis, Universitat Autonoma de Barcelona, Barcelona, Spain (2004)
2. Kamp, H., Reyle, U.: From Discourse to Logic. Kluwer Academic Publishers, London (1993)
3. Ma, M., Kevitt, P.M.: Visual semantics and ontology of eventive verbs. In: Proc. of the 1st International Joint Conference on NL Processing, pp. 278–285 (2004)
4. Schäfer, K., Brzoska, C.: F-Limette Fuzzy Logic Programming Integrating Metric Temporal Extensions. Journal of Symbolic Computation 22(5-6), 725–727 (1996)
5. Talmy, L.: Toward a Cognitive Semantics, vol. 1. Bradford Book (2000)

Conditional Random Fields for High-Level Part Correlation Analysis in Images

Giuseppe Passino and Ebroul Izquierdo

Queen Mary, University of London
Mile End Rd
London, E1 4NS, UK
{giuseppe.passino,ebroul.izquierdo}@elec.qmul.ac.uk

Abstract. A novel approach to model the semantic knowledge associated to objects detected in images is presented. The model is aimed at the classification of such objects according to contextual information combined to the extracted features. The system is based on Conditional Random Fields, a probabilistic graphical model used to model the conditional a-posteriori probability of the object classes, thus avoiding problems related to source modelling and features independence constraints. The novelty of the approach is in the addressing of the high-level, semantically rich objects interrelationships among image parts. This paper presents the application of the model to this new problem class and a first implementation of the system.

1 Introduction

The part-based image analysis is a promising approach for object detection and image classification systems, being based on the strong correlation among objects represented within images. A natural way to address this problem is represented by the probabilistic graphical models, in which a node in a graph is associated to each image part in order to perform inference on them (i.e., deduce a labelling). Such models make the relationships between nodes explicit, simplifying the inference and giving to the problem an immediate interpretation. The Conditional Random Fields (CRF) [1] represent a solution to model the part labels probabilities and the dependencies between the extracted features and the part labels configuration. The system presented in this paper is based on the idea that a CRF can be applied to study high-level relationships between semantically rich atomic objects. As the characteristic dimension of the parts considered as constituents of the image to be labelled grows, two effects are expected: the reduction of the number of elements to handle, and possibly the growth of the semantics associated to the elements. This application is conceptually rather different from the original problems solved using CRF models, and an analysis and reconsideration of major assumptions has to be made.

B. Falcidieno et al. (Eds.): SAMT 2007, LNCS 4816, pp. 264–267, 2007.

2   Related Research

The CRF, introduced by Lafferty et al. for problems related to text labelling [1], have been recently applied with promising results to the image classification problem [2,3,4,5]. The various approaches differ in the graph structure employed to model this bidimensional problem, and in the nature of the atomic elements chosen as parts (nodes in the graph).

While some systems [4,5] consider the pixel as atomic element, simpler graphical structures can be obtained by considering interest points [3] or fixed-size image blocks [2] as graph nodes. The pixel-based approaches usually lead to complex systems that take into account primarily local dependencies. However, multiscale strategies as in [5] can be devised to handle long-range dependencies. When structures other than pixels are used to build the graph and extract the features, the choice of these structures has to be done appropriately, in order not to obtain coarse level heterogeneous parts without any underlying semantics, or fine-grained interest points that lead to complex unstructured graphs that make the problem computationally intractable. Nonetheless, these systems are simpler than the pixel-based ones, and additional complexity can be allocated for other tasks, for example to infer the class of the image from the labelling of the parts that can be considered as latent hidden variables, as in [3].

3   Proposed System

The proposed system is aimed at the classification of the images based on a coarse-grain parts analysis. The system is composed of a cascade of two blocks: the first one is demanded to the preprocessing of the images in order to extract the parts (or patches) to be labelled and from them the feature vectors, while the second block is the real classifier.

3.1   Image Segmentation

The parts features are mainly related to simple colour information, in order to retain the complexity low and to focus on the capabilities of the model to discriminate between categories adequately described by their own colour.

The images have been segmented using the multidimensional extension of the anisotropic diffusion filter [6], a non-uniform blurring process which preserves the boundaries between homogeneous colour regions. Being an iterative approach, the filtering can be expressed as the solution for $t \to +\infty$ of the heat equation $\frac{\partial I(\mathbf{x},t)}{\partial t} = \nabla(c(\mathbf{x},t)\nabla I(\mathbf{x},t))$, in which $I(\mathbf{x},t)$ represents the image at different steps, $\mathbf{x}$ is a point on the image, t is the image scale, and c is the non-constant diffusion coefficient which determines the non-linear behaviour, which is actually a decreasing function of the image gradient magnitude. In the colour domain this equation is applied separately to the real-domain luminance and complex-domain chrominance image channels obtained working in the 1976 CIE Lu^*v^* colour space. A feature vector containing colour and size information is associated to each patch.

3.2 Inference Model

The CRF are undirected graphical models used for the classification of a set of correlated entities, that is, the association of a label from a predefined set to each entity, conditioned on some observation. The classification is performed by modelling the conditional a posteriori probability of each labelling configuration given the observed data. This probability is the only requirement for a Maximum A Posteriori (MAP) decision approach: unlike generative approaches, additional hypotheses on the dependence of the observation from the labels distribution are not required.

This work is based on a hidden-variables CRF, introduced in [3]. The part labels are not given in the training set, but they form a layer of latent variables. The training image class is provided instead. The learning of the parts traits is therefore unsupervised, the image class can be directly inferred, and a set of annotated images is not required for the training phase. The mathematical bases of this model are almost the same as for the original CRF presented in [1], but in this case is the joint probability of the image class y and the part labels $\mathbf{h}$ conditioned on the observation $\mathbf{x}$ to be modelled as $p(y, \mathbf{h}|\mathbf{x}) = \exp\{\Psi(\mathbf{x}, y, \mathbf{h}; \theta)\}/Z(x; \theta)$, Z being a normalisation factor, and θ the set of parameters of the model.

The form of the so-called local function Ψ summarises the probabilistic model used in order to describe the problem. The only constraint derived from the CRF framework is that it has to be a sum of functions depending only on the graph's cliques labels, so that the Markov property on the graph is satisfied. In this case, being the graph structured as a tree, the form of the local function is

$$\Psi(\mathbf{x}, y, \mathbf{h}; \theta) = \sum_{i=1}^{m} \sum_{k \in \mathcal{K}_i^1} \theta_k f_k^1(x_i, h_i) + \sum_{i=1}^{m} \sum_{k \in \mathcal{K}_i^2} \theta_k f_k^2(y, h_i) + \tag{1}$$
$$+ \sum_{i,j \in E} \sum_{k \in \mathcal{K}_{i,j}^3} \theta_k f_k^3(y, h_i, h_j) \ ,$$

where the functions f_k^1, f_k^2, f_k^3 encompass the dependencies between the feature vector $\mathbf{x}$ and the part label h_i, between the part labels h_i and the image category y, and between the part labels pairs h_i, h_j and the image category y, respectively. The functions are the product of indicator functions $\delta(\cdot)$ in respect to the discrete variables h, y, and selector functions applied to the feature vector $\mathbf{x}$ to extract its single components. The index i spans among the nodes in the graph and among the set of edges E respectively, while the sets $\mathcal{K}$ indicate the set of parameters and functions related to a specific node or edge in the graph.

The inference in the model can be done with a likelihood maximisation approach in a conceptually straightforward way. The form of the log-likelihood $L(\theta) = \sum_{i=1}^{M} \log\left(p(y_i|\mathbf{x}_i, \theta)\right) - \frac{\|\theta\|^2}{2\sigma^2}$, where the sum is on the training set and the second term is a Gaussian prior. The gradient can be calculated in closed-form, and a quasi-Newton method is used for the solution search.

4 Results and Conclusions

The experimental validation of the system is based on the Caltech 101 categories dataset, where images belonging to the "faces" category are discriminated from

Table 1. Comparison between our model ("our") and the reference one ("reference"), showing relative elapsed time (T_r), likelihood prior variance (σ^2) and classification accuracy (η).

Model	T_r	σ^2	η
our	0.13	10^4	77%
reference	1	0.1	90%

images randomly chosen from all the other categories. This data set is chosen because the skin colour can be a discriminant for the face category, so it makes sense to use colour-based features. The results are compared with the reference model [3], in which a similar inference algorithm is used. In Table 1 the performance comparison results are shown.

The reduced classification performances compared with the reference model can be explained with the low number of constraints accentuating the convergence problems of the likelihood maximisation, and with the poverty of the employed features. The segmentation algorithm can fail in isolating the face boundaries in presence of low-contrasted background, as well.

These results are however encouraging, because they show that the model is able to handle dependencies between nodes. Some drawbacks are highlighted as well, as excessive simplicity of the employed features, that makes the likelihood maximisation process unlikely to find the optimal solution. Further studies on the model can take advantage of the proposed framework to integrate different level of analysis for the classification of the images.

References

1. Lafferty, J., McCallum, A., Pereira, F.: Conditional random fields: Probabilistic models for segmenting and labeling sequence data. In: Proc. 18th International Conf. on Machine Learning, pp. 282–289. Morgan Kaufmann, San Francisco (2001)
2. Kumar, S., Hebert, M.: Discriminative random fields: A discriminative framework for contextual interaction in classification. In: ICCV 2003. Proceedings of the 2003 IEEE International Conference on Computer Vision, vol. 2, pp. 1150–1157 (2003)
3. Quattoni, A., Collins, M., Darrell, T.: Conditional random fields for object recognition. In: Neural Information Processing Systems Vision (2004)
4. Shotton, J., Winn, J., Rother, C., Criminisi, A.: Textonboost: Joint appearance, shape and context modeling for multi-class object recognition and segmentation. In: Leonardis, A., Bischof, H., Pinz, A. (eds.) ECCV 2006. LNCS, vol. 3951, Springer, Heidelberg (2006)
5. He, X., Zemel, R.S., Carreira-Perpinan, M.A.: Multiscale conditional random fields for image labeling. In: CVPR 2004. Computer Vision and Pattern Recognition, 2004. Proceedings of the 2004 IEEE Computer Society Conference on, vol. 2, pp. 695–702 (2004)
6. Lucchese, L., Mitra, S.: Colour segmentation based on separate anisotropic diffusion of chromatic and achromatic channels. Vision, Image and Signal Processing, IEE Proceedings 148(3), 141–150 (2001)

Document Layout Substructure Discovery*

Claudio Andreatta

Fondazione Bruno Kessler - IRST, 38050 Povo, Trento, Italy
andreatta@itc.it

Abstract. In this paper we present a system, DoLSuD, for the automatic discovery of relevant substructures in a document layout. DoLSuD, Document Layout Substructure Discovery, extracts, analyzes and describes the visual content of structured documents, such as catalogs, in order to discover repeating and distinctive substructures in the document layout and to establish relations between textual and image content. The paper presents the system along with experimental results and the web based service which utilizes the analysis results.

1 Introduction

Effective acquisition, organization, processing, use and sharing of the knowledge embedded in multimedia content play a major role for competitiveness in the modern information society and for the emerging knowledge economy. Hence a strong motivation for the research of methods that can analyze and index multimedia data and provide solutions for automatically acquiring and accessing knowledge. At the same time there is a considerable effort to achieve a practical integration of multimedia meta-data into the Semantic Web Framework [1].

In this paper we describe the system DoLSuD, Document Layout Substructure Discovery. DoLSuD is a system supporting document level information extraction and analysis from multimedia resources. The information conveyed by a multimedia document may be analyzed at two different levels: the document level, in which the geometrical layout is investigated considering both textual and visual features, and the content level, where the subject of text and images must be discovered. The DoLSuD system extracts the visual content in order to discover repeating and distinctive structures in the document layout and to establish relations between textual and image content. Such relations are useful to exploit the semantic annotation of the textual part so that we can annotate images and guide image processing algorithms; the purpose of which is to automatically derive concepts from the raw image data [2] and to detect and recognize objects [3].

The following sections describe the approach in detail applied to a practical scenario: the analysis of furniture catalogs. Section 2 describes the document layout analysis algorithms and Section 3 concludes with evaluations and some directions for future works.

* This work was supported by the European Union under the project **VIKEF**: Virtual Information and Knowledge Environment Framework (*http://vikef.net*). Image courtesy of IKEA Italia, copyright held by the IKEA group.

B. Falcidieno et al. (Eds.): SAMT 2007, LNCS 4816, pp. 268–271, 2007.

2   Document Analysis: Layout Substructure Discovery

The automatic information extraction from multimedia data is central for over-coming the so-called knowledge acquisition bottleneck. Multimedia sources of information such as documents contain text and images thus requiring infor-mation extraction approaches combining several different techniques, ranging from Natural Language Processing to Image Analysis and Understanding. An in-depth discussion of the multimedia annotation and information extraction methodology adopted is beyond the scope of this paper, further details may be found here [4].

In order to extract from the document image pictures of isolated items (high-lights) and scenes, it is necessary to correctly identify the page background color and then by means of connected component analysis proceed to the segmentation of the document page. The document structure is then represented as a directed graph $G(V, E)$ whose vertices are the structural elements in the page and the edges represent the spatial relationships among the elements on the page. The classes of structural elements considered are: textual paragraphs bounding boxes V_P and images V_I. Each element is identified on the page by a connected set of pixels: a region on the document image. Being interested in the text-image relationships, we consider only the directed edges with paragraphs V_P as source and images V_I as target. The spatial relationship is described by means of the two dimensional vector connecting the centers of mass of the elements c_i, the minimum $L2$ distance between the two sets and the ratio of the area of inter-section to the area of the union of the two regions. All vectors and distances are normalized, analyzing their distribution on the whole graph. Thus the edge $e(V_i, V_j) = e_{ij} \in E$ label is a triple $(\mathbf{v}_{ij}, d_{ij}, a_{ij})$.

One method for discovering knowledge in structured data is the identification of common, repeating and distinctive substructures within the data. The moti-vation behind this process is to find patterns capable of compressing the data and to simplify or summarize the data substituting the discovered patterns. Iter-ating this process it is possible to hierarchically re-organize the data and provide varying level of abstraction. The data mining algorithm upon which this work is based is SUBDUE [5,6]. SUBDUE is a system that discovers interesting sub-structures in structural data based on inexact graph matching techniques and a computationally constrained beam search algorithm guided by heuristics.

Given the input graph G and a substructure S, we want to associate a simi-larity measure to all the subgraphs of G in order to identify the one that most closely looks like S: the set I of instances of S in G. In the inexact graph match approach used in this paper, each distortion of a graph is assigned a cost and the matching cost is the sum of all the distortions. Possible distortions are node and edge insertion, deletion and substitution (e.g. $SubEdge(e_{ij}) = \|\mathbf{v_i} - \mathbf{v_j}\|$).

The score of an instance $i \subset G$ of S is defined as:

$$s(i, S, G) = 1 - d(i, S, G) = \max\left(0, 1 - \frac{matching\_cost(i, S)}{size(i)}\right) \qquad (1)$$

where $size(G) = \#nodes(G) + \#edges(G)$.

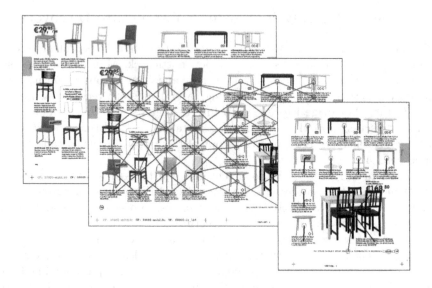

Fig. 1. The substructure discovery algorithm processing phases: the document page, the layout graph representation and the detected caption-figure relationships.

SUBDUE uses domain-independent heuristics to evaluate a substructure: cognitive saving, compactness, connectivity, coverage and minimum description length (see [5,6]). In this work the SUBDUE heuristics were adapted to the specific task. The heuristics considered to evaluate a substructure were: the minimum description length, the coverage computed for the graph nodes, and, in order to strengthen the most repeating substructures, a modified cognitive saving. The nodes coverage and the cognitive savings are defined as:

$$nodes\_cov(S, G) = \frac{\sum_{i \in I} s(i, S) * \#par\_nodes(i)}{\#par\_nodes(G)} + \frac{\sum_{i \in I} s(i, S) * \#img\_nodes(i)}{\#img\_nodes(G)}$$

$$cognitive\_sav(S, G) = \left(\sum_{i \in I} s(i, S)^\alpha * size(i)^\beta \right) - size(S)^\beta \qquad (2)$$

where α and β are used to weight the contribution of the complexity versus the number of instances, $\#par\_nodes(G)$ and $\#img\_nodes(G)$ are the number of paragraph and image nodes in the graph. The value of a substructure S is computed as a product of the three proposed heuristics.

After a substructure is discovered the graph is simplified by removing all of the instances of the structure and the procedure is repeated on the pruned graph in order to find new interesting subgraphs. When the most promising substructures are identified, an association score is assigned to all edges: $Score(e_{ij}) = \lambda_{ij} + \mu d_{ij} + \nu a_{ij}$, where λ_{ij} is the matching score obtained by the subgraph matching algorithm and the other contribution helps to handle the problem of document pages which lack a structured layout or have a mixed content.

3   Experiments and Conclusions

We present here some preliminary results of system performance. While qualitative experiments have been conducted on different furniture catalogs such as IKEA, BRF, ZANOTTA and SEMANAVERDE, the IKEA 2006 Italian catalog has been analyzed by the system and the image-caption links evaluated comparing the algorithm output with a manually annotated ground truth. The catalog has 185 pages, from which 1332 product captions and 1348 image regions have been extracted, the system detected 1446 image-caption relations. A relation is defined correctly if the product described by the caption is depicted by the related image region, otherwise it is defined as incorrect. The following table reports the obtained results.

Total relations	Correct	Incorrect
1446	1302	144
	90%	10%

In this paper we presented a system supporting document level information extraction and analysis from digital catalogs. It identifies recurrent structures in large complex documents supporting further fully or semi automatic processing stages and it can provide abstract structured components resulting in a hierarchical view of the document. Automatically extracting meaningful links between images and annotated text from the catalog structure of allows us to exploit the semantic annotation of the textual part to guide image processing algorithms, to recognize depicted objects, to infer correspondences between word and specific image parts and to enhance the performances of traditional content based image retrieval systems. As future work we plan to extend the experiments in order to evaluate the background detection and image segmentation stage.

References

1. van Ossenbruggen, J., Stamou, G., Pan, J.Z.: Multimedia Annotations and the Semantic Web. In: SWCASE. Proc. of the International Workshop on Semantic Web Case Studies and Best Practices for eBusiness (2005)
2. Barnard, K., Duygulu, P., Forsyth, D., de Freitas, N., Blei, D., Jordan, M.: Matching words and pictures. Journal of Machine Learning Research 3, 1107–1135 (2002)
3. Andreatta, C., Lecca, M., Messelodi, S.: Memory-based object recognition in digital images. In: VMV 2005. Proceedings of 10th International Fall Workshop - Vision, Modelling, and Visualization, Erlangen, Germany (November 16-28, 2005)
4. Bartolini, R., Giovannetti, E., Marchi, S., Montemagni, S., Andreatta, C., Brunelli, R., Stecher, R., Niedere, C.: Ontology learning in multimedia information extraction from product catalogues. In: Staab, S., Svátek, V. (eds.) EKAW 2006. LNCS (LNAI), vol. 4248, Springer, Heidelberg (2006)
5. Cook, D., Holder, L.: Substructure discovery using minimum description length and background knowledge. Journal of Artificial Intelligence Research 1, 231–255 (1994)
6. Coble, J., Rathi, R., Cook, D.J., Holder, L.B.: Iterative structure discovery in graph-based data. International Journal on Artificial Intelligence Tools 14(1-2), 101–124 (2005)

Recognition of JPEG Compressed Face Images Based on AdaBoost

Chunmei Qing and Jianming Jiang

School of Informatics, University of Bradford, UK
{C.Qing, J.Jiang1}@bradford.ac.uk

Abstract. This paper presents an advanced face recognition system based on AdaBoost algorithm in the JPEG compressed domain. First, the dimensionality is reduced by truncating some of the block-based DCT coefficients and the nonuniform illumination variations are alleviated by discarding the DC coefficient of each block. Next, an improved AdaBoost.M2 algorithm which uses Euclidean Distance(ED) to eliminate non-effective weak classifiers is proposed to select most discriminative DCT features from the truncated DCT coefficient vectors. At last, the LDA is used as the final classifier. Experiments on Yale face databases show that the proposed approach is superior to other methods in terms of recognition accuracy, efficiency, and illumination robustness.

Keywords: Discrete Cosine Transform(DCT); AdaBoost algorithm; Linear Discriminant Analysis(LDA); Euclidean Distance(ED); Yale database.

1 Introduction

The Discrete Cosine Transform (DCT) has been employed in face recognition due to its several advantages in feature extraction [1][2]. In this paper, facial features are first extracted by the block-based DCT embedded in all JPEG compressed face images, which greatly reduces dimensionality of the original face image as well as has low complexity in implementation. Besides, it turns out that by simply discarding the DC coefficient of each block, the proposed system is robust against nonuniform brightness variations of images.

Introduced by Freud and Schapire [3], the AdaBoost algorithm has been successfully used to select Haar-like features for face detection [4] and for learning the most discriminative Gabor features for recognition [5]. In this paper, we present an advanced face recognition system that is based on the use of the AdaBoost.M2 algorithm to extract the most discriminating JPEG DCT features of face images. The Euclidean Distance(ED) between the candidate weak classifier and the selected weak classifiers is examined to avoid redundant classifiers. Better performance has been observed when the selected JPEG DCT features are applied for face recognition, which uses Linear Discriminant Analysis(LDA)[6] as the final classifier.

B. Falcidieno et al. (Eds.): SAMT 2007, LNCS 4816, pp. 272–275, 2007.
© Springer-Verlag Berlin Heidelberg 2007

2 Face Feature Selection in JPEG Compressed Domain

2.1 The JPEG DCT Coefficient Feature

In JPEG standard, the original image $f(x, y)$ is initially partitioned into $p \times q$ rectangular nonoverlapping blocks (8×8 blocks): $b(i, j), i = 1, \cdots, p, j = 1, \cdots, q$, where each block is transformed independently using the 2D-DCT basis function. DCT coefficients are coded in a zigzag order and DCT coefficients with great magnitude are mainly located at the upper-left corner of each block. So some redundant information may be first removed by truncating the last $64 - k$ DCT coefficients of each block so that the dimensionality of the coefficient vectors can be reduced. The nonuniform illumination effect can be reduced just by discarding the DC coefficient of each block. After truncating some DCT coefficients, the JPEG DCT coefficient features will be

$$F = [g_{11}^2, \cdots, g_{11}^k, \cdots, g_{ij}^2, \cdots, g_{ij}^k, \cdots, g_{pq}^2 \cdots, g_{pq}^k]^T, \tag{1}$$

where $[g_{ij}^2, \cdots, g_{ij}^k]^T$ are the DCT coefficients of the subimage block $b(i, j)$.

2.2 Feature Redundancy Elimination: Improved AdaBoost.M2

AdaBoost.M2 is an adaptive algorithm to boost a sequence of weak classifiers to form a stronger classifer, in that the weights are updated dynamically according to the errors in previous learning [3]. As reported in [7] and [5], many classifiers selected by the AdaBoost algorithm might be similar, so they are redundant. In this paper we use the most common Euclidean Distance(ED) to eliminate non-effective weak classifiers. Before a weak classifier is selected, the ED between the new classifier and each of the selected ones is examined to make sure that the information carried by the new classifier has not been captured before. To determine whether the new classifier is redundant or not $d(h_u)$ is compared with a pre-defined Threshold Euclidean Distance(TED). If it is smaller than the TED, the information carried by the classifier has already been captured. Details of the improved AdaBoost.M2 algorithm are listed in Fig. 1.

3 Experimental Results

We use the Yale Face Database [8] to evaluate the proposed face recognition method. In order to be consistent with the experiments of [6], the closely cropped images, which scaled to an image size of 72×64, are used in our experiment. In addition, five images of each individual are randomly selected as the training set, and the other six images as the probe set in our experiments. To reduce the possible variation, we performed each experiment 10 times and all the result data given in this paper is an average of them. Furthermore, the decision stump as the weak classifier and the LDA as the final classifier are used.

In Table 1, all experimental results in recognition rates on the Yale face database are listed in comparison with a range of representative algorithms including:

PCA[6], LDA[6], IPCA_ICA[9], and DCT+FLD+RBF[10]. We can see from Table 1 that our proposed method achieves comparable performance. However, the main advantage of this method is its lower complexity in implementation since DCT is already embedded in all JPEG compressed face images, which is crucial for high-speed face recognition in large databases.

Input: learning set $\{(x_1, y_1), \cdots, (x_N, y_N); x_i \in \mathbb{X}, y_i \in \mathbb{Y}\}, \mathbb{Y} = \{1, \cdots, |\mathbb{Y}|\}$; weak classifier of the form $h : \mathbb{X} \times \mathbb{Y} \to [0, 1]$.

Initialization: $D_1(i) = \frac{1}{N}$, weight vector $w_{i,y}^1 = \frac{D_1(i)}{|\mathbb{Y}|-1}$

For $t = 1, \cdots, T$:

1. Set $W_i^t = \sum\limits_{y \neq y_i} w_{i,y}^t$, for $y \neq y_i$: $q_{i,y}^t = \frac{w_{i,y}^t}{W_i^t}$, and $D_t(i) = \frac{W_i^t}{\sum\limits_{i=1}^N W_i^t}$.

2. Call the candidate weak classifier h_j with distribution D_t and label weighting function q_t. Calculate the pseudo-loss

   $\epsilon_j = \frac{1}{2} \sum\limits_{i=1}^N D_t(i) \left(1 - h_j(x_i, y_i) + \sum_{y \neq y_i} q_{i,y}^t h_j(x_i, y) \right)$

   LoopCounter=0

   Do

   - Choose h_u with lowest error ϵ_u from the candidate classifiers, and calculate the minimum Euclidean Distance(ED) $d(h_u)$.
   - If $d(h_u) \geqslant TED$, the classifier is found, $h_t = h_u, \epsilon_t = \epsilon_u$. Go to Step 3.
   - Else, remove h_u from the candidate list.
   - End If

   LoopCounter=LoopCounter+1

3. Set $\beta_t = \frac{\epsilon_t}{1-\epsilon_t}$

4. For $i = 1, \cdots, N$ and $y \in \mathbb{Y}\backslash\{y_i\}$ set the new weight vectors to be
   $w_{i,y}^{t+1} = w_{i,y}^t \beta_t^{\frac{1}{2}(1+h_t(x_i,y_i)-h_t(x_i,y))}$

Output: final classifier H: $H(x) = arg\max_{y \in \mathbb{Y}} \sum_{t=1}^T \ln\left(\frac{1}{\beta_t}\right) h_t(x, y)$

Fig. 1. Details of the improved AdaBoost.M2 algorithm

Table 1. Comparison on the Yale database(closely cropped faces)

Methods	Recognition Rate(%)	Notes
Eigenface(PCA) [6]	74.5	pixel domain
Fisherface(LDA) [6]	93.9	pixel domain
IPCA_ICA [9]	98.2	pixel domain
DCT + FLD + RBF [10]	95.2	pixel domain
Proposed algotithm	96.67	JPEG compressed domain

4   Conclusion

A novel feature selection algorithm has been proposed in this paper. Facial features are first extracted by truncating the block-based DCT coefficients embedded in all JPEG compressed face images, which greatly reduces dimensionality of the original face image as well as has low complexity in implementation. Besides, we have explored another property of the block-based DCT. It turns out that by simply discarding the first DCT coefficient of each block, the proposed system is robust against nonuniform brightness variations of images. In order to obtain the most invariant and discriminating JPEG DCT features of faces, the AdaBoost.M2 which uses the decision stump as weak classifiers is further applied to the feature vectors. By introducing Euclidean distance measure into AdaBoost.M2, the improved algorithm can reduce redundancy among selected features. Based on this algorithm, JPEG DCT features and the LDA, a fast and robust method for face recognition has also been developed. The recognition rate was 96.67% for JPEG compressed domain recognition on the closely cropped Yale face database.

Acknowledgments. The authors wish to acknowledge the financial support under EU IST FP-6 Research Programme with the integrated project: LIVE (Contract No. IST-4027312).

References

1. Eickeler, S., Müller, S., Rigoll, G.: Recognition of jpeg compressed face images based on statistical methods. Image Vision Comput. 18, 279–287 (2000)
2. Hafed, Z.M., Levine, M.D.: Face recognition using the discrete cosine transform. Int. J. Comput. Vis. 43, 167–188 (2001)
3. Freund, Y., Schapire, R.E.: A decision-theoretic generalization of online learning and an application to boosting. J. Comp. & Sys. Sci. 55, 119–139 (1997)
4. Viola, P., Jones, M.J.: Rapid object detection using a boosted cascade of simple features. In: Proc. IEEE Conf. on Computer Vision and Pattern Recognition, Kauai, Hawaii, pp. 511–518 (2001)
5. Shen, L., Bai, L.: Mutualboost learning for selecting gabor features for face recognition. Pattern Recognition Letters 27, 1758–1767 (2006)
6. Belhumeur, P.N., Hespanha, J.P., Kriegman, D.J.: Eigenfaces versus fisherfaces: Recognition using class specific linear projection. IEEE Trans. Pattern Anal. Mach. Intell. 19, 711–720 (1997)
7. Li, S.Z., Zhang, Z.: Floatboost learning and statistical face detection. IEEE Trans. Pattern Anal. Machine Intell. 26, 1112–1123 (2004)
8. Yale: (University face database available [online],
   http://cvc.yale.edu/projects/yalefaces/yalefaces.html
9. Dagher, I., Nachar, R.: Face recognition using ipca-ica algorithm. IEEE Transactions on Pattern Analysis and Machine Intelligence 28, 996–1000 (2006)
10. Er, M.J., Chen, W., Wu, S.: High-speed face recognition based on discrete cosine transform and rbf neural networks. IEEE Trans. Neural Netw. 16, 679–691 (2005)

Camera Motion Analysis Towards Semantic-Based Video Retrieval in Compressed Domain

Ying Weng and Jianmin Jiang

School of Informatics, University of Bradford, BD7 1DP, UK
{Y.Weng,J.Jiang1}@bradford.ac.uk

Abstract. To reduce the semantic gap between low-level visual features and the richness of human semantics, this paper proposes new algorithms, by virtue of the combined camera motion descriptors with multi-threshold, to automatically retrieve the semantic concepts, i.e., close-up, and panorama, directly in MPEG compressed domain based on camera motion analysis. Extensive experiments illustrate that the proposed algorithms provide promising retrieval results under real-time application scenario and without human intervention.

Keywords: Camera motion analysis, MPEG compressed domain, semantic-based video retrieval.

1 Introduction

The ability to retrieve visual information by semantic terms in compressed domain will bring new opportunities for efficient utilisation of multimedia material in improved ways. While considerable research has been conducted in image/video retrieval [1]-[3], only few solutions have been given to this challenging task with limited decoding of MPEG videos. To this end, our work focuses on using low-level descriptors to automatically retrieve high-level semantic concepts, i.e., close-up, and panorama, directly in MPEG compressed domain based on camera motion analysis.

Previous research on close-up detection is mostly in sports video [5]. To improve the coverage of close-up retrieval and implement it extensively, zoom-in descriptor is extracted directly in compressed domain and when its value is larger than a certain two-level threshold, the frame is retrieved as a close-up.

Panorama is very common to everyone, carries more information and provides a larger field of view, thus panorama retrieval is significant for applications [4]. We investigate automatic panorama retrieval in compressed domain exploiting the combined descriptors, i.e., zoom-out and pan, with two-layered threshold.

2 Close-Up Retrieval

The camera motion descriptor is utilised, i.e., zoom-in, with two-level threshold to automatically retrieve close-ups. Let the class labels for thresholding be denoted as TH_S and TH_M ; the former represents the threshold for a single detected zoom-in

B. Falcidieno et al. (Eds.): SAMT 2007, LNCS 4816, pp. 276–279, 2007.
© Springer-Verlag Berlin Heidelberg 2007

frame, and the latter represents the threshold for continuous multiple detected zoom-in frames. Then the thresholding operation is as follows: (i) If only one single zoom-in frame is detected, then the threshold is set to TH_S. Moreover, on the condition that its zoom factor μ has a higher value than TH_S, this frame is retrieved as a close-up; (ii) If continuous multiple zoom-in frames are detected, then the threshold is set to TH_M. Meanwhile, the maximum zoom factor μ_{max} and the minimum zoom factor μ_{min} for these continuous frames are obtained. If the ratio of μ_{max} to μ_{min} is larger than TH_M, these frames are retrieved as close-ups.

The above can be summarised in (1) and Fig. 1:

$$C_F = \begin{cases} F_S, & if \ \mu > TH_S \\ F_M, & if \ \dfrac{\mu_{max}}{\mu_{min}} > TH_M \end{cases} . \tag{1}$$

where F_S denotes a single detected zoom-in frame, F_M denotes continuously multiple detected zoom-in frames, and C_F denotes retrieved close-ups.

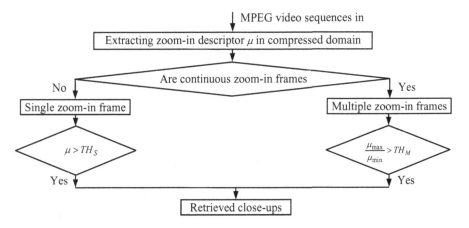

Fig. 1. Overview of the proposed automatic close-up retrieval in MPEG compressed domain

3 Panorama Retrieval

Automatic panorama retrieval exploits the combined camera motion descriptors, i.e., zoom-out and pan, with two-layered threshold is investigated. Let the class labels for thresholding be denoted as TH_Z and TH_P; the former represents the threshold for a detected zoom-out frame, and the latter represents the threshold for panning. The approach is divided into two steps: (i) On the condition that a zoom-out frame is detected, the first threshold is set to TH_Z, and while its zoom factor μ is less than TH_Z, this frame is selected as a candidate; (ii) If a candidate frame is further detected with panning, the second threshold is set to TH_P, and if its pan rate $f\alpha$ has a higher absolute value than TH_P, this frame is retrieved as a panorama frame.

The above can be summarised in (2) and Fig. 2:

$$F_C = F_Z, \quad if \ \mu < TH_Z;$$
$$P_F = F_C, \quad if \ |f\alpha| > TH_P. \tag{2}$$

where F_Z denotes a detected zoom-out frame, F_C denotes a candidate frame, and P_F denotes a retrieved panorama.

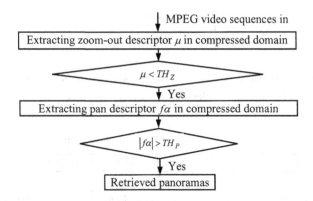

Fig. 2. Overview of the proposed automatic panorama retrieval in MPEG compressed domain

4 Experimental Results and Evaluations

The proposed automatic close-up and panorama retrieval algorithms are tested on a database with different MPEG video clips, including well-known TREC2001 documentary video sequences, movies, sports, and news. TH_S is set to 1.001, TH_M is set to 4, TH_Z is set to 0.998, and TH_P is set to 4. The recall and precision rates for close-up and panorama retrievals are listed in Table 1. The results demonstrate that the proposed algorithms are computationally efficient, also achieve superior performances in terms of both recall and precision rates. The samples of retrieval results from the test video clips are shown in Fig. 3 and 4, respectively.

Table 1. Summary of experimental results for close-up and panorama retrieval

	Videos	NAD32	NAD55	Movie1	Movie2	Sports1	News1	Average
Close-up	Recall	86.67%	95.00%	90.63%	92.86%	93.42%	94.74%	92.22%
	Precision	89.66%	87.69%	81.69%	78.00%	92.21%	94.74%	87.33%
	Videos	BOR08	BOR11	BOR19	Movie3	Sports2	Sports3	Average
Panorama	Recall	89.74%	90.20%	86.02%	88.46%	87.50%	91.84%	89.60%
	Precision	86.42%	85.19%	91.95%	82.14%	90.32%	93.75%	88.89%

(a) from NAD32 (b) from NAD55 (c) from Movie1 (d) from Sports1 (e) from News1

Fig. 3. Samples of close-up retrieval results from the test database

(a) from BOR08 (b) from BOR11 (c) from BOR19 (d) from Movie3 (e) from Sports2

Fig. 4. Samples of panorama retrieval results from the test database

5 Conclusions

In this paper, we build close-up retrieval via zoom-in descriptor with two-level threshold, and panorama retrieval via combined descriptors, i.e., zoom-out and pan, with two-layered threshold. The whole process is under real-time application scenario and without human intervention. Extensive experiments show that computational complexity and retrieval performance of the proposed algorithms are well balanced.

Acknowledgments. The authors would like to acknowledge the financial support under European IST FP6 Integrated Project: LIVE (Contract No. IST-4-027312).

References

1. Djordjevic, D., Izquierdo, E.: An object- and user-driven system for semantic-based image annotation and retrieval. IEEE Trans. Circuits Syst. Video Technol. 17(3), 313–323 (2007)
2. Jiang, J., Weng, Y., Guo, B., Feng, Y.: Robust-to-rotation texture descriptor for image retrieval in wavelets domain. Journal of Electronic Imaging 15(1), 1–14 (2006)
3. Hsieh, J.W., Yu, S.L., Chen, Y.S.: Motion-based video retrieval by trajectory matching. IEEE Trans. Circuits Syst. Video Technol. 16(3), 396–409 (2006)
4. Sun, X., Foote, J., Kimber, D., Manjunath, B.S.: Region of interest extraction and virtual camera control based on panoramic video capturing. IEEE Trans. Multimedia 7(5), 981–990 (2005)
5. Liu, L., Ye, X., Yao, M., Zhang, S.: A semantic description scheme of soccer video based on MPEG-7. In: Aizawa, K., Nakamura, Y., Satoh, S. (eds.) PCM 2004. LNCS, vol. 3332, pp. 298–305. Springer, Heidelberg (2004)
6. Tan, Y.P., Saur, D.D., Kulkarni, S.R., Ramadge, P.J.: Rapid estimation of camera motion from compressed video with application to video annotation. IEEE Trans. Circuits Syst. Video Technol. 10(1), 133–146 (2000)

Challenges in Supporting Faceted Semantic Browsing of Multimedia Collections

Daniel Alexander Smith, Alisdair Owens, m.c. schraefel, Patrick Sinclair,
Paul André, Max L. Wilson, Alistair Russell, Kirk Martinez, and Paul Lewis

IAM Group, School of Electronics and Computer Science, University of Southampton,
Southampton, United Kingdom
{das05r,ao,mc,pass,pa2,mlw05r,ar5,km,phl}@ecs.soton.ac.uk

Abstract. We discuss three approaches, 3store, D2R and MySQL, we have
explored to support efficient querying of multimedia data sources via mSpace, a
rich UI. Our results underline key research challenges facing the development
of high performance RDF query layers to support complex real-time UIs.

Keywords: triple stores, RDF browser, mSpace, multimedia indexing.

1 Introduction

Web based interfaces to multimedia collections traditionally enable collection search
by simple keyword queries which produce a list of links to be explored further. More
sophisticated user interfaces (UIs) also allow browsing by single
topics/categories/facets[1] which can then be sorted in a variety of ways.[2] Over the past
several years we have been looking at mechanisms to enable richer strategies for
exploring multimedia archives. The mSpace framework presents a multicolumn
faceted browsing interface that allows a person to select instances in a column/facet
on the data, and have the data in the columns to the right be filtered by that selection.
Each selection also populates information about that selection into an associated pane.
The columns are placed in a "slice" and can be moved around. Likewise different
facets or dimensions in the data space can be added or subtracted from that slice.
Information about any selected instance can also be saved for later reference. This
approach has been described in detail elsewhere [1].

In the original mSpace UI, each column was populated one at a time, much like the
Apple OS X Finder: the first column has a set of data; a selection in that set
determines the instances which populate the next column; a selection must be made in
that column to populate the one next to it. In user studies, we saw that people found
the interface more tractable if every column was fully populated in advance. We call
this approach "pre-pop." A feature added to complement pre-pop is "backward
highlighting." With backward highlighting, a selection means the data in the columns
to the right are filtered, showing every possible result within the restrictions applied
by the selection; second, the *possible* paths back from the initial selection in the

---

[1] http://flamenco.berkeley.edu/demos.html
[2] http://www.open-video.org/

B. Falcidieno et al. (Eds.): SAMT 2007, LNCS 4816, pp. 280–283, 2007.

columns to the left of the selection are highlighted. Thus, backwards highlighting provides additional cues for users to understand with what information their current selection is associated. While effective, these features have meant significantly increased query complexity and hence an increased hit on performance.

In the following sections we describe our experience with three approaches to optimize query performance and UI experience where our efforts have been motivated to maintain a Semantic Web deployment; we conclude with consideration of research challenges for future performance with heterogeneous data sources.

2   3Store: Performance Testing Semantic Web Querying

The mSpace framework was designed initially as a Semantic Web system in order to support the aggregation of heterogeneous data sources, and to deliver the above exploratory search experience [2] to users. In previous iterations of mSpace, we utilised 3store [3], an RDF triple store backed by MySQL as the storage and query layer. This was motivated by the attractiveness of RDF(S) as a data format: it allows for relatively simple integration of a wide variety of data sources, as well as basic inferential capabilities. Further, it uses a simple, standard query language (initially RDQL, now SPARQL), allowing people or agents to browse the dataset using mechanisms other than the mSpace browser. A forerunning project, CSAKTiveSpace [4], was based on the same technology and had coped with a substantial dataset (40 million+ triples), we were confident that 3store would offer sufficient performance to support mSpace. This was initially borne out: our early similarly sized datasets performed adequately.

As we implemented the new pre-pop and backward highlighting features on a dataset of approximately 100 times the size of CSAKTiveSpace's, query response times became unacceptably slow for real-time interactive use. Pre-pop in particular can result in many more queries being performed due to having to reload every column, rather than just one, some of which can be particularly demanding if the selections made do not restrict the search space very much. Performance issues were particularly noticeable on queries that returned a large result set, or less predictable results. For example, an issue in optimising 3store for high query performance is that it is a tool for storing generic RDF, and a given piece of RDF data does not have a predictable structure. This means that it is difficult to generate a representative, optimised schema for 3Store's MySQL backend. As it stands, 3store conceptually stores a triple in one table with columns including a hash of each of subject, predicate, and object. This table (as well as a few supporting tables) are extremely long, containing a row for triple, and data is retrieved via self-joins. This schema gives the MySQL query optimiser little opportunity to perform significant optimisation on queries, and gives us relatively little control over indexing. We needed to find a solution that would offer us the ability to define a more detailed schema that more closely represented the data while providing us with a SPARQL[3] interface to minimise the effort involved with a code transition, and to maintain Semantic Web compatibility. This motivation lead us to try D2R.

---

[3] http://www.w3.org/TR/rdf-sparql-query/

3   D2R Server: Best of Both Worlds - or Not Quite?

D2R Server[5] is an application layer that translates SPARQL queries into SQL, in order to provide a "semantic shim" on to a relational database. The key reason for changing the storage layer was that specific compound indexes across our schema could be created in a way that was not possible with the triple-storage method that 3store used. This approach also provides more structurally specific information to the SQL query optimiser, thus providing additional performance increases.

The subsystem that configures mSpace for a specific dataset is known as the mSpace Model [1]. Previously this was stored in the triplestore, and queried out as part of the processing. Due to the way D2R is configured as a layer between a relational database and a SPARQL query-point, asserting a small amount of arbitrarily structured RDF like the mSpace Model would have been problematic to implement with D2R, so the decision was made to instead alter the mSpace server such that the model was stored separately to the data. This is also attractive as it gives the additional benefit that remote SPARQL end-points that we do not control can be configured locally and explored using mSpace, without any agreement with, or alteration of, the remote data.

The solution looked promising, but ultimately was not suitable for several reasons. First, the SQL statements that the D2R Server created were inefficient, and considerable and timely effort would be needed to experiment with optimizing these queries. That D2R Server has been implemented using Java means that memory is filled much earlier than if the database was queried directly by the mSpace Server. It was also not possible to perform keyword matching efficiently across the data, due to a lack of an indexable query system for keyword string matching, with only the SPARQL REGEX syntax supported. D2R performs these in its local virtual machine memory space, rather than in the database itself, again causing a performance hit. At this point, we wanted to see if it were possible to achieve the kind of performance we needed by going directly to a highly indexable, tuneable system, so we tried MySQL.

4   SQL and MySQL: Losing the Semantic Web for Performance

SQL is a lower-level language than SPARQL. In SPARQL, one describes the pattern of data that is desired, and it is the responsibility of the triple store to decide how to retrieve the information. In SQL, it is possible to express the same query in a variety of different ways, depending on how we wish the database to achieve the result. This is both an advantage and a disadvantage: it offers the ability to tweak our queries manually to achieve optimal performance, but this is a laborious process. Ultimately it would be preferable to have the backing data store do the optimising work for us. Since the mSpace server now had the model separated from the data store, moving from a D2R storage to a straight SQL one was relatively cheap and offered considerable performance gains over a system using the same database, but with a D2R layer included. It also allowed the system to capitalise on the full-text indexing available in MySQL for scalable string searches of the data.

The focus of this approach is purely the performance of getting information to the UI in real time. While this means that the UI is responsive and scales well to large datasets, it means sacrificing the original Semantic Web aspects of this work. As such, future work will re-explore the possibilities of re-incorporating support for aggregated heterogeneous sources, looking at how best to translate the performance gains we do get from SQL while utilising Semantic Web tools for data aggregation.

5 Conclusions

Integration of new, more backend-intensive technologies, as well as working with larger collections of data has raised interesting research and technical challenges for mSpace in particular and Semantic Web technologies in general. The very reason for the Semantic Web is to bring together heterogeneous data to enable rich queries of it: our work has challenged Semantic Web technologies to support such interfaces for even a single data set. We discovered that existing triple storage solutions were unable to provide interactive-level real-time performance as our needs grew, and were forced to implement traditional relational database support. While this transition aided performance to the UI, it has meant that benefits gained from the use of Semantic Web technologies, such as ease of data aggregation, and the simple interface of SPARQL, are lost. Future optimisations of the mSpace server may yield some performance improvements, but in order to perform the desired move back to solely using Semantic technologies there is a requirement for triple stores with response times several orders of magnitude better than that offered by 3store. This will likely require stores that have moved beyond the long triple list format, and begun to involve indexing and query optimisations that cope with RDF data's unpredictable structure.

References

1. schraefel, m.c., Smith, D.A., Owens, A., Russell, A., Harris, C., Wilson, M.L.: The evolving mSpace platform: leveraging the Semantic Web on the Trail of the Memex. In: Proceedings of Hypertext, Salzburg (2005)
2. schraefel, m.c., Wilson, M., Russell, A., Smith, D.A.: mSpace: improving information access to multimedia domains with multimodal exploratory search. Commun. ACM 49(4), 47–49 (2006)
3. Harris, S., Gibbins, N.: 3store: Efficient Bulk RDF Storage. In: 1st International Workshop on Practical and Scalable Semantic Systems, Sanibel Island, Florida, pp. 1–15 (2003)
4. Shadbolt, N., Gibbins, N., Glaser, H., Harris, S., schraefel, m.c.: CS AKTive Space, or How We Learned to Stop Worrying and Love the Semantic Web. IEEE Intelligent Systems 19(3), 41–47 (2004)
5. Bizer, C., Cyganiak, R.: D2R server-publishing relational databases on the Semantic Web (poster). In: Fensel, D., Sycara, K.P., Mylopoulos, J. (eds.) ISWC 2003. LNCS, vol. 2870, Springer, Heidelberg (2003)

A Study of Vocabularies for Image Annotation

Allan Hanbury*

Pattern Recognition and Image Processing group (PRIP),
Institute of Computer Aided Automation, Vienna University of Technology,
Favoritenstraße 9/1832, A-1040 Vienna, Austria
hanbury@prip.tuwien.ac.at

Abstract. In order to evaluate image annotation and object categorisation algorithms, ground truth in the form of a set of images correctly
annotated with text describing each image is required. Statistics on the
WordNet categories of keywords collected from recent automated image
annotation and object categorisation publications and evaluation campaigns are presented. These statistics provide a snapshot of keywords
used to train and test current image annotation systems as well as information on the usefulness of WordNet for categorising them.

1 Introduction

Automated image annotation and object categorisation are currently important
research topics in the field of computer vision [1,2,3,4]. To measure progress
towards successfully carrying out this task, evaluation of algorithms which automatically extract this sort of metadata is required. For successful evaluation of
these algorithms, reliable ground truth is necessary. This ground truth is usually
in the form of a manual or computer-assisted annotation of images by keywords.

For *automated image annotation*, the aim is to automatically assign suitable
keywords to describe images or regions of images based on image features [1,2,3].
Object Categorisation is concerned with the identification of particular objects
[4,5]. Object categorisation can be seen as annotation of an image by keywords
describing the objects present. Due to the nature of the algorithms applied,
automated image annotation techniques have in general annotated the images
using a selection from a larger vocabulary of keywords than object categorisation.

In this paper we provide an overview of the keywords that have been used
for the annotation of images for evaluation purposes. We present statistics on
the distribution of the WordNet categories of keywords from datasets used in
automated image annotation and object categorisation research and evaluation
campaigns. This provides a snapshot of keywords used to train and test current
image annotation systems. The list of keywords created in this work could be
used as the start of a more comprehensive vocabulary for the annotation of
images, similarly to the way in which the vocabulary in [6] was begun.

---

* This work was partially supported by the European Union Network of Excellence
MUSCLE (FP6-507752).

B. Falcidieno et al. (Eds.): SAMT 2007, LNCS 4816, pp. 284–287, 2007.

Table 1. The sources of the keywords. The left column gives the source and references, and the right column gives the number of keywords obtained from the source.

Source	# Keywords
PASCAL VOC Challenge 2005 databases [5]	101
EU LAVA Project [7]	10
Chen and Wang [8]	20
Microsoft Research Cambridge Databases [4]	35
Fei-Fei et al. [9]	101
Carbonetto et al. [2]	55
Li and Wang [3]	433
Barnard et al. [1]	323
University of Washington Ground Truth Image Database	392

2   Analysis of Visual Keywords

We created a list combining all the keywords used in the papers and datasets listed in Table 1. These sources correspond to datasets which have been made available on-line. Each keyword was entered into the list only once, even if it occurred in more than one source list. Nouns in plural forms were converted to singular form. Keywords not present in WordNet were excluded. This led to a combined list containing 792 keywords.

We categorised these keywords using WordNet [10] categories from a higher level. Choosing a suitable level for all classes of words proved to be difficult, as the different branches of the WordNet hierarchy have different depths. For example taking the top level of the WordNet hierarchy (the word "entity") as level 0, the word "goalpost" is found at level 11, while the word "waterfall" is at level 4. We therefore decided to use a mixture of categories from levels 3 and 4 of the hierarchy, manually chosen to minimize the number of categories (e.g., each keyword should if possible not form its own category, as would happen if we chose level 4 for the keywords describing bodies of water) while preventing the creation of categories containing too many keywords. The "food" category occurs in both levels 3 and 4 in different branches of the hierarchy, referring to "solid food" and "nutrients". To prevent the creation of two very similar categories, the members of both categories have been fused. For proper nouns, we used the "instance of" relation to place them in the hierarchy. Verbs and adjectives were placed in their own categories. Some words occur in more than one branch of the WordNet hierarchy. In this case, we manually chose the "most visual" of the branches, or included more than one branch if the corresponding senses were applicable to images. The distribution of the number of keywords per category is shown in Table 2. The full categorised keyword list is available for download[1].

The *artefact* category contains the most keywords, followed by the *living thing* category, where the latter includes humans, plants and animals. This reflects the presence of objects and animals in both the Corel dataset and the datasets

---

[1] http://muscle.prip.tuwien.ac.at/keywords_with_wordnet_categories.txt

Table 2. The number (#, column 2) of keywords occurring in each category. The level ℓ at which each category occurs in the WordNet hierarchy is shown in the 3rd column. The rightmost 3 columns contain the names of the levels above the chosen category.

category	#	ℓ	level 1	level 2	level 3
artefact	264	4	physical entity	object	whole
living thing	153	3	physical entity	object	
location	114	3	physical entity	object	
event	40	4	abstract entity	abstraction	psychological feature
adjective	34				
food	22	3	physical entity	substance	
		4	physical entity	substance	solid
natural object	21	4	physical entity	object	whole
geological formation	18	3	physical entity	object	
cognition	14	4	abstract entity	abstraction	psychological feature
body of water	14	3	physical entity	thing	
material	12	3	physical entity	substance	
body part	11	4	physical entity	thing	part
fundamental quantity	10	4	abstract entity	abstraction	measure
natural phenomenon	10	4	physical entity	process	phenomenon
land	10	3	physical entity	object	
attribute	9	3	abstract entity	abstraction	
group	8	3	abstract entity	abstraction	
communication	8	3	abstract entity	abstraction	
gas	5	4	physical entity	substance	fluid
solid	5	3	physical entity	substance	
relation	4	3	abstract entity	abstraction	
drug	2	4	physical entity	causal agent	agent
suspension	1	4	physical entity	substance	mixture
chemical process	1	4	physical entity	process	natural process
verb	1				
bodily process	1	4	physical entity	process	organic process

collected for object categorisation tasks. The *location* keywords also contain a large number of proper nouns, such as "Mexico" and "British Columbia", which are present in the Corel annotations. It is less likely that these proper nouns can be successfully associated automatically with images. The *event* category contains keywords such as "war", "parade" and various types of sport.

The categories containing few keywords show which branches of the WordNet hierarchy are sparsely populated: the only *bodily process* is "dining", the only *chemical process* is "fire" and the only *suspension* is "steam". These sparsely populated branches could also be seen as a demonstration that the WordNet hierarchy is not ideally suited to intuitive categorisation of concepts found in images. An example of this is that "sky" is classified under *gas*. The reason that there are 5 keywords in this category is that the Washington Database contains the combined keywords "clear sky", "cloudy sky", "overcast sky", etc. These common keywords forming part of sparsely populated branches should also be considered when WordNet is used as the basis for a vocabulary, as in [11].

The advantage of using WordNet categories is that it is straightforward to examine categories obtained from other levels of the hierarchy. A solution to the problem of some of the sparsely populated categories would be to use the level 2 category *substance*. This would fuse the categories *food*, *material*, *gas*, *solid* and *suspension* into a single category containing 45 keywords. However, "sky" would now fall into the rather non-intuitive *substance* category.

3 Conclusion

We analyse the keywords that have been used to annotate images in a number of publications and evaluation campaigns. These keywords are placed into categories obtained from higher levels of the WordNet hierarchy. From this analysis one can see that the main automated annotation effort has been directed at images of everyday objects and of living things. This categorisation also reveals some disadvantages of using the WordNet hierarchy to create intuitive categories for an image annotation vocabulary. The investigation of a more intuitive categorisation of keywords for image annotation is an interesting topic to pursue.

References

1. Barnard, K., Duygulu, P., de Freitas, N., Forsyth, D., Blei, D., Jordan, M.I.: Matching words and pictures. Journal of Machine Learning Research 3, 1107–1135 (2003)
2. Carbonetto, P., de Freitas, N., Barnard, K.: A statistical model for general contextual object recognition. In: Pajdla, T., Matas, J(G.) (eds.) ECCV 2004. LNCS, vol. 3021, pp. 350–362. Springer, Heidelberg (2004)
3. Li, J., Wang, J.Z.: Automatic linguistic indexing of pictures by a statistical modeling approach. IEEE Trans. PAMI 25(9), 1075–1088 (2003)
4. Winn, J., Criminisi, A., Minka, T.: Object categorization by learned universal visual dictionary. In: Proc. ICCV, pp. 1800–1807 (2005)
5. Everingham, M., et al.: The 2005 PASCAL visual object classes challenge. In: Selected Proceedings of the First PASCAL Challenges Workshop (2006)
6. Jörgenson, C., Jörgenson, P.: Testing a vocabulary for image indexing and ground truthing. In: Proc. Internet Imaging III, pp. 207–215 (2002)
7. Perronnin, F., Dance, C., Csurka, G., Bressan, M.: Adapted vocabularies for generic visual categorization. In: Leonardis, A., Bischof, H., Pinz, A. (eds.) ECCV 2006. LNCS, vol. 3954, pp. 464–475. Springer, Heidelberg (2006)
8. Chen, Y., Wang, J.Z.: Image categorization by learning and reasoning with regions. Journal of Machine Learning Research 5, 913–939 (2004)
9. Fei-Fei, L., Fergus, R., Perona, P.: Learning generative visual models from few training examples an incremental bayesian approach tested on 101 object categories. In: Proc. Workshop on Generative-Model Based Vision (June 2004)
10. Miller, G.A., Beckwith, R., Fellbaum, C., Gross, D., Miller, K.: Introduction to WordNet: An on-line lexical database. International Journal of Lexicography 3(4), 235–244 (1990)
11. Zinger, S., Millet, C., Mathieu, B., Grefenstette, G., Hède, P., Moëllic, P.A.: Extracting an ontology of portrayable objects from WordNet. In: Proc. MUSCLE/ImageCLEF Workshop on Image & Video Retrieval Evaluation, pp. 17–23 (2005)

Towards a Cross-Media Analysis of Spatially Co-located Image and Text Regions in TV-News

Thierry Declerck[1] and Andreas Cobet[2]

[1] DFKI GmbH, Language Technology Lab,
Stuhlsatzenhausweg.3, 66123 Saarbrücken, Germany
declerck@dfki.de
[2] Technische Universität Berlin, Communication Systems Group,
Straße des 17. Juni, 10623 Berlin, Germany
cobet@nue.tu-berlin.de

Abstract. We describe in this poster/short paper on-going work on the extraction and semantic interpretation of text regions in television news programmes. We present some of the data we consider in this work, the actual technologies in use and where they have to be improved. Finally we briefly discuss a possible innovative and valuable approach to the establishment of a cross-media analysis framework.

1 Introduction

The European Network of Excellence "K-Space"[1], which started in 2006, is dealing with semantic inferences for semi-automatic annotation and retrieval of multimedia content. The aim of the project is to contribute in narrowing the gap between content descriptors that can be computed automatically by current machines and algorithms, and the richness and subjectivity of semantics in high-level human interpretations of audiovisual media: the so-called *Semantic Gap*.

The project deals with the integration of knowledge structures, as encoded in high-level representation languages, and low-level descriptors for audio-video content, taking also into account knowledge that can be extracted from sources that are complementary to the audio/video stream, mainly speech transcripts and text surrounding images or textual metadata describing a video or images, or even text included in the images. These complementary resources typically fall into two groups: primary resources, which are directly attached to multimedia, and secondary resources, which are more loosely coupled with the audio-video material.

We concentrate in this position paper on the primary complementary resources, and investigate the possible use of text extracted from images by means of detection of textual regions in images and optical character recognition (OCR) processes for adding semantics to images and also to support audio/video analysis. We describe a first experiment on text extraction from news videos (news programmes of the German broadcaster "ARD"), for which we identified a list of relevant patterns of

---

[1] "K-Space" stays for "Knowledge Space of Shared Technology and Integrative Research to Bridge the Semantic Gap"; see also www.k-space.eu.

B. Falcidieno et al. (Eds.): SAMT 2007, LNCS 4816, pp. 288–291, 2007.

textual information appearing on the TV screen during those news programmes, and their relations to displayed images. We show 2 examples of such patterns (out of 6 we identified), and explain what can be gained from those patterns, which are particular in the sense that the text belonging to one semantic unit might be distributed around an image. The use of the extracted text for supporting the semantic annotation of their containing images implies the applications of linguistic analysis of the extracted text and an appropriate detection images regions. We concentrate in this paper on a short description of such patterns, showing results of the detection of textual regions in images, of the OCR procedure and of linguistic analysis applied to the extracted text.

2   The Data

In the following we just present 2 typical examples of presenting news information to the TV public[2], showing how the broadcaster (here the German public broadcaster ARD) combines image and text to convey information.

Fig. 1. In this pattern, we can see the speaker and a background image, directly surrounded by two textual contributions (ignoring here the name of the News programme and the date)

In the first case above, we consider the two textual regions being close to the background image, just above and below it. The text analysis tool applied to the extracted text can detect the topic (decision of the Parliament about election) and also where it takes place (in Kiew). Interesting here: there is no linguistic hint, that this "decision" is being discussed in Kiew: We can infer this only on the base of heuristics applied to the distribution of words around the image. On the base of world

---

[2] As already mentioned above, they are more such patterns, which cannot be described here due to space limitation.

knowledge, we can also infer that the Parliament presented here is the Ukrainian one (this information being most probably given by the speaker). Other information we can recognize: the voice to be heard in the audio streaming is in this pattern belonging to the news speaker. In case we know her name, this information can help in improving speaker recognition. In other patterns of information display, we can assume that the voice being heard is not from the speaker, but from someone presented in the image (or video).

A more complex pattern, with respect to semantic interpretation is shown in Fig. 2.

Fig. 2. We can see above the background picture a short phrase and below the picture the name of a person. The text should be read as "Accusations against the son of Annan".

In Fig. 2 the person name below the image is pointing to the content of the image. But the information on the top of the image is mentioning: "complains against son". So here we need also some inferences to get the point that the accusations are addressed against the son of the person shown in the image, but that the son is not shown here.

3 First Results

The textual region detection tools used in our experiment perform quite good[3], but the OCR tools applied have still to be improved[4], and below we can see one error generated by the OCR procedure: the name of Kiew being represented as "Krew". But

---

[3] This information on the base of a small informal evaluation done on the data. A formal evaluation is still to be proposed.

[4] First steps have been done in this respects, which we will present in a next version of this short paper.

here when we know that we deal with Named Entities, a list (or gazetteer) of such Entities can be given to the OCR mechanisms for matching/correcting their results against it. An example of an output of the system (for the example given in Fig. 1) is:

Krew 4 78 452
Parlamentsbeschluss 4 84 102
Zur 4 84 150
Wahl 4 130 143

Here we have a good result, with only 1 error ("Krew" instead of "Kiew"). We can extract out of this data structure the different text contributions. We cannot propose for an analysis of the string "Krew" for the time being (unknown word). The (automatic) linguistic dependency analysis of the textual contribution at the top of the image is giving:

[NP Parlamentsbeschluss *(head_noun)* [PP zur Wahl] *(noun_modifier)*]

The identified head noun is the main topic. And in fact this corresponds to the image, showing a parliament. So the key frame is not about the „election", but about a parliament decision about the election. The dependency analysis allows thus to reduce considerably the number of key words that can be used for indexing, replacing them by structured textual fragment. We can map here the head noun onto an ontology as well, so that the image might be also annotated with concepts like "political institutions" or the like.

4 First Conclusions

On the base of the patterns described above, we start to provide a textual analysis, which has to take into account the non-sequential list of words distributed over the screen. For this, we have to take into account the region information of the text parts, also including information about the region of the image to which the text parts are "belonging".

The linguistic and semantic annotation resulting from the analysis of those textual parts can then support the semantic annotation of the image as well. Very interesting in this context, is the fact that the image itself contributes to the semantic content of the text, since the image plays sometimes the role of filling the gap of omitted textual elements in the image displayed in the actual news. So for examples we noticed that only short nominal phrases are used (very seldom verbs are used in this context), and that images are acting as linking information between the textual part. This is very well documented in the figure 2 above, where the "complains against son" and "Annan" are linked by the image of Annan. In normal natural text, we would expect the sentence "Complains against the son of Annan".

If this sentence would have been associated with an image, we would certainly expect to have not Annan himself shown in the image, but rather the son of Annan. In this sense the patterns we recognized can really give extended support to person detection in images, in a cross-media fashion.

Towards Person Google: Multimodal Person Search and Retrieval

Lutz Goldmann, Amjad Samour, and Thomas Sikora*

Communication Systems Group
Technical University of Berlin
Berlin, Germany

Abstract. Content based multimedia retrieval systems have been proposed to allow for automatic and efficient indexing and retrieval of the increasing amount of audiovisual data (image, video and audio clips). The search for specific persons within this data is an important subtopic due to its large range of applications. This article describes an original system for multimodal person search and provides some initial performance results that demonstrate the efficiency of the system.

1 Introduction

With the increasing amount of available multimedia data, efficient systems for searching and retrieving relevant AV documents are needed. Since keyword based indexing is very time consuming and inefficient due to linguistic and semantic ambiguities, content based multimedia retrieval systems have been proposed, that search and retrieve AV documents based on audio and visual features. While content based image retrieval has been a very active research field, only some work has been done in the field of person search and retrieval, where the goal is to find a AV document with a specific person present within the audio and the visual stream. An original system for multimodal person search and retrieval is proposed in this article which is based on audio and video analysis techniques combined by a multimodal fusion approach.

2 System Overview

Figure 1 gives an overview of the proposed system. The initial system is based on the query by example paradigm, where the user selects an AV document, the system compares it to the AV documents in the database and retrieves them ranked according to their similarity.

2.1 Audio Analysis

The goal of the audio analyis part is to retrieve audio segments based on the voice characteristics of a person without considering the spoken content.

---

* The work presented in this paper was supported by the European Commission under contract FP6-027026 K-Space.

B. Falcidieno et al. (Eds.): SAMT 2007, LNCS 4816, pp. 292–295, 2007.

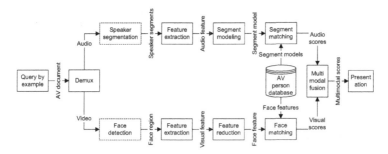

Fig. 1. Overview of the person search and retrieval system

Feature extraction. For describing the audio segments mel frequency cepstral coefficients (MFCC), introduced by Davis et al. [1], are applied. They are widely used in the area of speech recognition and audio classification since they provide a good description of audio characteristics under a wide range of conditions and with reasonable computational costs. A sliding window divides the audio segment into multiple overlapping frames with a length of 20 ms and an overlap of 10 ms. For each frame a feature vector consisting of 13 MFCC's and the log energy of the frame is created. Depending on the segment length, different number of feature vectors are extracted.

Segment modeling. In order to reduce the temporal characteristics of the audio data within a segment and to create a robust model of the spectral characteristics of the speaker's voice, each speaker segment is modeled as a multivariate Gaussian distribution.

Segment matching. The goal of the matching stage is to compare an audio segment with all audio segments within the database. Since each audio segment is described using a statistical model, model selection techniques are suitable for the comparison. The Bayesian information criterion (BIC), proposed by Schwarz et al. [2] is applied to compute the distance between two segments.

2.2   Visual Analysis

The goal of the visual analysis part is to retrieve persons based on their facial appearance.

Feature extraction. The face region is determined based on the pupil positions, obtained either manually or automatically, and an anthropometric face model [3]. In order to handle different face sizes, each face region is scaled to a common size. Since uneven illumination tends to change the appearance of faces, statistical normalization methods are applied globally and locally.

Feature reduction. In order to reduce the feature dimensionality and maintain the most relevant information, the principal component analysis (PCA) [4] is applied. This leads to a set of eigenfaces that form a reduced basis onto which the original feature vectors are projected.

Feature matching. The goal of the matching stage is to compare a query face with all faces in the database. Since each face is described by a feature vector, vector distances can be used for the comparison. For the initial experiments the Euclidean distance was chosen.

2.3   Multimodal Fusion

The general goal of multimodal fusion is to exploit the complementary character of multimodal sources to increase the robustness of the system with regard to the single modalities. More specifically, the idea here is to combine the voice and face characteristics to resolve ambiguities within an individual modality. For the proposed system score level fusion was chosen since it provides the best tradeoff in terms of information content and ease of fusion.

Score normalization. Scores of different modalities usually exhibit quite different characteristics (type, distribution, range) which make it very difficult to combine them in a suitable way. The goal of the score normalization step is to modify the location and variation of their distributions to transform them into a common domain. Out of the large number of possible normalization techniques the z-score normalization has been chosen for the initial system.

Score fusion. Different fusion rules (product, sum, min, max) are considered within the system. For each of the AV documents, these fusion rules combine the corresponding audio and visual scores into a multimodal score.

3   Experiments

The initial experiments are based on the VALID database that contains multimodal data (audio, video) of 106 persons (27 female, 79 male).

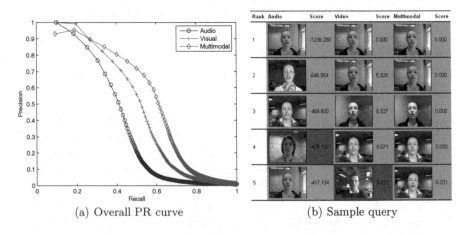

(a) Overall PR curve   (b) Sample query

Fig. 2. Retrieval performance of the different variants (audio, visual, multimodal)

Several evaluation measures have been proposed for evaluating search and retrieval systems [5]. They can be divided into precision/recall and rank measures. Both classes of measures have been considered.

Figure 2 shows a comparison of the audio only, visual only, and the multimodal system based on precision vs. recall (PR) curves and a sample query. In the current system, the visual modality outperforms the audio modality. Furthermore, it can be seen that the multimodal system shows a considerable performance improvement over the single modality systems. For 20 retrieved documents the mean recall values of the audio, visual, and multimodal system are $R = \{49; 58; 66\}\%$ while the corresponding precision values are $P = \{24; 29; 33\}\%$ respectively.

4   Conclusions

A system for multimodal person search and retrieval using voice and face characteristics has been developed. Initial experiments provide encouraging results and justify the proposed solution. While both modalities perform well individually, a performance improvement can be achieved with the multimodal system.

Future work will explore the different parts of the system in more detail and evaluate variants of the system. Furthermore, different query paradigms and relevance feedback techniques will be incorporated into the system. Another aspect is to analyze the influence of the detection steps (face detection, speaker segmentation) onto the retrieval performance.

References

1. Davis, S.B., Mermelstein, P.: Comparison of parametric representations for monosyllabic word recognition in continuously spoken sentences. IEEE Transactions on Acoustics, Speech, and Signal Processing 28(4), 357–366 (1980)
2. Schwarz, G.: Estimation the dimension of a model. Ann. Stat. 6, 461–464 (1978)
3. Farkas, L.G.: Anthropometry of the Head and Face. Raven Press (1994)
4. Jain, A.K., Duin, R.P.W., Mao, J.: Statistical pattern recognition: A review. IEEE Transactions on Pattern Analysis and Machine Intelligence 22, 4–37 (2000)
5. Mueller, H., Mueller, W., et al.: Performance evaluation in content based image retrieval: Overview and proposals. Technical report, University of Geneva (1999)

Event Detection in Pedestrian Detection and Tracking Applications

Philip Kelly, Noel E. O'Connor, and Alan F. Smeaton

Centre for Digital Video Processing, Adaptive Information Cluster,
Dublin City University, Ireland

Abstract. In this paper, we present a system framework for event detection in pedestrian and tracking applications. The system is built upon a robust computer vision approach to detecting and tracking pedestrians in unconstrained crowded scenes. Upon this framework we propose a pedestrian indexing scheme and suite of tools for detecting events or retrieving data from a given scenario.

Keywords: Pedestrian Detection, Tracking, Stereo, Event Detection.

1   Introduction

Robust detection and tracking of humans is a key enabling technology for many applications. It is key to knowing who is where in a scene *and* what their actions have been. It potentially allows other layers in an application's framework to infer beliefs about those people. However, depending upon the end-user application, a variety of event detectors may need to be defined. In this paper, we propose a framework for detecting user-defined events at run-time which may then be used in a variety of surveillance applications.

2   Pedestrian Detection and Tracking

Many of the person detection and tracking techniques described in the literature make assumptions about the environmental conditions, pedestrian and background colour intensity information [1], occlusions [2], the pedestrian flow, pedestrian movements, or that a person will exist within the scene for a significant number of frames un-occluded[3]. We have developed a robust pedestrian detection and tracking system [4] for a single stereo camera that makes *none* of the above assumptions and in addition requires *no* external training. In addition, it is camera viewpoint invariant and is able to handle a large variety of pedestrian appearances, making a single constraining assumption that a person is standing vertically with respect to the groundplane. During tracking, the system can obtain statistics about each pedestrian, such as colour appearance models, the duration of time spent in the camera's field of view and the path they have traversed through the scene. Since stereo information is used, both the 3D position, velocity and the height of a pedestrian can be accurately obtained.

B. Falcidieno et al. (Eds.): SAMT 2007, LNCS 4816, pp. 296–299, 2007.

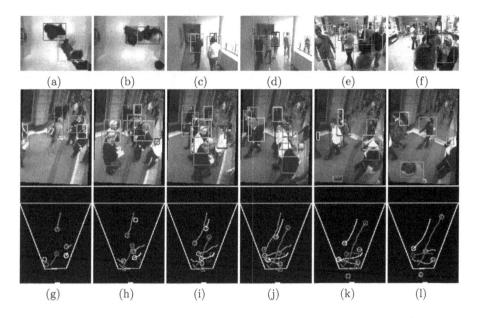

(a)   (b)   (c)   (d)   (e)   (f)

(g)   (h)   (i)   (j)   (k)   (l)

Fig. 1. Examples; (a)-(f) Pedestrian Detection; (g)-(l) Pedestrian Tracking

Evaluation of this system on 3,500 manually annotated ground truth pedestrians from sequences with varying camera height, camera orientation and environmental conditions reveal extremely accurate performance of our proposed approach. Figure 1(a)-(f) presents some illustrative results where each detected pedestrian is enclosed by a bounding box of a certain colour. An example of a tracking sequence can be seen in figure 1(g)-(l), where the second row of images depict the scene from a *plan-view* or *birds-eye view* orientation. In these plan-view images, the white lines indicate the bounds of the scene, the position of detected pedestrians in that frame are illustrated by a circle of the same colour as their bounding box, and tracks are depicted as "tails" from the centre of the circle to previous positions in the scene.

3   Event Detection

Beyond detection and tracking, many applications require a further level of processing whereby the *actions* of pedestrians must be interpreted. For some applications these event detectors can be hard-coded into the application framework, for example surveillance applications which determine pedestrian flow densities during specific time periods. For many applications, however, the exact event detector *cannot* be hard-coded into the system as; (1) the event definition is dependent on an undefined scene; or (2) the event is itself undefined. The first scenario is typical of many Ambient Intelligence (AmI) applications [5], as although the event required to be detected is known (e.g. detect pedestrians

in a designated area waiting to cross the road), information about the scene
(e.g. the exact designated area) is unknown. The second scenario is typical of
general purpose surveillance applications where events are determined by a user
for detection either at run-time or *after* the event has occurred. Take for example a typical CCTV surveillance system, if the surveillance video is augmented
with robust pedestrian tracking and statistical information it becomes possible
to quickly search the video for specific events via the augmented meta-data. For
example, if a lost child's appearance is known (e.g. the colour of their clothes and
their height) then all possible detections of the child can be subsequently flagged
from every camera in the system. Other examples include adding "watched" areas in a scene (e.g. to determine the number of pedestrians who entered a specific
shop) or flagging "unusual" events such as lingering pedestrians (e.g. to review
all footage of persons who were in the vicinity of an area that was vandalised
for a significant amount of time).

4   Proposed Approach

Building on our prior work, we have developed a pedestrian indexing scheme and
suite of tools for detecting events and retrieving data from a given scenario. One
such tool is the creation of 3D *hotspot* regions from 2D plan-view images - see
the 2D yellow coloured area in figure 2(a). In this figure, it should be noted that
a background colour model has been projected onto the image-plane to allow
the gauging of distance and orientation within the plan-view image. Using this
plan-view image, a hotspot is simply created in the proposed system by circling a
region of interest within the plan-view image. The resultant hotspot can be seen
as a 3D area of interest which can be incorporated into the definition of events
for some application scenarios. For example, in the framework of an automated
pedestrian traffic light system it can be used to define an area where pedestrians
tend to wait before crossing the road – see figures 2(b)-(f). A detected event,
which in turn leads to a changing of the traffic lights, could then be defined
using this hotspot region and an event (or query) syntax. An example of the
syntax is, $Event(cross) = hpt(1), p(g5), t(10s)$, which declares an event called
cross to be detected if on hotspot number 1 ($hpt(1)$) there are greater than 5
pedestrians ($p(g5)$) who have been waiting for more than 10 seconds ($t(10s)$).
The proposed system can be easily used across scenes with differing underlying
structure whilst the event syntax allows a number of user-defined events to be
created and searched for during run-time in content-based retrieval applications.

5   Future Work

These techniques will underpin a user-defined event detection system which can
be simply tailored for a suite of pedestrian detection and tracking applications.
Currently the event syntax consists of a number of features including; hotspots,
timings (such as the length of time an object is in the scene), and pedestrian
numbers. However, the syntax should be extended to incorporate other features,

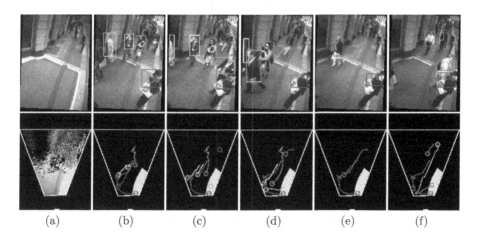

(a)   (b)   (c)   (d)   (e)   (f)

Fig. 2. Pedestrian crossing application; (a) Hotspot; (b)-(f) Waiting to cross the road

including; time of day, pedestrian interactions (such as pedestrians walking in a group or on their own) and pedestrian statistics such as colour, height, velocity and position).

Acknowledgements

This material is based on works supported by Science Foundation Ireland under Grant No. 03/IN.3/I361. The authors wish to acknowledge the support of the European Commission under the FP6-027026-K-SPACE contract. This paper was awarded a best poster prize at the 2007 K-Space Jamboree Workshop for PhD students.

References

1. Senior, A.W.: Tracking with probabilistic appearance models. In: IEEE International Workshop on Performance Evaluation of Tracking and Surveillance Systems, pp. 48–55 (June 2002)
2. Harville, M.: Stereo person tracking with adaptive plan-view templates of height and occupancy statistics. International Journal of Computer Vision 22, 127–142 (2004)
3. Elgammal, A.E, Davis, L.S.: Probabilistic framework for segmenting people under occlusion. In: IEEE International Conference on Computer Vision, vol. 2, pp. 145–152 (2001)
4. Kelly, P., Cooke, E., O'Connor, N.E., Smeaton, A.F.: Pedestrian detection using stereo and biometric information. In: International Conference on Image Analysis and Recognition, pp. 802–813 (September 2006)
5. Remagnino, P., Foresti, G.L.: Ambient intelligence: A new multidisciplinary paradigm. In: IEEE Transactions on Systems, Man and Cybernetics, vol. 35, pp. 1–6 (January 2005)

SAMMI
Semantic Affect-Enhanced MultiMedia Indexing

Marco Paleari[1], Benoit Huet[1], and Brian Duffy[2]

[1] Eurecom Institute, B.P. 194,
F-06904 Sophia Antipolis Cedex, France
paleari@eurecom.fr
[2] The SmartLab, University of East London
London, UK

Abstract. Multimedia indexing is about developing techniques allowing people to effectively find media. Content-based methods become necessary when dealing with big databases. Current technology allows exploring the emotional space which is known to carry very interesting semantic information. In this paper we state the need for an integrated method which extracts reliable affective information and attaches this semantic information to the medium itself. We present a list of possible applications and advantages that the emotional information can bring about together with a framework called SAMMI and the preliminary results of this newly initiated research work.

1 Introduction

Emotions have been demonstrated to influence many different human cerebral functions and in particular human memory [1]. In this paper we present few possible scenarios involving content-based indexing, retrieval, and summarization of media and we show how a coupled affect and semantic approach can improve results of such a kind of systems. We then detail an architecture combining emotion recognition through multimodal fusion and automatic semantic labeling/tagging of videos for content-based retrieval and summarization.

Even though studies from the indexing and retrieval community acknowledge that emotions are an important characteristic of media and that they might be used in many interesting ways as semantic tags only few efforts have been done to link emotions to content-based indexing and retrieval of multimedia [2,3,4,5,6]. [2,3] analyze the text associated to a film searching for occurrences of emotionally meaningful terms; [4] analyze pitch and energy of the speech signal of a film; [5] canalize features such as tempo, melody, mode, and rhythm to classify music and [6] uses information about textures and colors to extrapolate the emotional meaning of an image. The evaluation of these systems lack of completeness but when the algorithms are evaluated they allow to positively index as much as 85% of media showing the feasibility of this kind of approach.

State of the art algorithms for emotion recognition usually use the speech signal and/or the facial expression (see [7] for a thorough overview) approaching

B. Falcidieno et al. (Eds.): SAMT 2007, LNCS 4816, pp. 300–303, 2007.

a recognition score of 90%. Some limitations are nevertheless usually applied on the training and testing data which makes the data not realistic. Illumination, audio quality, database size, head movement and position, user (in)dependency, or distractions such as beard or glasses represent usually the main challenges for this kind of systems. Only few works have exploited the intrinsically multimodal nature of emotions by using two or more modalities, usually audio and video, and claiming interesting performances (around 90%).

2   Motivations and Case Studies

It seems, in many cases, very reasonable to use emotions for indexing and retrieval tasks. For example one could argue it is simpler to define music as "romantic" or "melancholic" than to define its genre, tempo or melody. Similarly film and book genres are strongly linked to emotions as can clearly be seen in the case of comedies or horrors. We argue emotions need to be coupled to other content-based semantic tags to build complete and flexible systems.

One example showing the importance of a multi-disciplinary approach could be where one is trying to summarize one action movie: one may look for scenes regarding gunfights and therefore looking for shootings. Supposing there are, in the film, scenes in a shooting range, we may not want to select them. Looking at the content alone would return these scenes together with the real gunfights while only looking for emotionally relevant scenes instead would result in finding scenes which do not contains shootings at all. The combination of the two, however, will be able to return scenes which are emotionally relevant and do contain shootings and that are, therefore, likely to belong to gunfights. The same principles can be applied to an indexing scenario: an action movie could be, for example, characterized by the fact of having an ongoing rotation of surprise, fear, and relief and for having explosion or shooting scenes.

We have seen, so far, how emotions can join other media content descriptors on order to improve upon the performance of content-based retrieval and semantic indexing systems. In the next section we describe SAMMI, a framework we are developing which allows creating such a kind of systems.

3   Semantic Affect-Enhanced MultiMedia Indexing

This section describes SAMMI, a framework explicitly designed for extracting reliable real-time emotional information through multimodal fusion of affective cues and to use it for emotion-enhanced indexing and retrieval of videos.

There are three main limitations of existing work on emotion-based indexing and retrieval that have been shown: 1) emotion estimation algorithms are very simple and not very reliable, 2) emotions are generally used without being coupled with any other content information and 3) the evaluation of the experiments is preliminary and quite incomplete;

SAMMI estimates emotions through a multimodal fusion paradigm. Speech is analyzed and different feature sets are extrapolated: pitch, speech

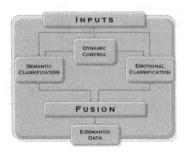

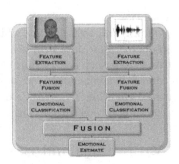

Fig. 1. SAMMI's architecture   **Fig. 2.** Bimodal emotion recognition

formants, energy, MFCC, and Rasta-PLP. Those feature sets are fed to different classification systems (e.g. HMM, GMM, and SVM) to have different emotion estimates to compare. Simultaneously a face is found in the video and the expression is analyzed through motion flow and feature point positions and movements; these features are also fed to different classification systems. Multimodal feature fusion will be experimented, leading to additional emotion estimates.

The different emotion appraisals are fused to extrapolate a single emotion estimate (see Fig. 2). Dynamic control (Fig. 1) is used to adapt the multimodal fusion according to the qualities of the various modalities at hand. Indeed if lighting is inadequate the use of color information should be limited and the emotion estimate should privilege the auditory modality.

SAMMI couples emotions and other semantic information (Fig. 1). The extraction of different feature sets from the same media, as well as the application of different classification techniques and the use of different modalities are all characteristics which assure good reliability; the use of dynamic control assure stability in presence of noise.

4   Preliminary Results and Concluding Remarks

We have currently developed the automatic and real-time extraction of the feature points from the video. When a face found (Haar classifier) the video is cropped, resized and equalized. Twelve facial zones are considered. For each zone some points are followed along the video (Lukas & Kanade algorithm). The trajectory of the center of mass of the 12 point sets is used as output (Fig. 3).

With the obtained data we trained two classifiers (a NN and a SVM) with different settings and we reached an average 48.4% recognition rate with a strong predominance of the anger and sadness emotions compared to the others (Fig. 4). The analysis of the temporal information reveals that different emotions are recognized according to different temporal patterns. Each point of the graphs in Fig. 4 represents the likelihood to recognize, in a video-shot at a specific time (x axis), the given emotion (column) known the expressed emotion (line).

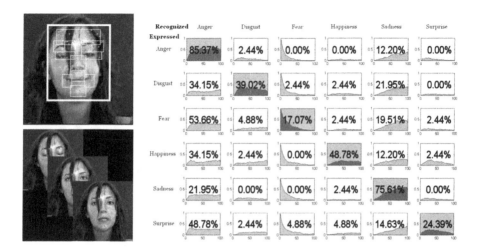

Fig. 3. Video processing   **Fig. 4.** Confusion Matrix Table

Future work will thus explore possibilities for exploiting this temporal pattern. Additionally, we think some improvements can be reached by using 6 different detectors (one for each emotion) instead of one classifier and by exploiting the multimodality, intrinsic in emotions, and therefore by processing audio.

We believe the examples we have exposed justify the need of such a multidisciplinary approach by making clear its positive impact on tomorrows multimedia indexing and retrieval systems. We argue that this is possible because of the very nature of emotions which facilitates bridging the semantic gap.

References

1. Damasio, A.R.: Descartes' Error: Emotion, Reason, and the Human Brain. Avon books, NY (1994)
2. Salway, A., Graham, M.: Extracting information about emotions in films. In: Proceedings of ACM Multimedia 2003, Berkeley, CA, USA, pp. 299–302 (2003)
3. Miyamori, H., Nakamura, S., Tanaka, K.: Generation of views of TV content using TV viewers' perspectives expressed in live chats on the web. In: Proceedings of ACM Multimedia 2005, Singapore, pp. 853–861 (2005)
4. Chan, C.H., Jones, G.J.F.: Affect-based indexing and retrieval of films. In: Proceedings of ACM Multimedia 2005, Singapore, pp. 427–430 (2005)
5. Kuo, F.F., Chiang, M.F., Shan, M.K., Lee, S.Y.: Emotion-based music recommendation by association discovery from film music. In: Proceedings of ACM Multimedia 2005, Singapore, pp. 507–510 (2005)
6. Kim, E.Y., Kim, S.J., Koo, H.J., Jeong, K., Kim, J.I.: Emotion-Based Textile Indexing Using Colors and Texture. In: Wang, L., Jin, Y. (eds.) FSKD 2005. LNCS (LNAI), vol. 3613, pp. 1077–1080. Springer, Heidelberg (2005)
7. Pantic, M., Rothkrantz, L.: Toward an Affect-Sensitive Multimodal Human-Computer Interaction. In: Proceedings of IEEE, vol. 91, pp. 1370–1390 (2003)

Author Index

Lecture Notes in Computer Science

Sublibrary 3: Information Systems and Application, incl. Internet/Web and HCI

For information about Vols. 1– 4469
please contact your bookseller or Springer

Vol. 4663: C. Baranauskas, P. Palanque, J. Abascal, S.D.J. Barbosa (Eds.), Human-Computer Interaction – INTERACT 2007, Part II. XXXIII, 735 pages. 2007.

Vol. 4662: C. Baranauskas, P. Palanque, J. Abascal, S.D.J. Barbosa (Eds.), Human-Computer Interaction – INTERACT 2007, Part I. XXXIII, 637 pages. 2007.

Vol. 4658: T. Enokido, L. Barolli, M. Takizawa (Eds.), Network-Based Information Systems. XIII, 544 pages. 2007.

Vol. 4656: M.A. Wimmer, J. Scholl, Å. Grönlund (Eds.), Electronic Government. XIV, 450 pages. 2007.

Vol. 4655: G. Psaila, R. Wagner (Eds.), E-Commerce and Web Technologies. VII, 229 pages. 2007.

Vol. 4654: I.-Y. Song, J. Eder, T.M. Nguyen (Eds.), Data Warehousing and Knowledge Discovery. XVI, 482 pages. 2007.

Vol. 4653: R. Wagner, N. Revell, G. Pernul (Eds.), Database and Expert Systems Applications. XXII, 907 pages. 2007.

Vol. 4636: G. Antoniou, U. Aßmann, C. Baroglio, S. Decker, N. Henze, P.-L. Patranjan, R. Tolksdorf (Eds.), Reasoning Web. IX, 345 pages. 2007.

Vol. 4611: J. Indulska, J. Ma, L.T. Yang, T. Ungerer, J. Cao (Eds.), Ubiquitous Intelligence and Computing. XXIII, 1257 pages. 2007.

Vol. 4607: L. Baresi, P. Fraternali, G.-J. Houben (Eds.), Web Engineering. XVI, 576 pages. 2007.

Vol. 4606: A. Pras, M. van Sinderen (Eds.), Dependable and Adaptable Networks and Services. XIV, 149 pages. 2007.

Vol. 4605: D. Papadias, D. Zhang, G. Kollios (Eds.), Advances in Spatial and Temporal Databases. X, 479 pages. 2007.

Vol. 4602: S. Barker, G.-J. Ahn (Eds.), Data and Applications Security XXI. X, 291 pages. 2007.

Vol. 4601: S. Spaccapietra, P. Atzeni, F. Fages, M.-S. Hacid, M. Kifer, J. Mylopoulos, B. Pernici, P. Shvaiko, J. Trujillo, I. Zaihrayeu (Eds.), Journal on Data Semantics IX. XV, 197 pages. 2007.

Vol. 4592: Z. Kedad, N. Lammari, E. Métais, F. Meziane, Y. Rezgui (Eds.), Natural Language Processing and Information Systems. XIV, 442 pages. 2007.

Vol. 4587: R. Cooper, J. Kennedy (Eds.), Data Management. XIII, 259 pages. 2007.

Vol. 4577: N. Sebe, Y. Liu, Y.-t. Zhuang, T.S. Huang (Eds.), Multimedia Content Analysis and Mining. XIII, 513 pages. 2007.

Vol. 4568: T. Ishida, S. R. Fussell, P. T. J. M. Vossen (Eds.), Intercultural Collaboration. XIII, 395 pages. 2007.

Vol. 4566: M.J. Dainoff (Ed.), Ergonomics and Health Aspects of Work with Computers. XVIII, 390 pages. 2007.

Vol. 4564: D. Schuler (Ed.), Online Communities and Social Computing. XVII, 520 pages. 2007.

Vol. 4563: R. Shumaker (Ed.), Virtual Reality. XXII, 762 pages. 2007.

Vol. 4561: V.G. Duffy (Ed.), Digital Human Modeling. XXIII, 1068 pages. 2007.

Vol. 4560: N. Aykin (Ed.), Usability and Internationalization, Part II. XVIII, 576 pages. 2007.

Vol. 4559: N. Aykin (Ed.), Usability and Internationalization, Part I. XVIII, 661 pages. 2007.

Vol. 4558: M.J. Smith, G. Salvendy (Eds.), Human Interface and the Management of Information, Part II. XXIII, 1162 pages. 2007.

Vol. 4557: M.J. Smith, G. Salvendy (Eds.), Human Interface and the Management of Information, Part I. XXII, 1030 pages. 2007.

Vol. 4541: T. Okadome, T. Yamazaki, M. Makhtari (Eds.), Pervasive Computing for Quality of Life Enhancement. IX, 248 pages. 2007.

Vol. 4537: K.C.-C. Chang, W. Wang, L. Chen, C.A. Ellis, C.-H. Hsu, A.C. Tsoi, H. Wang (Eds.), Advances in Web and Network Technologies, and Information Management. XXIII, 707 pages. 2007.

Vol. 4531: J. Indulska, K. Raymond (Eds.), Distributed Applications and Interoperable Systems. XI, 337 pages. 2007.

Vol. 4526: M. Malek, M. Reitenspieß, A. van Moorsel (Eds.), Service Availability. X, 155 pages. 2007.

Vol. 4524: M. Marchiori, J.Z. Pan, C.d.S. Marie (Eds.), Web Reasoning and Rule Systems. XI, 382 pages. 2007.

Vol. 4519: E. Franconi, M. Kifer, W. May (Eds.), The Semantic Web: Research and Applications. XVIII, 830 pages. 2007.

Vol. 4518: N. Fuhr, M. Lalmas, A. Trotman (Eds.), Comparative Evaluation of XML Information Retrieval Systems. XII, 554 pages. 2007.

Vol. 4508: M.-Y. Kao, X.-Y. Li (Eds.), Algorithmic Aspects in Information and Management. VIII, 428 pages. 2007.

Vol. 4506: D. Zeng, I. Gotham, K. Komatsu, C. Lynch, M. Thurmond, D. Madigan, B. Lober, J. Kvach, H. Chen (Eds.), Intelligence and Security Informatics: Biosurveillance. XI, 234 pages. 2007.

Vol. 4505: G. Dong, X. Lin, W. Wang, Y. Yang, J.X. Yu (Eds.), Advances in Data and Web Management. XXII, 896 pages. 2007.

Vol. 4504: J. Huang, R. Kowalczyk, Z. Maamar, D. Martin, I. Müller, S. Stoutenburg, K.P. Sycara (Eds.), Service-Oriented Computing: Agents, Semantics, and Engineering. X, 175 pages. 2007.

Vol. 4500: N.A. Streitz, A.D. Kameas, I. Mavrommati (Eds.), The Disappearing Computer. XVIII, 304 pages. 2007.

Vol. 4495: J. Krogstie, A. Opdahl, G. Sindre (Eds.), Advanced Information Systems Engineering. XVI, 606 pages. 2007.

Vol. 4480: A. LaMarca, M. Langheinrich, K.N. Truong (Eds.), Pervasive Computing. XIII, 369 pages. 2007.

Vol. 4473: D. Draheim, G. Weber (Eds.), Trends in Enterprise Application Architecture. X, 355 pages. 2007.

Vol. 4471: P. Cesar, K. Chorianopoulos, J.F. Jensen (Eds.), Interactive TV: A Shared Experience. XIII, 236 pages. 2007.